Bioinformatics and Beyond

This book is a comprehensive exploration of the dynamic interplay between bioinformatics and artificial intelligence (AI) within the healthcare landscape. This book introduces readers to the foundational principles of bioinformatics and AI, elucidating their integration and collaborative potential.

Bioinformatics and Beyond: AI Applications in Healthcare explores the transformative impact of data-driven insights, showcasing the applications of machine learning in diagnostics, personalized medicine, and genomic advancements. The book unveils the pivotal role AI plays in accelerating pharmaceutical research. Moreover, it addresses the practical implementation of AI in clinical decision support systems, while also critically examining challenges and ethical considerations associated with these technologies. Finally, the book looks toward the future, envisioning emerging trends and technologies that promise to reshape the future of healthcare. Aimed at professionals, researchers, and students across diverse disciplines, this book serves as an invaluable guide to understanding and navigating the evolving landscape of AI applications in healthcare.

This book is tailored to meet the needs of scientists, researchers, practitioners, professionals, and educators actively engaged in the realms of bioinformatics, artificial intelligence, and healthcare. It will be an indispensable resource for those seeking advanced strategies to address challenges and harness opportunities in the rapidly evolving fields of medical and biomedical research.

Artificial Intelligence for Sustainable Engineering and Management

Sachi Nandan Mohanty
School of Computer Science & Engineering (SCOPE),
VIT-AP University,
Amaravati, Andhra Pradesh, India
Deepak Gupta

Artificial intelligence is shaping the future of humanity across nearly every industry. It is already the main driver of emerging technologies like big data, robotics and IoT, and it will continue to act as a technological innovator for the foreseeable future. Artificial intelligence is the simulation of human intelligence processes by machines, especially computer systems. Specific applications of AI include expert systems, natural language processing, speech recognition and machine vision. The future of business intelligence combined with AI will see the analysis of huge quantities of contextual data in real-time to meet customer needs.

Green Metaverse for Greener Economies
Edited by Sukanta Kumar Baral, Richa Goel, Tilottama Singh, and Rakesh Kumar

Healthcare Analytics and Advanced Computational Intelligence
Edited by Sushruta Mishra, Meshal Alharbi, Hrudaya Kumar Tripathy, Biswajit Sahoo, and Ahmed Alkhayyat

AI in Agriculture for Sustainable and Economic Management
Edited by Sirisha Potluri, Suneeta Satpathy, Santi Swarup Basa, and Antonio Zuorro

Deep Learning in Biomedical Signal and Medical Imaging
Edited by Ngangbam Herojit Singh, Utku Kose, and Sarada Prasad Gochhayat

Sustainable Development Using Private AI: Security Models and Applications
Edited by Uma Maheswari V and Rajanikanth Aluvalu

Sustainable Farming through Machine Learning Enhancing Productivity and Efficiency
Suneeta Satpathy, Bijay Paikaray, Ming Yang, and Arun Balakrishnan

Bioinformatics and Beyond: AI Applications in Healthcare
Edited by Moolchand Sharma, Deepak Kumar Sharma, Deevyankar Agarwal, and Khoula Al Harthy

www.routledge.com/AI-for-Sustainable-Engineering-and-Management-series/book-series/AISEM

Bioinformatics and Beyond
AI Applications in Healthcare

Edited by
Moolchand Sharma, Deepak Kumar Sharma,
Deevyankar Agarwal, and Khoula Al Harthy

CRC Press is an imprint of the
Taylor & Francis Group, an **informa** business

Designed cover image: image credited to jijomathaidesigners [ShutterStock ID: 1662238171]

First edition published 2025
by CRC Press
2385 NW Executive Center Drive, Suite 320, Boca Raton FL 33431

and by CRC Press
4 Park Square, Milton Park, Abingdon, Oxon, OX14 4RN

CRC Press is an imprint of Taylor & Francis Group, LLC

© 2025 selection and editorial matter, Moolchand Sharma, Deepak Kumar Sharma, Deevyankar Agarwal, Khoula Al Harthy; individual chapters, the contributors

Reasonable efforts have been made to publish reliable data and information, but the author and publisher cannot assume responsibility for the validity of all materials or the consequences of their use. The authors and publishers have attempted to trace the copyright holders of all material reproduced in this publication and apologize to copyright holders if permission to publish in this form has not been obtained. If any copyright material has not been acknowledged please write and let us know so we may rectify in any future reprint.

Except as permitted under U.S. Copyright Law, no part of this book may be reprinted, reproduced, transmitted, or utilized in any form by any electronic, mechanical, or other means, now known or hereafter invented, including photocopying, microfilming, and recording, or in any information storage or retrieval system, without written permission from the publishers.

For permission to photocopy or use material electronically from this work, access www.copyright.com or contact the Copyright Clearance Center, Inc. (CCC), 222 Rosewood Drive, Danvers, MA 01923, 978-750-8400. For works that are not available on CCC please contact mpkbookspermissions@tandf.co.uk

Trademark notice: Product or corporate names may be trademarks or registered trademarks and are used only for identification and explanation without intent to infringe.

ISBN: 9781032832425 (hbk)
ISBN: 9781032832432 (pbk)
ISBN: 9781003508403 (ebk)

DOI: 10.1201/9781003508403

Typeset in Times
by Newgen Publishing UK

Dedication

Dr. Moolchand Sharma would like to dedicate this book to his father, Sh. Naresh Kumar Sharma and his mother, Smt. Rambati Sharma for their constant support and motivation, and his family members, including his wife, Ms. Pratibha Sharma, and son Dhairya Sharma. I also thank the publisher and my other co-editors for believing in my abilities.

Dr. Deepak Kumar Sharma would like to dedicate this book to my beloved parents, for their endless support and guidance; to my incredible wife, for her unwavering love and companionship; and to my wonderful kids, for inspiring me every day with their boundless joy and curiosity.

Dr. Deevyankar Agarwal would like to dedicate this book to his father, Sh. Anil Kumar Agarwal and his mother, Smt. Sunita Agarwal and his wife, Ms. Aparna Agarwal, and son Jai Agarwal for their constant support and motivation. I would also like to give my special thanks to the publisher and my other co-editors for having faith in my abilities.

Dr. Khoula Al Harthy would like to express my deepest gratitude to my family for their unwavering support throughout this journey. To my spouse, who endured late nights and countless revisions with a smile. Thanks to all book editors for their support and dedication to our work.

Contents

Preface ix
About the Book xi
About the Editors xiii
List of Contributors xv
Introduction and Scope xix

Chapter 1 AI and Machine Learning in Modern Healthcare 1

Kshatrapal Singh, Vijay Shukla, Dhiraj Gupta, and Yogesh Kumar Sharma

Chapter 2 Telemedicine and Remote Prenatal Care: A Soft Computing Approach 19

Pradnya S. Mehta, Pranav Bafna, Atharv Sawant, Pratik Patil, and Niraj Pandit

Chapter 3 Detection of Abnormality in Heart Rhythm Using a Machine Learning Approach 49

Prabhudutta Ray, Raj Rawal, and Ahsan Z. Rizvi

Chapter 4 AI in Drug Discovery and Development 66

Tahreem Shahzad, Arooj Fatima Tul Zahra, Ayesha Naeem, and Mujahid Tabassum

Chapter 5 Transforming Healthcare: Leveraging Machine Learning Algorithms for Diagnosis, Treatment, and Management 92

Indu Joseph Thoppil, K. Ashtalakshmi, and Ramesh Chundi

Chapter 6 Gestational Diabetes Prediction Using Hybrid Probabilistic Machine Learning Models 115

Lakshmi K., Umme Salma M., and Sangeetha Shathish

Chapter 7 AI in Diagnostics and Disease Prediction: Urolithiasis 130

Suneel Kumar and Dashrath Singh

Chapter 8	Unveiling Cheminformatics for Accelerated Drug Discovery and Development: A Computational-Guided Approach	161

Amanpreet Kaur and Debasish Mandal

Chapter 9	Transformative Applications of AI and Machine Learning in Bioinformatics for Healthcare Systems	178

Arooj Fatima Tul Zahra, Mujahid Tabassum, Nabiea Shehma, Moeza Anam, and Tripti Sharma

Chapter 10	Revolutionizing Drug Development: The Role of AI in Modern Pharmaceutical Research	206

Samridhi Agarwal and Amit Kumar Dutta

Chapter 11	Impact of AI on Healthcare from Diagnostics to Drug Discovery	228

Smrita Singh and Ashutosh Singh Chauhan

Chapter 12	Biomedical Imaging Techniques in AI Applications	242

T. Kalpana, R. Thamilselvan, and T. M. Saravanan

Chapter 13	Ethical Implications of AI in CRISPR: Responsible Genome Editing	256

Umesh Gupta, Ayushman Pranav, Rajesh Kumar Modi, and Ankit Dubey

Index 277

Preface

We are delighted to launch our book *Bioinformatics and Beyond: AI Applications in Healthcare* under the book series Artificial Intelligence for Sustainable Engineering and Management, CRC Press, Taylor & Francis Group. The value of the book lies in providing readers with a comprehensive understanding of how the synergy between bioinformatics and artificial intelligence (AI) is transforming healthcare. It emphasizes the practical applications, challenges, and ethical considerations associated with implementing AI in the healthcare domain, making it a valuable resource for researchers, healthcare professionals, and anyone interested in the intersection of technology and healthcare. Readers can anticipate future trends and emerging technologies, acquiring future-ready knowledge that positions them at the forefront of the evolving healthcare landscape. Essentially, this book not only imparts knowledge but also functions as a practical guide, empowering readers to apply advanced strategies and contribute meaningfully to the transformative advancements in medical and biomedical research.

This book's unique highlights include its practical applications, providing real-world solutions that empower professionals and researchers to navigate the intricate landscape of data management, machine learning algorithms, and cutting-edge advancements in genomics. It delves into the acceleration of drug discovery, the implementation of clinical decision support systems, and the ethical considerations crucial in the deployment of AI in healthcare. Specialized chapters on biomedical imaging, natural language processing, and discussions on telemedicine and public health informatics further amplify its practical relevance. Around 60 full-length chapters were received. Among these manuscripts, 13 chapters have been included in this volume. All the chapters submitted were peer reviewed by at least two independent reviewers and provided with a detailed review proforma. The comments from the reviewers were communicated to the authors, who incorporated the suggestions in their revised manuscripts. The recommendations from two reviewers were considered while selecting chapters for inclusion in the volume. The exhaustiveness of the review process is evident, given the large number of articles received addressing a wide range of research areas. The stringent review process ensured that each published chapter met the rigorous academic and scientific standards.

We would like to thank the authors of the published chapters for adhering to the schedule and incorporating the review comments. We extend heartfelt acknowledgment to the authors, peer reviewers, committee members, and production staff whose diligent work shaped this volume.

Moolchand Sharma
Deepak Kumar Sharma
Deevyankar Agarwal
Khoula Al Harthy
Editors

About the Book

The book is a comprehensive exploration of the dynamic interplay between bioinformatics and artificial intelligence (AI) within the healthcare landscape. This book introduces readers to the foundational principles of bioinformatics and AI, elucidating their integration and collaborative potential. It delves into the transformative impact of data-driven insights, showcasing the applications of machine learning in diagnostics, personalized medicine, and genomic advancements. With a focus on drug discovery, the book unveils the pivotal role AI plays in accelerating pharmaceutical research. Moreover, it addresses the practical implementation of AI in clinical decision support systems, while also critically examining challenges and ethical considerations associated with these technologies.

Finally, the book looks toward the future, envisioning emerging trends and technologies that promise to reshape the future of healthcare. Aimed at professionals, researchers, and students across diverse disciplines, this book serves as an invaluable guide to understanding and navigating the evolving landscape of AI applications in healthcare. This book is tailored to meet the needs of scientists, researchers, practitioners, professionals, and educators actively engaged in the realms of bioinformatics, AI, and healthcare. By offering targeted insights into cutting-edge developments, the book is an indispensable resource for those seeking advanced strategies to address challenges and harness opportunities in the rapidly evolving fields of medical and biomedical research.

About the Editors

Moolchand Sharma is currently an Assistant Professor in the Department of Computer Science and Engineering at the Maharaja Agrasen Institute of Technology, GGSIPU Delhi. He has published scientific research publications in reputed international journals and conferences, including SCI indexed and Scopus indexed journals such as *Cognitive Systems Research* (Elsevier), *Physical Communication* (Elsevier), *Intelligent Decision Technologies: An International Journal, Cyber-Physical Systems* (Taylor & Francis Group), *International Journal of Image & Graphics* (World Scientific), *International Journal of Innovative Computing and Applications* (Inderscience), and *Innovative Computing and Communication Journal* (Scientific Peer-reviewed Journal). He has authored/co-authored chapters with international publishers like Elsevier, Wiley, and De Gruyter. He has authored/edited four books with a national/international level publisher (CRC Press, Bhavya publications). His research areas include artificial intelligence, nature-inspired computing, security in cloud computing, machine learning, and search engine optimization. He is associated with various professional bodies like IEEE, ISTE, IAENG, ICSES, UACEE, Internet Society, and life membership of Universal Innovators research lab. He has taught fore more than nine years. He is the co-convener of the ICICC, DOSCI, ICDAM and ICCCN Springer Scopus Indexed conference series and ICCRDA-2020 Scopus Indexed IOP Material Science & Engineering conference series. He is also the organizer and co-convener of the International Conference on Innovations and Ideas towards Patents (ICIIP) series. He is the advisory and TPC committee member of the ICCIDS-2022 Elsevier SSRN Conference. He is also the reviewer of many reputed journals from Springer, Elsevier, IEEE, IEEE JBHI, Wiley, Taylor & Francis Group, IJEECS, and World Scientific Journal, and many Springer conferences. He has served as a session chair in many international Springer conferences. He received his PhD from DCR University of Science & Technology, Haryana in 2024. He completed his M. Tech (CSE)) in 2012 from SRM University, NCR Campus, Ghaziabad, and B. Tech (CSE) in 2010 from KNGD Modi Engg. College, GBTU.

Deepak Kumar Sharma is working as an Associate Professor in the Department of Information Technology, Indira Gandhi Delhi Technical University for Women (IGDTUW), Kashmere Gate, Delhi, India. Earlier, he worked as Assistant Professor at Netaji Subhas University of Technology (Formerly N.S.I.T.), Dwarka, Delhi. He obtained his Ph.D. in Computer Engineering from University of Delhi, India in 2016. His research interests include opportunistic networks, wireless ad hoc and sensor networks, software-defined networks, and IoT networks. He has over 17 years of experience in academics. He has published various research papers in reputed international journals like *ETT Wiley, IEEE Systems Journal, IEEE IoT Journal, Computer Communication Elsevier, IJCS Wiley,* etc., and conferences of repute like IEEE AINA, and GLOBECOM. He has also authored various book chapters in edited books published by IET, Wiley, Springer, and Elsevier. He has served as session chair in many conferences and is also a reviewer of various reputed journals like *ETT Wiley*, *AIHC Springer*, and *IJCS Wiley*.

Deevyankar Agarwal received his Ph.D. with outstanding grades in the applied artificial intelligence domain from the Department of Information Technology and Telecommunications, University of Valladolid, Spain. He has extensive experience in machine learning, having conducted research on applied deep learning and published papers on the automatic detection of COVID-19 and early detection of Alzheimer's by using state-of-the-art CNN architectures and medical images in Q1 journals (*Applied Soft Computing* (Elsevier) and *Journal of Medical Systems* (Springer Nature)). His Ph.D. thesis "DEEP-AD: The Deep Learning Model for Diagnostic Classification and Prognostic Prediction of Alzheimer's Disease", aimed to design a web platform based on 3D transfer learning, deep learning, and ensemble learning (for uncertainty quantification) to address the problem of early detection and prognostic progression of Alzheimer's. He has 22 years of experience as an educator, and has been working at the University of Technology and Applied Sciences, Muscat, since September 2013 as a lecturer in the Computer Engineering Department. Additionally, he also worked as a research coordinator, where his job was to keep track of the funding process for professors in the engineering department, to act as a go-between for funding agencies, enterprises, and the department, and to try to get as much funding as possible to develop products and projects that will benefit society.

Khoula Al Harthy is an esteemed researcher with a focus on information security and a keen interest in blockchain technology. Her career is dedicated to the empowerment of local expertise in cutting-edge technologies. As a recognized speaker, she contributes her insights at both national and international events. In a significant leadership role, Khoula has served as the President of the Oman Blockchain Club for two years. This organization is pivotal in disseminating knowledge and education about blockchain applications across Oman, fostering a deeper understanding of this revolutionary technology. In the academic realm, Khoula holds a Ph.D. in Risk Management. She is the head of the Computer Science and Creative Technologies department at GCET. Notably, she is the initiator of the Blockchain Research Thematic Group, driving forward the academic exploration and application of blockchain technology. Her work in this area highlights her commitment to integrating advanced tech into academia and beyond, shaping the future of information cybersecurity and AI in the region.

Contributors

Samridhi Agarwal
Amity Institute of Biotechnology,
Amity University Jharkhand,
Ranchi, India

Moeza Anam
Faculty of Human Dietitian and Nutrition,
University of Chenab, Gujrat, Punjab, Pakistan

K. Ashtalakshmi
Dayananda Sagar University,
Bengaluru, India

Pranav Bafna
Vishwakarma Institute of Information Technology,
Pune, India

Ashutosh Singh Chauhan
Archaeological Survey of India, Agra,
Uttar Pradesh, India

Ramesh Chundi
Dayananda Sagar University,
Bengaluru, India

Ankit Dubey
SCSET, Bennett University,
Gr. Noida-201310, U.P., India

Amit Kumar Dutta
Amity Institute of Biotechnology, Amity University,
Jharkhand, Ranchi, India

Dhiraj Gupta
Greater Noida Institute of Technology, Greater Noida
Uttar Pradesh, India

Umesh Gupta
SCSET, Bennett University,
Gr. Noida-201310, U.P., India

Lakshmi K.
CHRIST (Deemed to be University),
Bengaluru, India

T. Kalpana
Kongu Engineering College,
Erode, Tamilnadu, India

Amanpreet Kaur
Department of Chemistry and Biochemistry, Thapar Institute of Engineering and Technology, Patiala, Punjab, India.

Suneel Kumar
Mangalayatan University, Aligarh,
Uttar Pradesh, India

Umme Salma M.
CHRIST (Deemed to be University),
Bengaluru, India

Debasish Mandal
Department of Chemistry and Biochemistry, Thapar Institute of Engineering and Technology, Patiala, Punjab, India

Pradnya S. Mehta
Vishwakarma Institute of Information Technology,
Pune, India

Rajesh Kumar Modi
SCSET, Bennett University,
Gr. Noida-201310, U.P., India

Ayesha Naeem
Forman Christian College University,
Lahore, Pakistan

Niraj Pandit
Vishwakarma Institute of Information
  Technology,
Pune, India

Pratik Patil
Vishwakarma Institute of Information
  Technology,
Pune, India

Ayushman Pranav
SCSET, Bennett University,
Gr. Noida-201310, U.P., India

Raj Rawal
Gujarat Pulmonary and Critical
  Care,
Medicine, SHALL Hospital,
Gandhinagar, India

Prabhudutta Ray
Institute of Advanced Research,
Department of Computer Science
  & Engg.,
Gandhinagar, India

Ahsan Z. Rizvi
Institute of Advanced Research,
Department of Computer Science
  & Engg.,
Gandhinagar, India

T. M. Saravanan
Kongu Engineering College,
Erode, Tamilnadu, India

Atharv Sawant
Vishwakarma Institute of Information
  Technology,
Pune, India

Tahreem Shahzad
Department of Chemistry,
University of Narowal,
Narowal, Pakistan

Tripti Sharma
IT Department,
University of Technology and Applied
  Sciences,
Muscat, Oman

Yogesh Kumar Sharma
Greater Noida Institute of Technology,
  Greater Noida,
Uttar Pradesh, India

Sangeetha Shathish
CHRIST (Deemed to be University),
Bengaluru, India

Nabiea Shehma
School of Medicine, Yangtze University,
  Jingzhou,
Hubei, China

Vijay Shukla
Greater Noida Institute of Technology,
  Greater Noida,
Uttar Pradesh, India

Dashrath Singh
Krishna Pharmacy College, Bijnor,
Uttar Pradesh, India

Kshatrapal Singh
Greater Noida Institute of Technology,
  Greater Noida,
Uttar Pradesh, India

Smrita Singh
Creative Bioinformatics and
  Science,
Morna, Uttar Pradesh, India

Mujahid Tabassum
South East Technological University,
X91 HE36 Waterford, Ireland

R. Thamilselvan
Kongu Engineering College,
Erode, Tamilnadu, India

Indu Joseph Thoppil
Dayananda Sagar University,
Bengaluru, India

Arooj Fatima Tul Zahra
Swinburne University of Technology Sarawak,
Kuching, Malaysia

Introduction and Scope

The book *Bioinformatics and Beyond: AI Applications in Healthcare* serves as an all-encompassing guide, delving into the dynamic confluence of bioinformatics and artificial intelligence (AI) within the realm of healthcare. The book opens with a comprehensive introduction, outlining the foundational principles of bioinformatics and AI, setting the stage for an interdisciplinary exploration of their collaborative potential. The book's scope extends beyond mere theoretical discussions, offering readers practical insights into real-world applications. From the intricacies of data management and processing to the transformative impact of machine learning in diagnostics, personalized medicine, and genomic advancements, the book navigates through the multifaceted landscape of healthcare. It unfolds the pivotal role of AI in accelerating drug discovery processes, addresses the implementation of clinical decision support systems, and critically examines the ethical considerations inherent in these technologies. With an eye on the future, the book anticipates emerging trends and technologies, making it an invaluable resource for professionals, researchers, and students navigating the evolving landscape of AI applications in healthcare.

The future trajectory of the book unfolds as a dynamic exploration into the evolving landscape of bioinformatics, AI, and healthcare. With a commitment to staying at the forefront of advancements, the book envisions delving into emerging technologies that redefine healthcare paradigms. Anticipating a surge in multi-omics data availability, the future scope encompasses a deeper dive into integrating and analyzing diverse datasets, fostering a more holistic comprehension of intricate biological systems. Precision medicine remains a focal point, with the book set to unravel how AI further tailors medical treatments based on individual genetic profiles. As healthcare demands real-time insights, the future chapters are poised to expand into real-time analytics, spotlighting AI's role in enhancing the decision-making processes in clinical settings. The ethical considerations surrounding AI in healthcare, alongside evolving regulatory frameworks, will be thoroughly explored to guide readers in navigating responsible implementation. Additionally, the book envisions a broader global health perspective, spotlighting AI's potential in addressing global health challenges and contributing to healthcare equity. Embracing a continuous learning approach, the book will provide updates on technological trends, ensuring that readers remain well-versed in the latest methodologies shaping the future of bioinformatics and AI in the dynamic healthcare landscape. The future of the book lies in its adaptive and comprehensive approach, offering a guiding beacon for those keen on navigating the transformative waves of technology in healthcare.

SCOPE OF THE BOOK

The scope of *Bioinformatics and Beyond: AI Applications in Healthcare* is broad and forward-thinking, encompassing a comprehensive exploration of the intersection between bioinformatics, AI, and healthcare. The book is designed to cover a wide range of topics and provide valuable insights into the transformative applications of these technologies in the healthcare domain. The key areas of scope include:

- Foundational Concepts
- Practical Applications
- Data Management and Processing
- Machine Learning Algorithms
- Genomics and Personalized Medicine
- Drug Discovery and Development
- Clinical Decision Support Systems (CDSSs)
- Biomedical Imaging and AI
- Challenges and Ethical Considerations
- Future Trends and Innovations

The scope of the book is ambitious, aiming to provide a comprehensive and accessible resource for scientists, researchers, practitioners, professionals, and educators interested in the transformative applications of bioinformatics and AI in the dynamic and evolving field of healthcare.

1 AI and Machine Learning in Modern Healthcare

Kshatrapal Singh, Vijay Shukla, Dhiraj Gupta, and Yogesh Kumar Sharma

1.1 PREVENTIVE HEALTHCARE

Measures performed to prevent the spread of disease are considered preventive healthcare. Disabilities and diseases are important processes that begin long before a person realizes they are impacted. Diseases result from lifestyle choices, transmission factors for diseases, genetics, and environmental influences.

1.1.1 What Is Preventive Healthcare?

Periodic medical care, such as screenings, examinations, and patient advice, is known as preventive healthcare. Its goal is to stop diseases, illnesses, and additional health issues.

By avoiding illness, preventive healthcare aims to lessen the effects of disease and associated risk variables. Prevention strategies can be utilized at any stage of an individual's life and across the illness spectrum to halt further decline over time. The greatest reason for death and disabilities around the globe is chronic disease, which is also connected with increasing medical bills. A public health model that enhances preventative care calls for both clinical preventative services and screening exams [1, 2]. One tactic for reducing use and enhancing health outcomes is to recognize and stop possible issues upstream. Figure 1.1 shows the role of preventive healthcare in a person's life via primary care, laboratory, hospital, pharmacy, and medical experts.

1.1.2 Preventive Healthcare in India

About 11% of India's total healthcare spending in 2019 went toward preventative care. The marketplace for preventive healthcare in India, which had a 2019 market price of INR 3.71 Tn, is projected to rise at a compound annual growth rate (CAGR) of 27.30% between 2020 and 2025 to touch INR 14.58 [3].

The healthcare sector is expanding due to the availability of state-of-the-art medical equipment and services as well as the rise in non-communicable and chronic illnesses such cancer, diabetes, osteoarthritis, and heart disease. Additionally, preventive healthcare is more accessible for Indians because of how affordable it is.

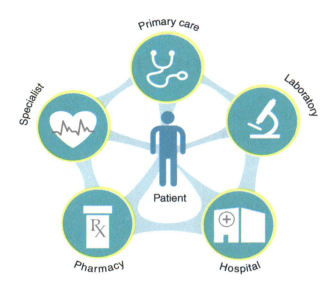

FIGURE 1.1 Preventive healthcare.

Moreover, limited healthcare insurance for preventive therapies and a lack of public awareness of preventive medical facilities are limiting market expansion.

1.1.3 Why Preventive Healthcare Is Important?

The ultimate goal of living a healthy lifestyle is preventive healthcare. Early diagnosis is the foundation of preventative healthcare's significance. It aids in early detection of potential problems and their early treatment to prevent worsening. Preventive healthcare is also significantly less expensive than paying a lot of money for therapy. Ultimately, developing a community with the best potential health condition depends heavily on preventive healthcare [4]. Figure 1.2 shows the importance of preventive healthcare in people's daily life in detail.

To encourage a healthier lifestyle and reduce the lethal effects of numerous diseases, healthcare reforms need to greatly increase the availability of preventive healthcare education. Children should be taught the value of preventative healthcare and how to live a healthy lifestyle by participating in exercise and dietary programs [5, 6]. They will know how to prioritize their health and welfare through their lives if these principles are instilled in them. Similar to children, adults must be more knowledgeable about recognizing early signs and symptoms of illness and taking steps to reduce stress in their everyday lives. Individuals are sometimes disinterested in using preventative healthcare services because they are unaware of how important they are.

AI and Machine Learning in Modern Healthcare

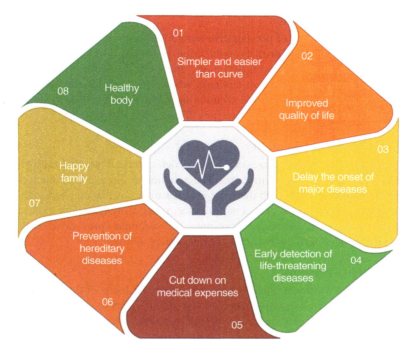

FIGURE 1.2 Importance of preventive healthcare.

1.1.4 Types of Preventive Healthcare

The five phases of illness progression are classified as follows: recovery/disability/death, subtle, clinical, susceptible, and underlying. Similar preventive health measures have been divided into comparable stages in order to concentrate on preventing different stages of an illness. These stages of control are called primordial, primary, secondary, and tertiary prevention. When combined, these strategies aim to lower risk in order to prevent both the onset of the disease as well as its upstream issues.

1.1.4.1 Primordial Prevention

Primordial prevention, the closest thing to preventative measures, was first reported in 1978. Putting a focus on social and environmental variables, it involves minimizing risk factors for the general public. Typically, these kinds of activities are promoted by national policy and legislation. Since primordial prevention is the most ancient type of prevention, it is often directed towards children in an attempt to minimize exposure to risks. In order to treat the early phases of natural illness, primordial prevention emphasizes the fundamental socioeconomic conditions that contribute to the onset of disease. Improving accessibility to safe sidewalks in urban areas is one example of how to promote physical activity, which lowers risk factors for conditions like type II diabetes, heart disease, obesity, and other conditions.

1.1.4.2 Primary Prevention

The fundamental objective of preventative methods is to protect a vulnerable person or population. The goal of primary prevention is to stop a disease before it ever begins. Therefore, those in good health make up its target audience. In order to prevent a disease from manifesting in a person who is sensitive to subclinical disease, risk-reduction measures or immune-stimulating treatments are usually used. As an example, the first line of defense is vaccines [7].

1.1.4.3 Secondary Prevention

Second-line prevention emphasizes early illness detection and targets healthy individuals with preclinical disease symptoms. The subclinical disease is characterized by pathologic changes, even though there are no outward signs that can be found during an appointment with a physician. One popular method of secondary prevention is screening. For example, a Papanicolaou smear is a secondary preventive method that finds cervical cancer early on before it becomes more serious.

1.1.4.4 Tertiary Prevention

The diagnosis and treatment phases of the illness are the focus of third-party prevention. It is administered to symptomatic persons in an effort to mitigate the severity of the sickness and any possible side effects. Although secondary prevention tries to stop illness before it starts, tertiary prevention tries to decrease the effects of an illness once it has already taken hold in an individual. Initiatives for recovery are often employed as tertiary preventative measures.

1.1.4.5 Quaternary Prevention

Quaternary prevention is defined as "activity performed to detect persons at danger of over-medicalization, to shield him from further medical intrusion, and to offer to him solutions, that are ethically appropriate," by the Wonca International Dictionary for General/Family Practice. The primary focus of Marc Jamoulle's initial introduction of this concept was on sick people without a condition [8]. The recently changed idea is the "Decision taken to protect individuals from medical therapies that have the potential to do more harm than good."

1.2 HEALTHCARE-RELATED BIG-DATA ANALYSIS

Since we routinely gather, store, process, and analyze enormous volumes of data, the idea of data has permeated every aspect of our daily lives. This characteristic is transdisciplinary and useful in a variety of sectors, such as technology, economics, medical care, and machine learning (ML). The potential utility of these enormous data sets, known as big data, in enhancing care services, clinical care, and public health has achieved popularity over the last few years.

The term "big data" in the context of medical care refers to the huge amounts of information generated through the application of digital technology for the collection of health records and the management of healthcare quality that would otherwise be too big and complicated for traditional systems [9].

Big-data analytics appear to have many positive, maybe life-saving, consequences in the healthcare industry. In its most basic form, big-style data refers to the massive volumes of data produced by automating activities and processes, which are subsequently gathered and handled by technologies. When employed in medical care, big data refers to precise health data on a population (or about an individual) and may help treat illnesses, prevent epidemics, lower costs, etc.

Treatment models have altered as a result of our extended lifespans, and a lot of these improvements are primarily the result of data. Since addressing any disease in its earliest stages is significantly less expensive, doctors would like to learn the most they can about a patient and as early in their life as possible. This allows them to see warning symptoms of serious illness as they emerge [10, 11]. Treatment is not as good as prevention, and having a full picture of a person will enable insurance companies to provide tailored patient health plans. The application of important factors in healthcare and healthcare data analytics serves to reinforce these ideas. This is an attempt by the industry to address the problems of data silos that patients face: fragments of their data being scattered around and stored in various locations such as hospitals, health care facilities, and offices, making effective interaction unfeasible.

However, there are always more resources available to medical practitioners to help them understand their patients better. The fact that this data usually comes in a variety of sizes and formats makes it challenging for the user to obtain it. The intelligence with which the data is managed holds greater significance than its bulk. With the right technology, data can be gathered from the many big-data sources in the healthcare industry rapidly and effectively.

In actuality, gathering large volumes of data for therapeutic applications has been costly and time-consuming for decades. It is now easier to collect this data, create comprehensive healthcare examinations, and turn it into relevant, crucial insights that can be used to provide high-quality healthcare thanks to today's continuously improving advancements [12]. The goal of healthcare data analysis is to use data-driven outcomes to predict and address issues before they become critical, as well as to assess medical procedures and therapies more rapidly, keep accurate inventory records, encourage patient participation in their own well-being, and give them the tools to do so.

1.2.1 Linguamatics

This is an active text analysis technique that is based on natural language processing (NLP) (I2E). I2E has the capability to extract and analyze a variety of data. Findings acquired with this method are ten times more rapid than with other methods, and interpretation of the data does not call for specialized knowledge. This method can extract information and facts about genetic links from unstructured data [13, 14]. For ML to produce clear, refined outputs, well-curated data must be used as the input. NLP, however, permits the extraction of clean and structured data that frequently remains concealed in unstructured input data when it is incorporated into electronic health records (EHRs) or clinical records in general as shown in Figure 1.3.

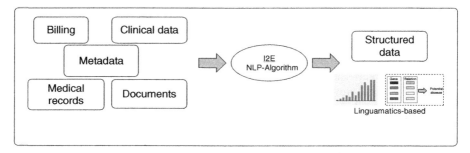

FIGURE 1.3 NLP-centric AI model applied in massive data retention and analysis.

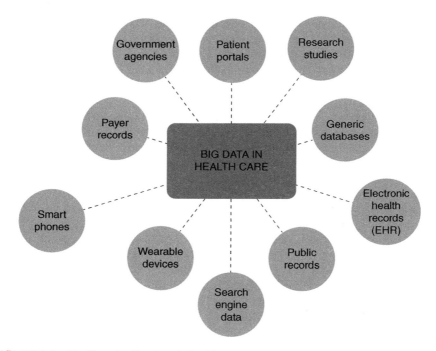

FIGURE 1.4 Big Data Applications in healthcare

1.2.2 Big-data Applications in Healthcare

Figure 1.4 shows the 21 big-data applications in the real world for the healthcare industries and depicts how an analytical technique may increase healthcare, improve systems, and finally save lives.

1.3 AI AND ML TOOLS FOR EARLY DETECTION AND DIAGNOSIS OF DISEASES

Artificial intelligence (AI) is the ability of a machine to simulate human learning functions in difficult scenarios, such as pattern and recognition of different images

AI and Machine Learning in Modern Healthcare

in the medical field. The use of AI in the healthcare industry changes how data is compiled, processed, and generated for medical safety. System planning is the term for the framework's core conceptual development. It includes the basic strategies, points of view, and how the framework performs under particular conditions. With a solid understanding of the framework's design, the customer can become knowledgeable of pits bounds and constraints. Figure 1.5 depicts a visual representation of the disease recognition model employing practical machines and deep learning classification techniques [15, 16].

Data processing involves several key steps that transform raw information into usable insights. The first step, **data preprocessing**, includes collecting raw data, fixing any issues such as errors or missing values, and cleaning the data to ensure consistency. Next, **data transformation** standardizes the data into a uniform format, applying techniques like normalization or encoding to prepare it for analysis. In **data reduction**, the focus shifts to refining the dataset by selecting only the most relevant features, often through dimensionality reduction, to make analysis more efficient. **Data collection and validation** follow, where data is gathered in organized phases, allowing for thorough testing to confirm that it accurately reflects real-world conditions. Finally, **analytical modeling** uses predictive models to estimate the likelihood of specific outcomes based on chosen variables. This systematic approach is particularly useful in fields like healthcare, where predictive models help assess disease risk. Each stage in this process contributes to building effective, data-driven models that aid in reliable decision-making.

Regular results of the tests could be coupled with more pertinent patient information, like age, gender, etc., in order to build disease-specific prediction, by incorporating AI into the test results workflow. Laboratories could be equipped to provide disease-specific patient's possible ratings by combining this data, which could aid doctors in learning about possible patient risks, diagnoses, or topics of interest. A number of healthcare organizations are currently utilizing ML and automated reasoning to create

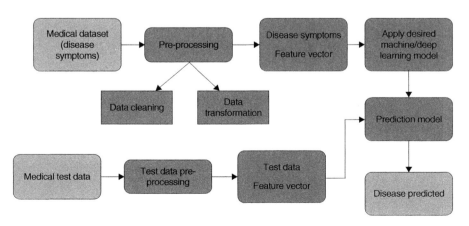

FIGURE 1.5 AI model for detection of diseases.

intelligent clinical decision support systems that may be incorporated into the current workflow for ordering tests and reviewing the results [17].

1.3.1 LIVER DISEASE SEVERITY ALGORITHM

Liver disease can be caused by a number of things. An overly inflamed and gradually fibrotic liver may result from excessive alcohol intake, obesity, diabetes, hepatitis infections, and medication use as shown in Figure 1.6. If detected promptly and treated effectively, liver disease may be cured. But frequently, this disease goes undiagnosed until a liver transplant is the only cure. Early detection of liver cancer may greatly benefit from the use of an AI-based predictive model to help determine patients at risk of developing serious liver illness [18, 19]. A predictive model may be used in conjunction with other clinically significant data to allow early diagnosis, preventing the advancement of cirrhosis, liver failure, the requirement for a liver transplant, and even fatality.

1.3.2 CANCER PREDICTIVE ALGORITHM

AI-based predictive analytics employing regular blood tests have the ability to assist doctors in making quicker diagnoses and providing effective therapy for people with cancer via early diagnosis of a patient's cancer risk in combination with other medically relevant information. Many scientists are currently attempting to develop predictive analytics and effectively create a future free of cancer worry.

1.3.3 AI AND ML IN COVID-19 DETECTION AND DIAGNOSIS

The identification and treatment of COVID-19 is crucial in the battle against the virus. The majority of today's diagnostic testing techniques are non-invasive, such as viral

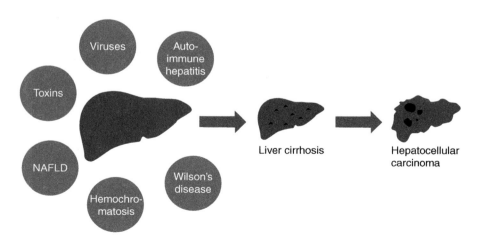

FIGURE 1.6 Factors for liver diseases.

throat swab testing, nucleic acid, serology, and chest X-ray and CT imaging. Quick and early diagnosis of infectious patients is essential for containing the pandemic's transmission and isolating the virus, and there is unquestionably a need for innovation in this field. Many AI tools have been invented for the identification and diagnosis of COVID-19 and SARS-CoV-2.

There is an ML framework that can discriminate among patients with and without COVID-19 using 11 important blood variables. With a total of 253 sample data, 105 of which come from patients whose COVID-19 disease was proved by using RT-PCR test, the model was built using the random forest ML approach and 49 clinically relevant blood test parameters (consisting of 24 routine hematological and 25 biochemical parameters) from 169 patients [20, 21].

AI is also used in medical imaging. High-dimensional imaging data can be gleaned from computed tomography (CT), magnetic resonance imaging (MRI), and X-ray. These tools offer a wealth of data that can be leveraged to create AI applications. Numerous important graphics phenotypes that are acquired through the qualitative and quantitative evaluation of structural alterations can be produced using imaging data. For example, one or more AI models use image segmentation models and image classification as part of COVID-19 diagnostics [22–24]. While an image classification process gathers properties from the region of interest and utilizes those elements as a framework for classification of (diagnosing) the images, image segmentation is employed to designate and classify the region of interest.

1.3.4 DIABETES DETECTION

The main factor causing high blood sugar levels is diabetes. AI is a cost-effective way to reduce diabetes-related ocular problems. Blood glucose prognosis has been divided into three categories, physiology-based, information-driven, and hybrid-based, as illustrated in Figure 1.7. Woldaregy et al. (2019) developed an integrated system and a concise ML handbook with the aim of evaluating the blood glucose level in individuals with type 1 diabetes. They discovered a number of ML strategies that are crucial for managing artificial pancreas functions and blood glucose alerts. They also proved that they understood the blood glucose simulator, which offered information for tracking and predicting blood glucose levels. Sugar levels can be affected by a variety of factors, including BMI, stress, illness, medications, amount of sleep, etc. Therefore, blood glucose prediction offers a prognosis of a person's blood sugar levels depending on the patient's previous and present histories, giving an indication of possible future related problems [25].

1.3.5 HEART DISEASES DIAGNOSIS

Studies assert that AI can predict when people will die of heart disease. Therefore, a variety of methods have been utilized to diagnose and forecast the intensity of the heart rate. Escamila et al. (2019) suggested a dimensionality reduction technique to use the highlights determination procedure to identify the highlights

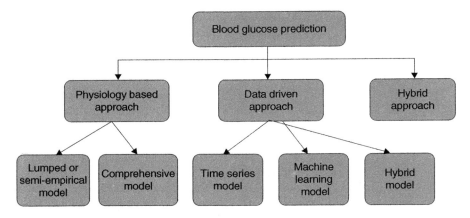

FIGURE 1.7 Blood glucose prediction techniques.

of cardiac disease. The sample used was the coronary sickness AI vault from UC Irvine, which has 74 highlights. The chi-square and head segments inquiry coupled with the irregular woods classifier achieved astounding precision incorporating deep learning in edge smart devices and in the practical implementation of heart disease detection. Tuli et al. (2019) suggested the Health Fog framework which included a body area sensor network, gateways, fogbus module, data cleaning, preprocessing, strategic planning, deep learning component, and assembling module among other software and hardware parts for their study could achieve excellent results.

1.3.6 Hypertension Disease Detection

Researchers have shown that by using input data about blood pressure, demographics, etc., AI has been capable of recognizing hypertension. A review of recent developments in the fields of computer science and medicine was given by Krittanawong et al. (2018). They showed how cutting-edge AI can be used for predicting the beginning stages of hypertension. Additionally, they claimed that AI is crucial in determining the causes of hypertension.

1.3.7 Skin Disease Diagnosis

At Sahlgrenska University's Department of Dermatology, Zaar et al. (2020) applied AI algorithms to classify clinical photographs of skin conditions, achieving a 56.4% improvement in diagnostic accuracy for the five primary recommended conditions. Similarly, Kumar et al. (2019) developed a two-stage approach combining computer vision and machine learning to assess and identify skin disorders. This method demonstrates the potential of AI-driven techniques in enhancing the accuracy and efficiency of dermatological diagnostics [26, 27].

1.4 AI AND ML IN PHARMA

In the pharmaceutical industry, AI refers to the use of automated algorithms for functions that normally require human intellect. The application of AI and ML in the pharmaceutical and biotech sectors has revolutionized how researchers create new medicines, treat diseases, and much more during the last 5 years.

The use of AI and ML in the pharmaceutical and biomedical industries has advanced from science fiction to reality in recent years. Pharma and biotech firms are increasingly using more automation via processes that combine data-driven choices with the usage of predictive analytics solutions. This kind of sophisticated data analysis will eventually combine ML and AI.

AI and ML can be applied to almost every aspect of the pharmaceutical industry, from drug discovery and production to marketing. Pharma companies may employ AI technology to increase cost-effectiveness, efficiency, and user-friendliness to all business activities by incorporating it into their main operations. The best part is that since AI technologies are built to continually acquire knowledge from new data and experience, they have the potential to be a powerful tool in the pharmaceutical industry's research and development divisions [28, 29].

Let us focus on a couple of the pharmaceutical industry's top AI and ML implementations.

1.4.1 Research and Development

Pharma companies use AI-enabled tools and advanced ML algorithms to expedite drug research worldwide. Since these intelligence technologies are designed to identify fine-grained patterns in large datasets, they can be used to solve issues pertaining to intricate biological networks. This ability is quite useful for looking at the trends in various diseases and figuring out which medicine combinations might work best to address specific symptoms of a certain ailment. Pharmaceutical companies can thus invest in the research and development of therapies that have the best chance of healing a disease or other medical condition.

1.4.2 Drug Discovery and Design

AI is helping with multi-target drug discoveries, medicinal recycling, biomarker identification, priority, morphological, and drug recognition and validation. This involves creating new substances and figuring out inventive biological objectives. The biggest benefit for pharmaceutical companies is the potential for AI to expedite the approval and market entry process for medications, especially when applied in drug trials. This could result in large cost savings, which could give patients access to more treatment alternatives and more reasonably priced prescription drugs [30].

1.4.3 Rare Diseases and Personalized Medicine

AI is being utilized in many different ways to identify diseases like cancer and even forecast health difficulties persons might encounter depending on their DNA by combining data from scanners, patient biology, and analysis. One illustration is IBM Watson for Oncology that combines patient medical data and histories to provide a customized treatment strategy. On the basis of a person's test findings, responses to prior medications, and previous client records for drug reactions, AI is also being employed to create individualized medicine therapies.

1.4.4 Identifying Clinical Trial Candidates

Selecting patients to take part in the trials is another application of AI in the pharmaceutical sector, in addition to aiding in the interpretation of clinical trial data. AI can scan genetic data to find the right patient population for an experiment and choose the best sample level employing strong predictive analytics. Free-form language patients enter into clinical trial tools and unstructured data like medical reports and intake forms can both be read by AI technology.

1.4.5 Predicting Treatment Results

One of the numerous time and cost-saving applications of AI is its ability to match drug therapies with individual patients, minimizing effort that typically involves trial and error. ML algorithms have the capacity to anticipate a person's response to potential drug therapies by drawing possible connections among variables that might be influencing the results, such as the body's ability to absorb the drugs, their distribution all over the body, and an individual's metabolic activity [31, 32].

1.4.6 Drug Dosage

Drug companies face a significant challenge in assuring that willing participants in clinical research adhere to a pharmaceutical study procedure. Volunteers in a drug study must be withdrawn from the trial if they do not follow the trial guidelines, or the results of the observation may be tainted. Making sure that participants take the required dosage of the researched medicine at the scheduled times is one of the critical components of a successful clinical study. Having a method to guarantee drug compliance is crucial for this reason. AI can separate the excellent apples from the bad using both techniques for analysing testing results and monitoring systems.

1.4.7 Marketing

Given that the pharmaceutical industry generates income, AI has the potential to be a useful tool in pharmaceutical marketing. Pharma companies can use AI to research and develop unique marketing strategies that ensure high sales and brand awareness. With the help of AI, organizations can map the user experience and identify which lead generation technique drove visitors to their website and ultimately persuaded them to make a purchase. As a result, pharmaceutical companies may focus more on the marketing strategies that generate the most leads and revenue expansion.

1.5 AI AND ML FOR ROBOT-ASSISTED SURGERIES

Who would have thought that one day, technology would advance to the point where robots might perform surgery? Robotic surgical assistants (SRAs) have entered the operating room (OR) thanks to technological advancements and AI, helping surgeons while they perform operations. Robot-assisted surgery is a groundbreaking idea that effectively combines the knowledge of experienced surgeons with cutting-edge digital technology (see Figure 1.8) [33].

1.5.1 The Role of Robotic Surgery

Success in the operating theatre depends on accuracy, prompt help, and the surgeon's skill. That is where robotic surgery is useful. A robotic surgery is a computer-controlled tool that assists in manipulating and directing surgical equipment so the surgeon can focus entirely on the delicate aspects of the operation. Robotic surgery provides unmatched speed and motion control, reducing hand trembling and preventing unintentional hand motions [34]. This enhances the surgeon's ability to perform with more flexibility, accuracy, mobility, and control, particularly throughout physically difficult surgeries as shown in Figure 1.8.

1.5.2 Robotic Surgery and AI

Modern surgery is changing the way we think about it because to AI-based procedures and the accuracy and command of robotic devices. With the aid of deep ML data, it has the ability to make the medium of communication among surgical robots and surgeons simpler. For example, it can detect a surgeon's actions and behaviours throughout surgery and translate those trends into commands that the robot can follow. AI therefore gathers data over time by observing operations performed by surgeons.

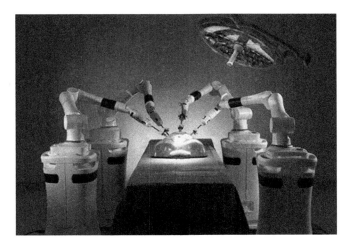

FIGURE 1.8 AI enabled robotic surgery.

AI helps surgical robots analyze and carry out cognitive tasks including judgement, problem-solving, voice recognition, and many others with the aid of all the gathered data and techniques. Additionally, AI and robotic surgery assist in equipment positioning, cancer detection, and scanning and surgery analysis.

1.5.3 Applications of Robotic Surgery

Over time, the uses for robotic surgery have progressively increased. Robotic surgery can more precisely mimic the patterns and movements of the surgeon. In hair transplant treatments, for example, surgical robots may remove hair follicles and implant them into the balding area on the scalp while retaining the proper force and pace. The use of robotic surgery during laparoscopy during kidney surgery is rapidly gaining popularity. Robotic laparoscopic surgery is less intrusive than traditional open surgery, which leads to less loss of blood and a faster recovery [35].

Robot-assisted surgery is the preferable choice over traditional cardiac surgery for the treatment of cardiac illness and disease. Robotic surgery, as opposed to traditional surgery, can be done using tiny incisions made between the ribs without requiring the rib cage to be split.

1.5.4 Potential Effects of AI in Surgeries

As discussed, AI innovations created by computer researchers will have a huge impact on medical treatment, in particular surgeries. Examples such as cognitive robotics, context-aware assistance, and predictive modeling illustrate the diverse applications of AI in healthcare. However, decision support systems that gather medical data to provide clinical recommendations still have limited impact on actual clinical practice. Although these systems can offer valuable insights, challenges in integration, user trust, and clinical workflow adaptation often restrict their effectiveness in routine medical settings [36–38].

In a similar vein, AI might enhance methods of predictive analytics that forecast survival following pancreatic cancer or post-operative complications like incisional hernia. These extra details can be made available to the correct patient at the correct time by perspective help systems. The surgical workflows may then be enhanced by analyzing the procedure to identify key operating room steps and categorize it employing ML. AI may shift surgical robotics practices toward cognitive surgical robots. Currently, surgical robots utilized in clinics are merely tele-manipulators with no autonomously function. Robotic devices for scenario automatic needle insertion have been explored in research.

1.6 VIRTUAL NURSING ASSISTANTS

One way to meet the increasing demands on the healthcare delivery system, longer life expectancies, aging of the population, and the world's increases emphasis on

AI and Machine Learning in Modern Healthcare 15

FIGURE 1.9 Virtual nursing assistants.

overall health is the use of virtual nurse assistants (VNAs). These are automated analytical and information systems designed to improve the standard and effectiveness of care while streamlining the work of medical staff in medical facilities, nursing homes, hospitals, and rehabilitation centers. Figure 1.9 shows a prototype of VNAs.

Patients can receive individualized experiences from a virtual assistant driven by AI. They can use it to diagnose their condition according to the symptoms, keep track of their health, make doctor's visits, and much more. VNAs can facilitate via online mode to any patient rather than looking for the origins of the problems/physical you are experiencing [39–41]. If you have common illnesses or concerns, the VNA will not just offer your health assistance but also give you the option to make an appointment to see a physician or an expert virtually.

Furthermore, VNAs are available 24/7. The advancement of chronic illnesses can be halted by using AI technology to improve patient engagement and self-management skills.

1.6.1 WHAT DO VNAs DO?

Virtual nursing assistants elevate expertise beyond past trends like telenurses and teledocs. Apps like MedWhat offer a chatbot environment in which users may register their moderate exercise and fitness routines and express inquiries, such as "What are the indications of the flu?" rather than speaking with a healthcare expert over the phone or via video conference.

Similar to GreatCall, several applications use wearable technology with AI to better understand the health of their target population. Some apps also offer telenurses and teledocs as part of their services, along with guidance on how to handle medical emergencies such as giving CPR [42–44].

1.7 CONCLUSION

To increase the efficiency of administering healthcare services and making medical decisions, AI integration is necessary. We examined some of the moral dilemmas that AI use cases face while encouraging quick adoption and ongoing integration into the healthcare system. The next ten years will be spent considering the value and foresight society can gain from these information assets, how AI can be used to interpret them to improve clinical outcomes, and the ongoing creation of new digital assets and tools. It is clear that we are at a turning point in the development of technology and medical practice. Although there is a lot of promise, there are also a lot of challenges that need to be overcome in terms of the real world and the reach of innovative applications. Achieving this goal will require a rise in integrated research in the field of AI applications in health care. In addition, we need to concentrate on developing the skills of healthcare professionals and fresh talent so that they can be proficient in the digital age and embrace rather than fear an AI-enhanced health system.

REFERENCES

1. Davahli MR, Karwowski W, Fiok K, Wan T, Parsaei HR. Controlling safety of artificial intelligence-based systems in healthcare. *Symmetry* 2021; 13:102.
2. Haque A, Milstein A, Fei-Fei L. Illuminating the dark spaces of healthcare with ambient intelligence. *Nature* 2020; 585:193–202.
3. Nachev P, Herron D, McNally N, Rees G, Williams B. Redefining the research hospital. *NPJ Digit Med* 2019; 2:119.
4. Wang A, Nguyen D, Sridhar AR, Gollakota S. Using smart speakers to contactlessly monitor heart rhythms. *Commun Biol* 2021; 4:319.
5. Muehlematter UJ, Daniore P, Vokinger KN. Approval of artificial intelligence and machine learning-based medical devices in the USA and Europe (2015–20): a comparative analysis. *Lancet Digital Health* 2021; 3:e195–203.
6. Esteva A, Robicquet A, Ramsundar B, Kuleshov V, DePristo M, Chou K, Cui C, Corrado G, Thrun S, Dean J. A guide to deep learning in healthcare. *Nat Med* 2019; 25:24–9.
7. Strodthoff N, Strodthoff C. Detecting and interpreting myocardial infarction using fully convolutional neural networks. *Physiological Measurement* 2019; 40:015001.
8. Bellemo V, Lim ZW, Lim G, Nguyen QD, Xie Y, Yip MY, Hamzah H, Ho J, Lee XQ, Hsu W, Lee ML. Artificial intelligence using deep learning to screen for referable and vision-threatening diabetic retinopathy in Africa: a clinical validation study. *Lancet Digit Health* 2019; 1:e35–44.
9. Ting DSW, Pasquale LR, Peng L, Campbell JP, Lee AY, Raman R, Tan GS, Schmetterer L, Keane PA, Wong TY. Artificial intelligence and deep learning in ophthalmology. *Br J Ophthalmol* 2019; 103:167–75.
10. Raumviboonsuk P, Krause J, Chotcomwongse P, Sayres R, Raman R, Widner K, Campana BJ, Phene S, Hemarat K, Tadarati M, Silpa-Archa S. Deep learning versus human graders for classifying diabetic retinopathy severity in a nationwide screening program. *NPJ Digit Med* 2019; 2:25.
11. Xie Y, Nguyen QD, Hamzah H, Lim G, Bellemo V, Gunasekeran DV, Yip MY, Lee XQ, Hsu W, Lee ML, Tan CS. Artificial intelligence for teleophthalmology-based diabetic retinopathy screening in a national programme: an economic analysis modelling study. *Lancet Digit Health* 2020; 2:e240–9.

12. Oktay O, Nanavati J, Schwaighofer A, Carter D, Bristow M, Tanno R, Jena R, Barnett G, Noble D, Rimmer Y, Glocker B. Evaluation of deep learning to augment image-guided radiotherapy for head and neck and prostate cancers. *JAMA Netw Open* 2020; 3:e2027426.
13. Senior AW, Evans R, Jumper J, Kirkpatrick J, Sifre L, Green T, Qin C, Žídek A, Nelson AW, Bridgland A, Penedones H. Improved protein structure prediction using potentials from deep learning. *Nature* 2020; 577:706–10.
14. Char DS, Shah NH, Magnus D. Implementing machine learning in health care – addressing ethical challenges. *N Engl J Med* 2018; 378:981–3.
15. Aicha AN, Englebienne G, van Schooten KS, Pijnappels M, Kröse B. Deep learning to predict falls in older adults based on daily-Life trunk accelerometry. *Sensors* 2018; 18:1654.
16. Shimabukuro D, Barton CW, Feldman MD, Mataraso SJ, Das R. Effect of a machine learning-based severe sepsis prediction algorithm on patient survival and hospital length of stay: a randomised clinical trial. *BMJ Open Respir Res* 2017; 4:e000234.
17. Schmidt-Erfurth U, Bogunovic H, Sadeghipour A, Schlegl T, Langs G, Gerendas BS, Osborne A, Waldstein SM. Machine learning to analyze the prognostic value of current imaging biomarkers in neovascular age-related macular degeneration. *Opthamology Retina* 2018; 2:24–30.
18. Vial A, Stirling D, Field M, Ros M, Ritz C, Carolan M, Holloway L, Miller AA. The role of deep learning and radiomic feature extraction in cancer-specific predictive modelling: a review. *Transl Cancer Res* 2018; 7:803–16.
19. Lee SI, Celik S, Logsdon BA, Lundberg SM, Martins TJ, Oehler VG, Estey EH, Miller CP, Chien S, Dai J, Saxena A, Blau CA, Becker PS. A machine learning approach to integrate big data for precision medicine in acute myeloid leukemia. *Nat Commun* 2018; 9:42.
20. Kaplan A, Haenlein M. Siri, Siri, in my hand: Who's the fairest in the land? On the interpretations, illustrations, and implications of artificial intelligence. *Bus Horiz* 2019; 62(1):15–25.
21. Singh D, Singh D, Manju, Gupta U. Smart Healthcare: A Breakthrough in the Growth of Technologies. In *Artificial Intelligence-based Healthcare Systems* 2023 Oct 27 (pp. 73–85). Cham: Springer Nature Switzerland.
22. Coombs C, Hislop D, Taneva SK, Barnard S. The strategic impacts of intelligent automation for knowledge and service work: an interdisciplinary review. *J Strateg Inf Syst* 2020; 29:101600.
23. Dreyer K, Allen B. Artificial intelligence in health care: brave new world or golden opportunity? *J Am Coll Radiol* 2018; 15(4):655–7.
24. Laï M-C, Brian M, Mamzer M-F. Perceptions of artificial intelligence in healthcare: findings from a qualitative survey study among actors in France. *J Transl Med* 2020; 18(1):1–13.
25. Esteva A, Robicquet A, Ramsundar B, Kuleshov V, DePristo M, Chou K, Cui C, Corrado G, Thrun S, Dean J. A guide to deep learning in healthcare. *Nat Med.* 2019; 25(1):24–9.
26. Turja T, Aaltonen I, Taipale S, Oksanen A. Robot acceptance model for care (RAM-care): a principled approach to the intention to use care robots. *Inf Manage* 2019; 57(5):103220.
27. Sohn K, Kwon O. Technology acceptance theories and factors influencing artificial intelligence-based intelligent products. *Telematics Inform* 2020; 47:101324.
28. Lu L, Cai R, Gursoy D. Developing and validating a service robot integration willingness scale. *Int J Hosp Manag* 2019; 80:36–51.

29. Reddy S, Allan S, Coghlan S, Cooper P. A governance model for the application of AI in health care. *J Am Med Inform Assoc.* 2020; 27(3):491–7.
30. Dwivedi YK, Hughes L, Ismagilova E, Aarts G, Coombs C, Crick T, Duan Y, Dwivedi R, Edwards J, Eirug A, Galanos V. Artificial intelligence (AI): multidisciplinary perspectives on emerging challenges, opportunities, and agenda for research, practice and policy. *Int J Inf Manag.* 2019; 57:101994.
31. Zandi D, Reis A, Vayena E, Goodman K. New ethical challenges of digital technologies, machine learning and artificial intelligence in public health: a call for papers. *Bull World Health Organ* 2019; 97(1):2.
32. Edwards SD. The HeartMath coherence model: implications and challenges for artificial intelligence and robotics. *AI Soc* 2019; 34(4):899–905.
33. Waring J, Lindvall C, Umeton R. Automated machine learning: review of the state-of-the-art and opportunities for healthcare. *Artif Intell Med* 2020; 104:101822.
34. Parikh RB, Obermeyer Z, Navathe AS. Regulation of predictive analytics in medicine. *Science* 2019; 363(6429):810–2.
35. Beregi J, Zins M, Masson J, Cart P, Bartoli J, Silberman B, Boudghene F, Meder JF. Radiology and artificial intelligence: an opportunity for our specialty. *Diagn Interv Imaging* 2018; 99(11):677.
36. Ain QU, Aleksandrova A, Roessler FD, Ballester PJ. Machine-learning scoring functions to improve structure-based binding affinity prediction and virtual screening. *WIREs Comput Mol Sci* 2015; 5: 405–424.
37. Gupta U, Pranav A, Kohli A, Ghosh S, Singh D. The Contribution of Artificial Intelligence to Drug Discovery: Current Progress and Prospects for the Future, *Microbial Data Intelligence and Computational Techniques for Sustainable Computing.* 2024 Mar 1:1–23.
38. DiMasi JA, Grabowski HG, Hansen RW. Innovation in the pharmaceutical industry: New estimates of R&D costs. *J Health Econo* 2016; 47: 20–33.
39. Du T, Liao L, Wu CH, Sun B. Prediction of residue-residue contact matrix foprotein-protein interaction with Fisher score features and deep learning. *Methods* 2016; 110: 97–105.
40. Sharma M, Deswal S, Gupta U, Tabassum M, Lawal I, editors. *Soft Computing Techniques in Connected Healthcare Systems.* CRC Press; 2023 Dec 20.
41. Gómez-Bombarelli R, Wei JN, Duvenaud D, Hernández-Lobato JM, Sánchez-Lengeling B, Sheberla D, Aguilera-Iparraguirre J, Hirzel TD, Adams RP, Aspuru-Guzik A. Automatic chemical design using a data-driven continuous representation of molecules. *ACS Cent Sci* 2018; 4: 268–276.
42. Gupta M, Srivastava D, Pantola D, Gupta U. Brain tumor detection using improved Otsu's thresholding method and supervised learning techniques at early stage. In *Proceedings of Emerging Trends and Technologies on Intelligent Systems: ETTIS 2022* 2022 Nov 16 (pp. 271–281). Singapore: Springer Nature Singapore.
43. Mishra S, Ahmed T, Sayeed MA, Gupta U. Artificial Neural Network Model for Automated Medical Diagnosis. In *Soft Computing Techniques in Connected Healthcare Systems* (pp. 34–54). CRC Press.
44. Malik K, Sadawarti H, Sharma M, Gupta U, Tiwari P, editors. *Computational Techniques in Neuroscience.* CRC Press; 2023 Nov 14.

2 Telemedicine and Remote Prenatal Care
A Soft Computing Approach

Pradnya S. Mehta, Pranav Bafna, Atharv Sawant, Pratik Patil, and Niraj Pandit

ORGANIZATION OF THE CHAPTER

The chapter begins with an introduction to telemedicine in Section 2.1, followed by an analysis of soft computing in telemedicine in Section 2.2. Section 2.3 explores the application of deep learning in remote prenatal monitoring, while Section 2.4 contains challenges and future directions as well as case study discussions on legal and ethical implications.

2.1 INTRODUCTION TO TELEMEDICINE

Telemedicine, or telehealth, is an innovative healthcare approach utilizing technology to facilitate the connection between patients and healthcare providers, eliminating the need for physical presence. This progressive method enables the secure electronic exchange of medical records, negating the necessity for in-person visits. Encompassing a broad spectrum of healthcare services, including video conferencing, remote patient care, and mobile health (mHealth), telehealth provides patients with the convenience of accessing timely medical services, irrespective of their location. Through telehealth, patients can remotely receive consultation, diagnosis, treatment, and follow-up care, empowering them to actively engage in their health management(Nguyen, 2021). The emphasis on patient-centred care within telehealth contributes to more effective management of chronic conditions and improved overall health outcomes. Importantly, it breaks down barriers to healthcare access for individuals in remote or underserved areas, addressing challenges such as transportation issues and prolonged waiting times. The patient-centric telehealth approaches represent a transformative force in healthcare, empowering individuals to play an active role in their health, enhancing access to care, and promoting seamless continuity of care.

2.1.1 Evolution of Telemedicine

The historical development of telemedicine in the realm of prenatal care has undergone a transformative journey marked by significant milestones and technological advancements as shown in Figure 2.1.

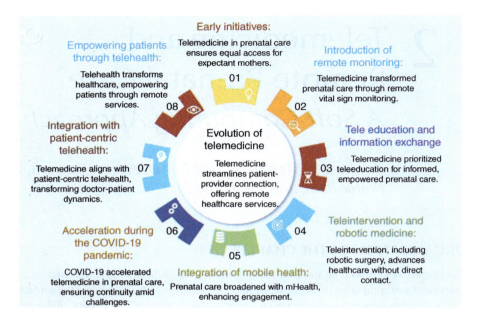

FIGURE 2.1 Progression in telemedicine.

Technological developments and major turning points have shaped the transformative path of telemedicine's historical development in the field of prenatal care. Telemedicine has consistently changed to address the evolving demands of pregnant women and healthcare professionals, starting with its early initiatives and continuing with the recent acceleration during the COVID-19 epidemic. The major phases of this development are listed below, with an emphasis on how each one has influenced the condition of telemedicine in prenatal care today:

1. **Early Initiatives**
   Telemedicine origins in prenatal care trace back to a visionary era marked by earnest efforts to connect healthcare providers with pregnant women in distant locales. The advent of telecommunication technologies not only shattered geographical barriers but also laid the groundwork for profound virtual connections. This trailblazing approach aimed to ensure that expectant mothers, regardless of their geographical location, had equal access to essential healthcare resources and expertise, forging a path toward inclusive prenatal care.
2. **Remote Monitoring**
   A pivotal leap in telemedicine's journey in prenatal care occurred with the introduction of remote monitoring technologies. This transformative innovation empowered healthcare providers to remotely track a spectrum of vital signs, from blood pressure (BP) to oxygen saturation (SpO2). This shift elevated the precision of maternal health monitoring. The ability to monitor

these vital signs remotely not only enhanced patient convenience but also underscored a commitment to proactive and personalized maternal care.

3. **Teleeducation and Information Exchange**
As telemedicine in prenatal care evolved, a critical emphasis was placed on teleeducation, facilitating seamless information exchange between healthcare professionals and expectant mothers. This phase aimed to make relevant and empowering information widely accessible, fostering transparent communication and consequently, enhancing the overall quality of prenatal care through informed decision-making. The goal was not just medical literacy but also ensuring that pregnant women felt empowered and actively engaged in their healthcare journey.

4. **Teleintervention and Robotic Medicine**
Telemedicine continued to evolve with the integration of teleintervention, enabling medical procedures, including surgeries, without direct patient-doctor contact. Robotic medicine emerged as a pioneering approach, showcasing the potential for advanced interventions while maintaining a physical distance. This era represented a convergence of technological innovation and a patient-centric approach to healthcare delivery as indicated in Figure 2.1, where cutting-edge procedures became more accessible, ensuring optimal outcomes while prioritizing patient safety and comfort (Nguyen, 2021).

5. **Integration of Mobile Health (mHealth)**
The landscape of telemedicine in prenatal care expanded significantly with the ascendancy of mobile health (mHealth) technologies. Mobile applications and devices seamlessly integrated into the prenatal care framework, empowering pregnant women to actively participate in their healthcare journey. These technologies not only facilitated convenient and frequent engagements but also promoted a holistic approach to prenatal well-being, transforming smartphones into personalized healthcare companions and enhancing patient agency in the management of their maternal health.

6. **Acceleration During the COVID-19 Pandemic**
The global upheaval brought about by the COVID-19 pandemic acted as a pivotal catalyst, hastening the expansion of telemedicine in prenatal care. In response to the imperative to minimize exposure, adhere to social distancing measures, and preserve healthcare resources, virtual healthcare practices became ubiquitous. Telemedicine emerged as an indispensable tool, ensuring uninterrupted access to prenatal care amid unprecedented challenges. This acceleration underscored the resilience of telemedicine and its capacity to adapt swiftly to ensure the continuity of care, proving to be a lifeline for expectant mothers during uncertain times.

7. **Integration with Patient-Centric Telehealth**
In the broader healthcare context, telemedicine seamlessly aligned with patient-centric telehealth approaches, revolutionizing the doctor-patient dynamic. This innovative healthcare model harnessed technology to create a seamless connection between patients and healthcare providers, eliminating the need for in-person visits. Services such as video conferencing, remote patient care, and mobile health (mHealth) collectively ushered in a new era of

timely, convenient, and patient-friendly medical services, transcending geographical boundaries. This was a shift from the traditional healthcare model to one that prioritized patient comfort, accessibility, and active participation in their healthcare decisions.
8. **Empowering Patients through Telehealth**
Telehealth emerged as a transformative force, placing the reins of healthcare management firmly in the hands of patients. This patient-centric model enabled remote consultation, diagnosis, treatment, and follow-up care, actively engaging individuals in their health journey. Particularly effective in managing chronic conditions, telehealth contributed to improve overall health outcomes. This patient empowerment not only enhanced health outcomes but also cultivated a sense of partnership between healthcare providers and patients, placing the focus on collaborative decision-making and personalized care plans (Heřman et.al, 2022).

2.1.2 Importance of Prenatal Care

Prenatal care is crucial for safeguarding the health of expectant mothers and their unborn children, significantly contributing to positive maternal and fatal outcomes. Research findings offer valuable insights into the multifaceted impact of adequate prenatal care on maternal and fatal health.

1. **Early Identification and Mitigation of Risks**
Adequate prenatal care acts proactively to identify and mitigate potential risks associated with pregnancy. Regular check-ups and screenings empower healthcare providers to monitor maternal health, promptly detecting complications or risk factors. Timely interventions during prenatal care can prevent or manage health issues, reducing the likelihood of adverse outcomes for both mother and child. By prioritizing early identification and intervention, prenatal care serves as a cornerstone for a healthier pregnancy journey.
2. **Optimal Fatal Development**
Consistent prenatal care is highlighted in research for its contribution to optimal fatal development. Monitoring fatal growth through routine check-ups allows healthcare professionals to address any abnormalities promptly. Adequate prenatal care ensures expectant mothers receive guidance on nutrition, lifestyle, and prenatal behaviours, promoting a healthy environment for fatal growth. The focus on optimal fatal development underscores the critical role that consistent and comprehensive prenatal care plays in ensuring the well-being of the unborn child (Sword, 2012).
3. **Emotional and Psychological Support**
Beyond physical aspects, prenatal care addresses the emotional and psychological well-being of expectant mothers. Supportive environments during prenatal check-ups reduce stress, anxiety, and mental health challenges, fostering a positive pregnancy experience. Recognizing the holistic nature of maternal health, prenatal care endeavours to create a supportive space that acknowledges and addresses emotional and psychological aspects.

2.1.2.1 Challenges in Accessing Prenatal Care, Especially in Remote or Underserved Areas

1. **Societal Barriers**
   Cultural norms, financial constraints, and transportation issues pose significant barriers to accessing prenatal care. Lack of awareness and cultural beliefs may lead to delayed or limited utilization of available healthcare services, particularly in vulnerable populations. Addressing societal barriers requires community-specific strategies, acknowledging and respecting cultural diversity to ensure inclusive prenatal care.

2. **Maternal Perspectives**
   Personal factors like poor motivation, fear of medical procedures, and concerns about societal judgments can hinder timely and comprehensive prenatal care. Addressing these barriers is crucial to ensure that maternal perspectives do not impede access to essential healthcare services during pregnancy. Recognizing and addressing maternal perspectives fosters a patient-centred approach, promoting a collaborative and supportive healthcare journey.

3. **Structural Hurdles**
   Healthcare facilities' locations, operational hours, and financial considerations present structural challenges to accessing prenatal care. Long wait times, a lack of child-friendly facilities, and language barriers exacerbate difficulties in accessing care, especially in remote or underserved areas. Overcoming structural hurdles necessitates a systemic approach, involving healthcare infrastructure improvements, accessibility enhancements, and tailored solutions for diverse communities.

4. **Provider-Patient Dynamics**
   Interactions with healthcare providers can either facilitate or hinder access to prenatal care. Improving communication, cultural competence, and sensitivity is essential to ensure the continuity of care, particularly in areas with limited healthcare resources. Fostering positive provider-patient dynamics contributes to a more supportive and inclusive prenatal care experience, addressing potential barriers associated with interpersonal interactions (Phillippi, 2009).

2.1.3 ROLE OF TELEMEDICINE IN PRENATAL CARE

Embarking on the enchanting journey of motherhood is a thrilling yet challenging adventure, and the advent of telemedicine is redefining the landscape of prenatal care. It explores the role played by telemedicine, addressing geographical and logistical hurdles while highlighting its potential for early detection through remote monitoring.

2.1.3.1 Overcoming Geographical Barriers

Traditional prenatal care with various areas often faced insurmountable challenges due to distance and limited access to specialized services. Recent studies shed light on the experiences of pregnant mothers in remote areas during the COVID-19 era.

Telemedicine emerges as a transformative solution, breaking down geographical barriers and connecting mothers with quality healthcare, irrespective of their physical location. This digital approach ensures that no mother feels isolated in her journey and empowers healthcare providers to extend their expertise universally. It is more than just a telehealth visit; it is a revolutionary approach to making quality prenatal care universally accessible.

2.1.3.2 Logistical Hurdles, Meet Telemedicine's Flexibility

The journey of pregnancy often involves juggling appointments with doctor visits, managing childcare, or dealing with connectivity issues – logistical hurdles familiar to many mothers. Telemedicine addresses these struggles, including poor internet connectivity and a lack of privacy during in-person visits. With its innovative audio and video solutions, telemedicine becomes the flexible ally every expectant mom deserves. Surprisingly, a significant chunk of respondents found telemedicine visits not only accessible but also remarkably easy. (Morgan, 2022)

2.1.3.3 Remote Monitoring: A Peek into the Future of Prenatal Care

Delving into the captivating world of high-risk pregnancies, conditions like anxiety, asthma, or carrying twins were found to influence preferences toward telehealth visits. This opens the door to the exciting realm of early detection and intervention through remote monitoring. Envision having vital signs monitored remotely, allowing healthcare providers to spot potential issues early in the pregnancy journey. The integration of cutting-edge technology propels telemedicine into the future, offering a comprehensive look into the health of both mom and baby.

2.1.3.4 Case Study of Telemedicine Evolution in Prenatal Care

Telemedicine's evolution in prenatal care is not just about the present moment, it is about creating a legacy of accessible, tech-driven healthcare for future generations. The telemedicine revolution is just beginning, and its potential is vast. The landscape of prenatal care is set to transform, embracing technology to nurture healthier beginnings for both mothers and their little ones. In this digital era, telemedicine is not just a tool, it is guiding towards a more accessible, equitable, and proactive approach to prenatal wellness (Furlepa, 2022).

2.1.3.4.1 Case Study

In rural areas, two pregnant women used telemedicine for prenatal care. One received medical advice, medication management, and emotional support via video conferencing. The other transmitted blood glucose data for monitoring and treatment adjustments. Both benefited from remote care, overcoming travel challenges and ensuring timely, high-quality care. Telemedicine empowered them to manage pregnancies effectively despite geographical barriers and health conditions, showcasing its potential in expanding healthcare access and improving quality of care during pregnancy.

2.2 SOFT COMPUTING IN HEALTHCARE

2.2.1 Soft Computing Overview

1. **The Essence of Soft Computing in Healthcare: A Revolutionary Approach**
   Soft computing has transformed problem-solving by embracing approximations to address complex issues with imprecise solutions. Unlike conventional computing methods, soft computing techniques exhibit tolerance towards imprecision, uncertainty, partial truth, and approximations. Recognized for their tractability and low solution cost, these techniques find widespread application across various fields. In the healthcare domain, soft computing stands out for its impact on medical diagnosis and prediction, introducing innovative solutions to elevate patient care and outcomes (Gambhir, 2022).

2. **The Role of Soft Computing in Healthcare**
   The healthcare industry, dedicated to the detection, treatment, analysis, prediction, and prevention of diseases, significantly benefits from the application of soft computing techniques. Soft computing plays a crucial role in addressing challenges such as healthcare resource shortages, insufficient diagnoses, and the need for cost-effective solutions. Using soft computing methods, healthcare organizations can extract valuable insights from vast data stores, enabling the development of efficient and effective healthcare systems.

3. **Soft Computing usage in Medical for Decision Making**
   Medical decision-making stands as a critical facet of healthcare, demanding accurate diagnoses and well-crafted treatment plans. Soft computing techniques emerge as intelligent tools supporting medical decision-making processes, surpassing traditional rule-based reasoning systems. These techniques empower healthcare practitioners to navigate inherently vague observations and draw meaningful inferences (Diwaker, 2022).

2.2.2 Understanding Software Informatics in Healthcare

The integration of procedures that allow the human brain to reason and learn from ambiguous or inaccurate information is the focus of software informatics in the healthcare industry as depicted in Figure 2.2. In contrast to sophisticated computational techniques that depend on concepts and accurate outcomes, fuzzy concepts can handle unclear, partial, and even contradictory data and still produce useful results as shown in Figure 2.2.

1. **Fuzzy Logic**
   The interpretation of diagnostic data obtained remotely, such as foetal monitoring data, maternal biomarkers, and ultrasound images, can be aided by fuzzy logic. It makes it possible to evaluate the health of the mother and the fetus by integrating a variety of factors and expert knowledge. Compared to typical binary techniques, fuzzy logic can provide more nuanced insights by handling the inherent ambiguity in diagnostic data. Personalized care

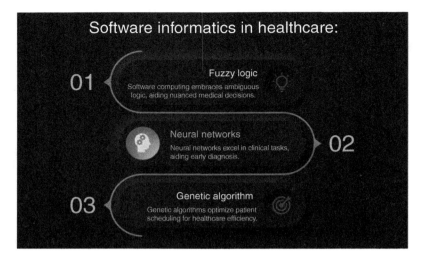

FIGURE 2.2 The flow of software informatics in healthcare.

plans that are suited to each patient's unique requirements and features are made possible by fuzzy logic. Fuzzy logic models take into account linguistic characteristics and fuzzy rules to produce customized suggestions for prenatal care that take into account medical history, cultural considerations, and maternal preferences. This tailored strategy optimizes health outcomes while increasing patient satisfaction and participation (de Silva et.al, 2003).

2. **Neural Networks**
   Neural networks, which draw inspiration from the architecture and operations of the human brain, are adept at extracting patterns from data. These sophisticated algorithms can carry out crucial tasks for clinical analysis, diagnosis, and prognosis because of their exceptional ability to recognize intricate patterns in data. Physicians can use a multitude of data to their advantage in order to identify illnesses early, create treatment programs, and eventually enhance patient outcomes as seen in Figure 2.2. These strong networks are valuable instruments in today's healthcare because they keep learning and developing.

3. **Genetic Algorithms**
   The natural selection process serves as the inspiration for the genetic algorithm, an optimization method used to solve difficult problems. The biological evolutionary process in which the most productive individuals will survive and procreate is the model for these algorithms. In order to minimize wait times, maximize capacity, and deliver care on time, it can be used to create a patient scheduling procedure. Various appointments are rated by the algorithm according to factors like patient urgency, provider location, and geographic distance. The algorithm finds efficient solutions that improve patient satisfaction and healthcare quality by making the most use of available time (de Silva et.al, 2014).

2.2.2.1 Example

In a hospital, a young patient awaits her diagnosis as her doctor utilizes fuzzy logic, a revolutionary tool, to decipher her complex symptoms. She presents with ambiguous symptoms like fatigue, mild pain, and intermittent nausea, challenging the diagnostic process. The fuzzy logic system mimics human cognitive processes, interpreting imprecise information to make informed decisions. It assigns degrees of membership to various diagnostic categories based on the patient's symptoms, allowing for nuanced analysis. Drawing from a vast medical database, the system generates possible diagnoses and their probabilities, enabling informed discussions between him and her doctor about treatment options. Fuzzy logic offers advantages in dealing with uncertain symptoms, analysing multiple symptoms comprehensively, and incorporating expert knowledge into diagnostics. Its use empowers healthcare professionals to personalize treatment plans and improve patient outcomes, showcasing the transformative potential of software informatics in healthcare delivery. This patient's experience highlights the impact of software informatics on enhancing healthcare quality.

2.2.3 APPLICATIONS AND TECHNIQUES OF SOFT COMPUTING IN HEALTHCARE

1. **Harnessing Adaptability and Handling Uncertainty**
   The integration of soft computing approaches has become pivotal in addressing complex challenges. Soft computing, inspired by the human mind, has made significant strides in healthcare, offering adaptable solutions that effectively handle uncertainty and imprecision.
2. **Soft Computing Approaches in Healthcare**
   Soft computing approaches have gained prominence in healthcare, proving their effectiveness in diagnosing and predicting diseases. Notable techniques include Particle Swarm Optimization (PSO), Genetic Algorithms (GAs), Artificial Neural Networks (ANNs), and Support Vector Machines (SVMs). These computational intelligence tools analyse healthcare data, offering accurate diagnoses (Sharma, 2021).
3. **Particle Swarm Optimization (PSO)**
   PSO, inspired by collective behaviour in nature, navigates a multidimensional search space to find optimal solutions. In healthcare, PSO enhances feature selection, parameter optimization, and medical image analysis, improving model performance and accuracy.
4. **Support Vector Machines (SVMs)**
   SVMs classify data by finding optimal hyperplanes in high-dimensional space. In healthcare, they aid in disease diagnosis, risk assessment, and treatment prediction, particularly beneficial for complex datasets and handling nonlinear relationships (Devi et.al, 2021; Gupta & Gupta, 2023).

2.2.4 THE IMPACT OF TELEMEDICINE ON PREGNANCY AND POSTPARTUM CARE

1. **Remote Consultation: Bridging Gaps in Maternal Healthcare Access**
   Telemedicine is catalysing a transformative shift in pregnancy and postpartum care, driven by a compelling need for innovation. This paradigmatic

change is explored through various dimensions, starting with the integral use of telemedicine in both prenatal and postnatal care. The subsequent sections illuminate the multifaceted advantages, including home monitoring of high-risk pregnancies, remote communication with doctors and breastfeeding support, and the ability to connect with experts regardless of geographical constraints.

2. **Use of telemedicine in prenatal and postnatal care**
   Prenatal and postnatal care is changing dramatically thanks in large part to the role of telemedicine. Moms can use telemedicine to arrange virtual visits with their doctors from the comfort of their homes. Long wait times, travel time, and the possibility of cross-contamination in medical facilities are all eliminated as a result. Additionally, by making it easier to monitor foetal vital signs and health at home, telemedicine empowers women to take an active role in health management. In addition, women can receive the best possible care no matter where they are by scheduling an immediate consultation with a remote specialist who possesses the utmost expertise in the field. During this crucial stage of life, telemedicine integration continues to support moms' and their partners' mental health. All pregnancy services are optimized and the quality of care is raised when these services are integrated (Arias, 2022).

3. **Reduce the need to travel for prenatal care**
   Prenatal care in hospitals can be less expensive and more convenient when provided virtually through telemedicine, which also provides in-person visits. This invention makes childcare more convenient and appropriate for hectic schedules by removing the burden of managing children, taking time off from work, and traveling. According to research, telemedicine programs for prenatal care can offer safe and affordable alternatives to protect moms' and babies' health while also producing pregnancy outcomes that are comparable to in-person care.

4. **Home monitoring of high-risk pregnancies**
   Healthcare has been transformed by telemedicine, particularly for those with high-risk pregnancies. By enabling patients to take part in their care from the convenience of their homes, it empowers them. Telemedicine enables patients to better manage their health and keep an eye on conditions like diabetes and hypertension through home care. This collaboration advances both health and self-efficacy. Telemedicine-assisted home care has shown to be especially beneficial in cases of pregnancy risk; it enables timely intervention and regular monitoring, ultimately improving the mother and the foetus (Gambhir S, 2016).

5. **Communication with Doctors and Breastfeeding Support**
   The incorporation of digital platforms and applications has transformed the way patients and physicians communicate, facilitating the acquisition and retention of crucial data. Breastfeeding is a special kind of telemedicine that has grown to be the mainstay of breastfeeding assistance, offering mothers vital direction and encouragement throughout their breastfeeding journey. Through

Telemedicine and Remote Prenatal Care   29

telefeeding, moms can overcome time and location barriers by interacting with lactation consultants and other healthcare professionals from the comfort of their own homes. These services offer timely assistance, prompt problem-solving, suitable guidance, and constructive support to ensure a successful breastfeeding experience (Galle, 2021).

2.2.4.1 Soft Computing in Remote Prenatal Care

Soft computing refers to a set of computational techniques that mimic human-like decision-making processes. In the context of remote prenatal care, these techniques enable healthcare providers to analyse vast amounts of data collected from various sources including wearable devices and telehealth platforms. Exploiting artificial intelligence (AI) algorithms and machine learning models, soft computing allows for real-time monitoring and analysis of vital signs, foetal development, and maternal health indicators. The use of soft computing in remote prenatal care promotes patient engagement and empowerment. Expectant mothers can actively participate in their own care by accessing real-time data about their health status through user-friendly interfaces or mobile applications as shown in Figure 2.3. This not only fosters a sense of control but also encourages adherence to recommended lifestyle modifications or treatment plans.

FIGURE 2.3   Pregnant women actively engaging in self-care.

2.2.5 REMOTE MONITORING WITH SOFT COMPUTING

2.2.5.1 Data Collection and Preprocessing

The acquisition of health data is a sophisticated process that blends information from diverse sources, such as wearable devices, medical records, and real-time sensors. Careful collection and preprocessing are pivotal for effectively leveraging this data in soft computing models. Postcollection, meticulous preprocessing is essential. Handling missing values involves techniques like mean imputation or regression-based imputation. Identifying and managing outliers utilizes statistical methods like Z-scores or interquartile range (IQR), eliminating or transforming outliers to align within an acceptable range. Additionally, ensuring proper data formatting for soft computing models is paramount. This often includes converting raw data into numeric representations, enabling algorithms to process information efficiently as seen in Figure 2.4. Techniques like scaling, normalization, or one-hot encoding in feature engineering further optimize model performance. This rigorous data preparation establishes a robust foundation for the subsequent application of advanced soft computing techniques in health data analysis (Bautista, 2020).

2.2.5.2 Fuzzy Logic for Interpretation

Fuzzy logic stands out as a powerful methodology for effectively handling uncertainty and imprecision in health data. By defining fuzzy sets and rules, it provides a framework for interpreting sensor readings and assessing the health conditions of both the mother and the foetus comprehensively. Fuzzy sets categorize health conditions, introducing distinctions like "healthy," "potentially at risk," or "critical condition," accommodating the inherent ambiguity in health data through membership degrees rather than rigid binary classifications. Illustratively, fuzzy inference systems establish rules correlating sensor readings with specific health conditions, derived from expert knowledge or data-driven approaches. The significant advantage

FIGURE 2.4 Data preprocessing for advanced computing analysis.

lies in delivering interpretable results that account for subjective terms, quantifying degrees of certainty or uncertainty associated with diverse health conditions instead of a binary yes/no response. Fuzzy inference systems contribute significantly to healthcare decision-making by quantifying uncertainty and imprecision, enhancing the overall decision-making landscape in healthcare settings.

2.2.5.3 Neural Networks for Pattern Recognition

Training neural networks for health data pattern recognition demands a well-designed architecture incorporating vital signs, historical records, and relevant information. Neural networks excel in health data analysis, requiring careful consideration of factors like hidden layer count, neurons per layer, and activation functions. Experimentation and fine-tuning based on performance metrics are vital. The architecture provides insights from vast datasets, balancing complexity with efficiency. While simpler tasks may suffice with a single hidden layer, intricate patterns may require multiple layers.

The number of neurons impacts representation and learning capacity. Activation functions like sigmoid, tanh, or ReLU introduce non-linearities crucial for capturing complex patterns. Thoughtful implementation empowers neural networks to discern intricate health data patterns, enhancing understanding of vital signs and historical records (Kaur et.al, 2023).

2.2.5.4 Genetic Algorithms for Optimization

Genetic algorithms offer a robust strategy for optimizing parameters in monitoring systems, including refining alert thresholds, feature selection, and neural network weights. Central to this approach is defining a fitness function aligned with performance objectives such as sensitivity, specificity, and predictive accuracy. Sensitivity measures true positive detection, specificity assesses true negative identification, and predictive accuracy signifies outcome prediction precision. Crafting a well-defined fitness function initiates the optimization process, guiding the genetic algorithm to maximize sensitivity and specificity while maintaining predictive accuracy. The iterative nature of genetic algorithms, evolving generations based on fitness scores, facilitates convergence towards an optimal solution, enhancing overall system performance. Genetic algorithms effectively adapt neural network weights, ensuring superior system performance tailored to specific requirements. Their versatility and adaptability make genetic algorithms valuable tools in optimizing health data monitoring systems, meeting dynamic demands with precision.

2.2.5.5 Data Fusion and Integration

Data fusion techniques play a pivotal role in integrating information from diverse sources to comprehensively evaluate maternal and foetal health. This involves combining data from wearable devices, electronic health records, and real-time sensor readings to create a holistic understanding. Wearable devices provide insights into vital signs, activity levels, and sleep patterns, while electronic health records offer medical history and diagnoses. Real-time sensor readings monitor vital signs like blood pressure and glucose levels. Intelligent fusion of these datasets enables accurate diagnoses, personalized care plans, and early risk identification during pregnancy.

Integrating physiological data with contextual information aids in understanding how environmental factors impact maternal-foetal health, facilitating proactive interventions for healthier pregnancies and optimal baby development. Leveraging data fusion techniques allows for effective integration of information, leading to informed decisions and improved outcomes for both mother and baby.

2.2.6 BENEFITS OF ADAPTIVE AND LEARNING ALGORITHMS IN MANAGING DYNAMIC HEALTH DATA

In the ever-evolving field of healthcare, the effective management of dynamic health data is crucial for providing quality care and improving patient outcomes. With the advent of adaptive and learning algorithms, healthcare professionals can harness the power of AI to analyse and make sense of large volumes of data in real-time.

Several benefits of using adaptive and learning algorithms to manage dynamic health data are changing the way healthcare is delivered in a number of ways as follows:

1. **Enhancing Predictive Analytics:** The synergy of adaptive algorithms with historical and real-time health data empowers healthcare providers to foresee specific health events. By identifying at-risk individuals early on, preventive measures can be promptly implemented, ushering in a new era of proactive and personalized healthcare.

FIGURE 2.5 Adaptive algorithms for remote patient monitoring.

2. **Enabling Remote Patient Monitoring:** Through the analysis of real-time data from wearable devices, adaptive algorithms facilitate remote patient monitoring. This breakthrough allows healthcare professionals to detect changes in a patient's health status swiftly, enabling timely interventions and reducing the dependence on hospitalization or emergency care indicated in Figure 2.5.
3. **Improving Diagnosis and Treatment:** The pivotal role of adaptive algorithms in improving diagnosis and treatment is underscored by their ability to examine vast datasets. Uncovering patterns and correlations not immediately apparent to human professionals lead to more accurate and timely diagnoses. This personalized approach optimizes treatment outcomes and minimizes the risk of adverse events (Adhikari, 2021).
4. **Improving Medication Management:** Significantly contributing to medication management, these algorithms dissect patient data and medication information. They play a crucial role in identifying potential drug interactions, adverse effects, or medication errors. This support aids healthcare providers in making informed decisions and enhances overall medication safety (Gupta et al., 2022).
5. **Enhancing Clinical Decision Support:** The collaboration between technology and healthcare expertise is evident in how adaptive algorithms enhance clinical decision support systems. By providing real-time insights and recommendations based on dynamic health data analysis, these algorithms enable more accurate and personalized care delivery. As AI technologies continue to advance, the future of healthcare holds great promise. Anticipated developments in sophisticated algorithms offer healthcare professionals the ability to unlock new insights, improve patient outcomes, and revolutionize the delivery and experience of healthcare. The journey toward a smarter and more responsive healthcare system fueled by adaptive algorithms to reshape the future of medicine (Singh et al., 2023).

2.2.7 Decision Support Systems

2.2.7.1 What Are Decision Support Systems for Prenatal Care?

Decision support systems for prenatal care using soft computing refer to the integration of AI and machine learning algorithms into healthcare platforms to assist healthcare professionals in making informed decisions for expectant mothers. These systems analyse various data points, including medical records, genetic information, and maternal health factors, to provide personalized recommendations and predictions for prenatal care.

2.2.7.2 Enhancing Informed Decision-Making: The Role of Systems in Supporting Healthcare Professionals with Deep Learning Various Techniques

The evolution of decision support systems (DSSs) for prenatal care, particularly those integrating soft computing techniques, marks a significant advancement in enhancing maternal and foetal health. Soft computing, encompassing methodologies such as fuzzy logic, neural networks, and genetic algorithms, introduces a flexible and adaptive approach to navigate the complexities inherent in prenatal care decision-making.

The following developments in deep learning systems have greatly enhanced healthcare practitioners' ability to make well-informed decisions, particularly in prenatal care:

1. **Early Stages**
   In the nascent stages of DSS development for prenatal care, traditional rule-based systems and statistical models were prevalent. However, the inherent uncertainty in prenatal data spurred researchers to explore soft computing approaches, paving the way for a more nuanced and adaptable decision support framework (Gogineni, 2012).
2. **Introduction of Fuzzy Logic**
   Fuzzy logic, adept at handling imprecise and uncertain information, found application in modelling various facets of prenatal care. From interpreting ultrasound results to assessing maternal health indicators, fuzzy logic transcended the limitations of traditional logic, providing a more realistic representation of complex medical data.
3. **Incorporation of Neural Networks**
   Inspired by the human brain, neural networks were seamlessly integrated into prenatal care DSSs. These networks proved invaluable in recognizing intricate patterns and relationships within diverse datasets, enhancing the predictive capabilities of the system. Historical patient data became a wellspring for predicting outcomes and identifying potential risk factors.
4. **Genetic Algorithms for Optimization**
   Mimicking natural selection, genetic algorithms were introduced to optimize parameters and enhance the efficiency of prenatal care decision models. This evolutionary approach played a pivotal role in fine-tuning algorithms, fostering better accuracy and reliability in decision support for healthcare professionals.
5. **Adaptive Learning Systems**
   Soft computing ushered in adaptive learning systems within prenatal care DSSs. These systems dynamically updated their knowledge based on incoming data, adapting to changing medical conditions and incorporating the latest research findings. This adaptive capability elevated the relevance and effectiveness of decision support (Davies, 2019).

The development of decision support systems for prenatal care has evolved into sophisticated, adaptive systems that empower healthcare professionals with enhanced capabilities for informed decision-making, ultimately contributing to improved maternal and foetal health outcomes.

2.2.8 Predictive Modelling and Risk Assessment

2.2.8.1 Predictive Modelling Using Soft Computing

Soft computing techniques exhibit remarkable flexibility, generating predictions from imprecise or incomplete data. They excel in deciphering the intricate interplay of factors influencing pregnancy complications, providing accurate predictions even in situations where underlying mechanisms are not fully understood. As conditions change throughout pregnancy and new knowledge emerges, soft computing techniques showcase their capability to update models in response to new data, ensuring predictive accuracy and relevance. This adaptability enhances the efficiency and responsiveness of prenatal care, making it a formidable asset in addressing the evolving dynamics of maternal health.

1. **Handling Imprecise and Uncertain Data**
   Soft computing techniques, including fuzzy logic, neural networks, and genetic algorithms, excel in addressing the challenges presented by imprecise and uncertain pregnancy-related data. The data, often fraught with missing values, inconsistencies, and subjective assessments, poses difficulties for traditional statistical methods. Soft computing techniques showcase their adaptability by effectively managing these uncertainties, allowing them to make predictions even in the presence of incomplete or imprecise data (Das, 2023).
2. **Insights from Complex Relationships**
   In the domain of pregnancy-related complications, characterized by intricate interactions among various factors, soft computing techniques demonstrate their ability to learn from complex relationships. Fuzzy logic, neural networks, and genetic algorithms adeptly capture these intricate interactions, enabling accurate predictions even when the underlying mechanisms are not fully understood.
3. **Adapting to Dynamic Conditions**
   Soft computing techniques offer a dynamic solution to the evolving nature of pregnancy-related risks. As conditions change throughout the pregnancy and new information emerges, these techniques continuously update their models based on new data. This adaptability ensures more accurate predictions as the pregnancy progresses, providing a responsive approach to changing conditions and enhancing the overall effectiveness of pregnancy-related care.

2.2.8.2 Early Risk Assessment in Prenatal Care

Soft computing DSSs not only aid in early risk assessment but also contribute to the development of personalized care plans tailored to the unique needs of each expectant mother. Here is how they facilitate personalized care plans:

1. **Comprehensive Data Analysis:** Soft computing DSSs analyse a broad range of maternal health indicators, genetic factors, and historical patient data. This comprehensive analysis allows for the early identification of potential risk factors that might contribute to complications during pregnancy.
2. **Dynamic Risk Scoring:** By utilizing adaptive learning systems, these DSSs continuously update their risk scoring mechanisms based on real-time data. This dynamic approach ensures that any changes in the patient's health status are promptly reflected in the risk assessment, allowing for timely interventions.
3. **Individualized Parameter Analysis**: These systems consider a multitude of factors, including genetic information, lifestyle choices, and historical health data, to create a detailed profile of the patient. This individualized approach ensures that the care plan is uniquely suited to address the specific health requirements of each expectant mother.

4. **Fuzzy Logic for Individual Sensitivity**: Fuzzy logic allows for the incorporation of imprecise and uncertain information, reflecting the individual sensitivity of patients to certain interventions or treatments. This ensures that the personalized care plan considers the nuances of each patient's health condition (Gupta & Gupta, 2022).
5. **Neural Networks for Treatment Optimization:** Neural networks (Gupta et al., 2024) assist in optimizing treatment plans based on the individual responses of patients to medications and interventions. This adaptive approach ensures that the care plan evolves in response to the patient's changing health dynamics.
6. **Real-Time Monitoring for Adjustments:** The integration of real-time monitoring and alert systems allows for continuous adjustments to the personalized care plan. If deviations or new information emerge, healthcare professionals are promptly notified, enabling them to adapt the care plan in real-time (Sharma, D., 2021).

2.3 CHALLENGES AND FUTURE DIRECTIONS

2.3.1 Ethical and Legal Considerations

1. **Privacy and Data Security**
   One of the primary ethical concerns surrounding remote prenatal care and soft computing applications is the risk of privacy breaches. With the collection and analysis of extensive personal health data, there is a potential for unauthorized access or use of this information (Singh et al., 2023). This could result in breaches of privacy, confidentiality, and even discrimination or stigmatization based on the health data collected. It is imperative for healthcare providers and technology companies to prioritize robust security measures to safeguard sensitive information and restrict access to authorized individuals only (Nittari, 2023).
2. **Reliability and Accuracy**
   Another ethical concern involves the potential biases in algorithms used in soft computing applications. These algorithms rely on large datasets, which may not fully represent the diverse population of pregnant women. If these algorithms are not thoroughly validated and tested for biases, they could generate inaccurate or unfair recommendations. It may not provide accurate recommendations for women from other racial or ethnic backgrounds. It is essential for researchers and developers to prioritize diversity and inclusivity in the datasets used to train these algorithms. Additionally, continuous monitoring and addressing of biases are crucial to ensure fair and accurate outcomes for all pregnant women.

2.3.2 Examine Legal Frameworks and Guidelines Governing Telemedicine Practices

1. **Licensing in Telemedicine Practices**
   Many jurisdictions require healthcare providers to be licensed in the state or country where the patient is located. Telemedicine providers must comply

with these licensing requirements, ensuring that practitioners are qualified to provide services within the patient's jurisdiction.

United States

Telemedicine licensing in the United States is governed at the state level. Healthcare providers offering telemedicine services typically need to be licensed in the state where the patient is located. Some states have reciprocity agreements that facilitate cross-border telemedicine practice.

United Kingdom

In the UK, healthcare professionals providing telemedicine services must be registered with the appropriate regulatory body, such as the General Medical Council (GMC) for doctors or the Nursing and Midwifery Council (NMC) for nurses. The licensing requirements may vary for different healthcare professions.

Australia

Australia's licensing requirements for telemedicine practitioners depend on the specific state or territory. The Australian Health Practitioner Regulation Agency (AHPRA) regulates the registration and licensing of healthcare professionals. Licensing may be required in the state or territory where the patient is located.

India

Telemedicine guidelines in India are governed by the Ministry of Health and Family Welfare. While there is a national framework for telemedicine, licensing requirements for healthcare providers may still vary by state. The Telemedicine Practice Guidelines provide a foundation for telemedicine practice.

Germany

In Germany, healthcare providers offering telemedicine services are subject to the licensing requirements of the relevant medical chamber. Licensing requirements may vary, and healthcare professionals must adhere to local regulations (Fields, 2020).

2. **Technology Standards**
   1. **HL7 (Health Level Seven International)**

   HL7 is an extensively espoused set of norms for the exchange, integration, sharing, and reclamation of electronic health information. It includes norms like HL7 v2, HL7 v3, and Fast Healthcare Interoperability coffers (FHIR). These norms grease the interoperability of health information systems, supporting flawless data exchange between different healthcare operations.

   2. **DICOM (Digital Imaging and Dispatches in Medicine)**

   DICOM is a standard for the communication and operation of medical imaging information and related data. It ensures interoperability between different medical imaging devices and systems, allowing the sharing and viewing of medical images in a harmonious manner. IHE (Integrating the Healthcare Enterprise) is an action that promotes the coordinated use of established standards, similar to HL7 and DICOM, to ameliorate the interoperability of healthcare information systems. IHE biographies give

guidelines for enforcing specific use cases, including those applicable to telemedicine (Sharma et al., 2023).

3. **ISO 13485 (Medical Devices)**
   ISO 13485 is a transnational standard specific to the quality operation system of medical bias. Telemedicine technologies, particularly those involving medical bias, should cleave to this standard to ensure the safety and effectiveness of the bias.

4. **HIPAA (Health Insurance Portability and Responsibility Act)**
   In the United States, telemedicine platforms must misbehave with HIPAA regulations to cover the sequestration and security of cases' health information. Adherence to HIPAA norms is pivotal for the secure transmission and storehouse of electronic defended health information (ePHI).

5. **IEEE 11073 (Health informatics-particular health device communication)**
   IEEE 11073 is a set of standards for personal health device communication. It defines protocols for interoperable communication between personal health devices and other systems, supporting the integration of data from various telehealth bias.

3. **Standard Guidelines for Telemedicine**

There are seven elements to consider before any telemedicine consultation as seen in Figure 2.6: identification of RMP and patient, mode of communication, consent, type of consultation, patient evaluation, and patient management.

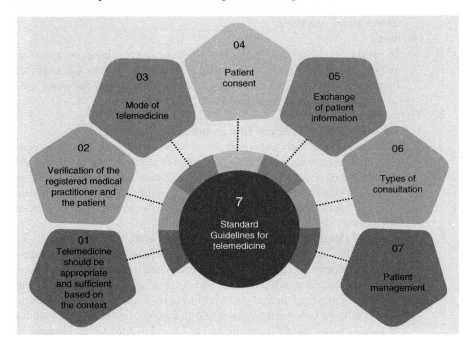

FIGURE 2.6 Telemedicine standard guidelines.

1. **Telemedicine should be appropriate and sufficient based on the context**
   Registered Medical Practitioners (RMPs) should exercise their professional judgment to determine whether a telemedicine consultation is suitable for a given situation or if an in-person consultation is necessary for the best interest of the patient. RMPs play a crucial role in ensuring the well-being of their patients. By using their professional judgment, they can determine whether a telemedicine consultation is appropriate or if an in-person consultation is more beneficial. Their expertise allows them to make informed decisions that prioritize the best interest of the patient.

2. **Verification of the registered medical practitioner and the patient**
   A RMP should verify and confirm the patient's identity using their name, age, address, email ID, phone number, registered ID, or any other appropriate form of identification. It is imperative for a registered medical practitioner to ensure the accurate identification of patients. This verification process helps maintain patient safety and confidentiality. To verify and confirm the patient's identity, the medical practitioner should gather information such as their name, age, address, email ID, phone number, registered ID, or any other suitable form of identification. By diligently following these procedures, healthcare professionals can ensure the correct and secure provision of medical services to patients.

3. **Mode of Telemedicine**
   Various technologies facilitate telemedicine consultations, each with distinct strengths, weaknesses, and contexts. Three primary modes include video, audio, and text. Medical practitioners must consider these aspects when selecting the appropriate mode for consultations. Video consultations offer real-time interaction, resembling in-person visits, ideal for visual assessments and rapport-building. However, stable internet and video conferencing software are prerequisites, and some patients may have privacy concerns when conferencing in public. Audio consultations rely on verbal communication, suitable when visual assessment is unnecessary. Text-based modes like chat and messaging provide asynchronous communication, offering flexibility in timing. They facilitate convenient exchanges without requiring simultaneous availability.

4. **Patient Consent**
   Patient consent is essential for any telemedicine consultation. Informed consent is a fundamental legal and ethical requirement. Patients must receive adequate information about the nature of telemedicine services, potential risks and benefits, as well as alternative options.

5. **Exchange of Patient Information**
   If the RMP feels that the information received is inadequate, he or she can request additional information from the patient. If a physical examination is crucial for consultation, the RMP should not proceed until an in-person consultation can be arranged to conduct the physical examination.

6. **Types of Consultation**
   First Consult means that the patient is consulting with the RMP for the first time.

Follow-Up Consult(s) means that the patient is consulting with the same RMP within 6 months of their previous in-person consultation, and this is for the continuation of care for the same health condition.

7. **Patient Management**
   An RMP may provide health promotion and disease prevention messages. These messages could be related to various topics such as diet, physical activity, smoking cessation, contagious infections, and more. Additionally, an RMP may offer advice on immunizations, exercises, hygiene practices, and mosquito control. Counselling is also a part of their role. This advice is specific to patients and may include various guidelines such as food restrictions, dos and don'ts for patients on anticancer drugs, proper use of a hearing aid, home physiotherapy, and other measures to mitigate the underlying condition. This may also include advice for new investigations that need to be carried out before the next consult (Loane, 2002).

2.3.3 Technological Challenges

2.3.3.1 Address Technical Challenges Associated with the Implementation of Soft Computing in Prenatal Care

1. **Data Quality and Availability**
   Soft computing algorithms heavily rely on high-quality, comprehensive data. In the context of prenatal care, obtaining accurate and sufficient data is challenging due to various factors such as the dynamic nature of pregnancy, limited monitoring resources, and variations in individual health profiles. Ensuring the availability and quality of diverse datasets is crucial for training robust models.
2. **Interoperability and Integration**
   Healthcare systems often use diverse electronic health record (EHR) systems, medical devices, and data formats. Integrating soft computing applications seamlessly into existing healthcare infrastructure is a significant challenge. Ensuring interoperability with different data sources and formats is essential for the effective implementation of soft computing in prenatal care.
3. **Algorithm Interpretability**
   Soft computing models, particularly complex machine learning algorithms, are often considered as "black boxes" due to their complexity. Interpreting the decisions made by these models in the context of prenatal care is critical for gaining the trust of healthcare providers and patients. Developing models with explainable and interpretable results is an ongoing challenge.
4. **Bias and Fairness**
   Soft computing models can inadvertently inherit biases present in the training data, leading to unfair or discriminatory outcomes. In prenatal care, biased predictions could have serious consequences. Addressing bias in the data and the algorithms themselves is crucial to ensure fair and equitable healthcare outcomes.

Telemedicine and Remote Prenatal Care 41

5. **Clinical Validation and Regulation**
   Soft computing applications used in prenatal care must undergo rigorous clinical validation to ensure their accuracy and reliability. Regulatory approval processes need to be established to evaluate the safety and efficacy of these technologies before widespread clinical adoption (Hossain, 2023).
6. **Human-Machine Interaction**
   Soft computing technologies should complement and enhance the capabilities of healthcare professionals rather than replace them. Ensuring effective human-machine interaction and understanding the role of healthcare providers in interpreting and acting upon soft computing outputs is a challenge that needs attention.
7. **Continuous Monitoring and Adaptation**
   Prenatal care is dynamic, and the health status of both the mother and the foetus can change rapidly. Soft computing models should be capable of continuous monitoring and adaptation to evolving conditions. Developing systems that can update and refine their predictions over time is an ongoing technical challenge.
8. **Limited Data for Rare Events**
   Prenatal complications and adverse events are relatively rare, making it challenging to train models effectively. Developing models that can handle imbalanced datasets and rare events is crucial for accurate prediction and early intervention (Malik et al., 2023).

2.3.3.2 Propose Potential Solutions and Devices

1. **Advanced Monitoring Devices**
   Develop and implement advanced, non-invasive monitoring devices that provide real-time data on maternal and foetal health. These devices could include wearables, sensors, and smart home devices to track vital signs, foetal movements, and other relevant parameters.
2. **Telemedicine Platforms with Integrated Soft Computing**
   Build telemedicine platforms that integrate soft computing applications, such as machine learning algorithms, to assist in data analysis, risk prediction, and decision support. These platforms should provide interpretable results to healthcare providers and include mechanisms for continuous learning and improvement.
3. **Enhanced Connectivity and Bandwidth**
   Improve internet connectivity and bandwidth in underserved areas to ensure that remote prenatal care is accessible to a wider population. This includes investing in infrastructure to support video consultations, remote monitoring, and data transmission.
4. **Interoperability Standards**
   Establish and adhere to interoperability standards for EHRs and medical devices. This would enable seamless integration of data from different sources, providing a comprehensive view of a patient's health history and facilitating more accurate assessments.

5. **User-Friendly Telehealth Platforms**
   Design telehealth platforms with user-friendly interfaces that are accessible to individuals with varying levels of technological literacy. This includes providing clear instructions, multilingual support, and easy navigation to enhance the user experience for pregnant individuals.
6. **Personalized Care Plans**
   Implement soft computing algorithms to analyse patient data and generate personalized care plans. These plans can take into account individual risk factors, preferences, and cultural considerations, ensuring that remote prenatal care is tailored to the specific needs of each patient.
7. **Remote Ultrasound and Imaging Technologies**
   Develop and deploy advanced remote ultrasound and imaging technologies that can be operated by healthcare professionals or even by pregnant individuals themselves under guidance. This would enhance the ability to remotely assess foetal development and identify potential issues. (Sahoo & Kumar, 2014)
8. **Patient Education and Engagement**
   Implement telehealth platforms that include educational resources and engagement features to empower pregnant individuals with knowledge about prenatal care. This could include video content, interactive modules, and regular communication to enhance health literacy.
9. **Regulatory Frameworks and Reimbursement Policies**
   Work with regulatory bodies to establish clear frameworks for the practice of remote prenatal care, ensuring that it meets legal and ethical standards. Additionally, advocate for reimbursement policies that recognize and support telehealth services to incentivise healthcare providers to offer remote prenatal care.

2.3.4 FUTURE PROSPECTS

2.3.4.1 Explore the Future Prospects of Telemedicine and Soft Computing in Prenatal Care

1. **Remote Monitoring and Wearable Technology**
   Increasing use of wearable devices and remote monitoring technology will allow pregnant individuals to track vital signs, foetal movements, and other relevant parameters from the comfort of their homes. This data can be continuously transmitted to healthcare providers, enabling proactive interventions and personalized care plans.
2. **Integration of Artificial Intelligence (AI) and Soft Computing**
   Further integration of AI and soft computing techniques, including machine learning and fuzzy logic, will enhance the accuracy and efficiency of prenatal care. Advanced algorithms will assist in risk prediction, early detection of complications, and personalized decision support for healthcare providers. (Kaur et al., 2023)

3. **Virtual Reality (VR) and Augmented Reality (AR) Applications**
   VR and AR technologies may play a role in enhancing the prenatal care experience. Virtual tours of labour and delivery rooms, educational simulations, and immersive experiences could help pregnant individuals and their families better understand the birthing process and make informed decisions. (Sahoo, 2014)
4. **Telegenetic Counselling**
   Genetic counselling services through telemedicine platforms will become more sophisticated, allowing for comprehensive discussions about genetic testing, family history, and potential risks. Soft computing applications may assist in analysing genetic data and providing tailored counselling based on individual risk factors.
5. **Predictive Analytics for Early Intervention**
   Advanced predictive analytics models will continue to evolve, allowing healthcare providers to identify and intervene in potential complications at an earlier stage. Soft computing applications can assist in analysing large datasets to predict risks and recommend timely interventions.
6. **Global Expansion of Telemedicine Services**
   Telemedicine will increasingly break down geographical barriers, providing access to prenatal care for individuals in remote or underserved areas. This global expansion will contribute to more equitable healthcare access and reduce disparities in maternal care.
7. **Regulatory Frameworks and Standardization**
   Governments and regulatory bodies will likely play a more active role in establishing clear guidelines and standards for telemedicine and soft computing applications in prenatal care. Standardization will contribute to increased trust among healthcare providers, patients, and regulatory agencies.
8. **Remote Postpartum Care**
   Beyond pregnancy, telemedicine and soft computing can extend to postpartum care, offering ongoing support and monitoring during the critical weeks and months after childbirth. This approach can enhance the continuity of care and address postpartum health concerns remotely (Su, 2022).
9. **Patient-Centred and Culturally Sensitive Platforms**
   Future platforms will be designed with a strong emphasis on patient-centred care and cultural sensitivity. Customizable interfaces, multilingual support, and culturally tailored content will ensure that telemedicine in prenatal care meets the diverse needs of pregnant individuals worldwide.

2.3.4.2 Emerging Technologies and Potential Areas for Further Research in Prenatal Care

Figure 2.7 shows the emerging technologies and potential areas for further research in prenatal care.

1. **Biometric Monitoring Wearables**
   Explore the integration of biometric monitoring wearables that can continuously track maternal and foetal health parameters, providing real-time data

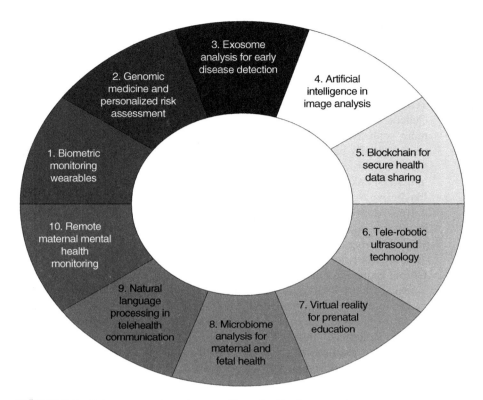

FIGURE 2.7 Future prospects and areas of investigation in prenatal care.

for healthcare providers. Investigate the accuracy, usability, and acceptance of such wearables in diverse populations.
2. **Genomic Medicine and Personalized Risk Assessment**
   Further research on the application of genomic medicine to prenatal care can lead to personalized risk assessments for genetic disorders. Investigate the ethical implications, patient education strategies, and integration of genomic data into decision-making processes. (Alexander & Kotelchuck, 2001)
3. **Exosome Analysis for Early Disease Detection**
   Examine the potential of exosome analysis for early detection of foetal abnormalities and pregnancy-related complications. Explore the use of exosomes in liquid biopsies for non-invasive monitoring and diagnostics, addressing challenges related to sensitivity and specificity. (Seth Kwabena Amponsah et al., 2023)
4. **AI in Image Analysis**
   Research the application of AI in image analysis for ultrasound and MRI scans. Develop algorithms that can automatically identify anomalies, assess foetal development, and provide quantitative data to augment clinical decision-making.
5. **Blockchain for Secure Health Data Sharing**
   Investigate the use of blockchain for secure and interoperable health data sharing in prenatal care. Explore decentralized platforms that allow seamless sharing of patient information while maintaining privacy and data integrity.

6. **Telerobotic Ultrasound Technology**
   Explore the potential of telerobotic ultrasound technology, allowing healthcare providers to remotely conduct ultrasound examinations. Assess the feasibility, accuracy, and patient satisfaction with telerobotic ultrasound systems in various healthcare settings.
7. **Virtual Reality for Prenatal Education**
   Conduct research on the effectiveness of virtual reality applications in prenatal education. Evaluate the impact of VR simulations on maternal understanding, preparedness for childbirth, and decision-making regarding prenatal interventions.
8. **Microbiome Analysis for Maternal and Foetal Health**
   Investigate the role of the maternal and foetal microbiome in pregnancy outcomes. Explore microbiome analysis techniques to understand how variations in microbial communities may impact maternal health, foetal development, and long-term health outcomes [28].
9. **Natural Language Processing in Telehealth Communication**
   Research the application of natural language processing in telehealth communication. Develop tools that can extract meaningful insights from patient-provider interactions during virtual consultations, aiding in the assessment of patient well-being and satisfaction.
10. **Remote Maternal Mental Health Monitoring**
    Explore technologies for remote maternal mental health monitoring. Develop tools that use sensor data, behavioural patterns, and self-reported information to assess and support maternal mental well-being during pregnancy and the postpartum period.

2.4 CONCLUSION

Telemedicine is revolutionizing prenatal care, transcending geographical constraints and providing accessible, high-quality healthcare to pregnant women worldwide. Soft computing techniques, including fuzzy logic, neural networks, and genetic algorithms, play a central role in analysing vast datasets and enhancing decision-making for healthcare providers. Through remote monitoring with soft computing, expectant mothers benefit from real-time tracking of vital signs, foetal development, and maternal health indicators, empowering them with self-care capabilities and promoting adherence to treatment plans. Soft computing systems for decision support further aid healthcare professionals by offering personalized care plans, early risk assessment, and predictive modelling.

Despite the challenges associated with implementing soft computing in prenatal care, such as data quality, interoperability, and technological limitations, ongoing efforts focus on addressing these issues through advanced monitoring devices, improved connectivity, and regulatory frameworks. The future outlook for telemedicine and soft computing in prenatal care is promising, with emerging technologies and areas of further research holding potential for enhanced accuracy, personalized risk assessment, and secure data sharing. As telemedicine continues to evolve, it presents a transformative paradigm in prenatal care, advancing maternal health and ensuring optimal outcomes for both mothers and their unborn children.

REFERENCES

Adhikari, S., Thapa, S., & Ghimire, A. (2021). Soft Computing Techniques for Medical Diagnosis, Prognosis and Treatment. In: Dash, S., Pani, S.K., Abraham, A., Liang, Y. (eds) *Advanced Soft Computing Techniques in Data Science, IoT and Cloud Computing. Studies in Big Data*, vol 89. Springer, Cham. https://doi.org/10.1007/978-3-030-75657-4_17

Alexander, G. R., & Kotelchuck, M. (2001). Assessing the role and effectiveness of prenatal care: history, challenges, and directions for future research. *Public Health Reports*, 116(4), 306.

Arias, M. P., Wang. E., Leitner, K., Sannah, T., Keegan, M., Delferro, J., Iluore, C., Arimoro, F., Streaty, T., & Hamm, R. F. (2022). The impact on postpartum care by telehealth: a retrospective cohort study. *American Journal of Obstetrics & Gynecology*, 4(3), 100611. doi: 10.1016/j.ajogmf.2022.100611. Epub 2022 Mar 22. PMID: 35331971; PMCID: PMC10134102.

Bautista, J. M., Quiwa, Q. A. I., & Reyes, R. S. (2020, November). Machine learning analysis for remote prenatal care. In *2020 IEEE Region 10 Conference (TENCON)* (pp. 397–402). IEEE.

Das, S., Mukherjee, H., Roy, K., & Saha, C. K. (2023). Fetal Health Classification from Cardiotocograph for Both Stages of Labor—A Soft-Computing-Based Approach. *Diagnostics*, 13(5), 858.

Davies, C., Fattori, F., O'Donnell, D., Donnelly, S., Ní Shé, É., O. Shea, M., ... & Kroll, T. (2019). What are the mechanisms that support healthcare professionals to adopt assisted decision-making practice? A rapid realist review. *BMC Health Services Research*, 19, 1–14.

de Silva, C. W. (2003). The role of soft computing in intelligent machines. *Philosophical Transactions of the Royal Society of London. Series A: Mathematical, Physical and Engineering Sciences*, 361(1809), 1749–1780.

Devi, M., Singh, S., Tiwari, S., Chandra Patel, S., & Ayana, M. T. (2021). A survey of soft computing approaches in biomedical imaging. *Journal of Healthcare Engineering*, 2021, 1563844.

Diwaker, C., Tomar, P., Poonia, R. C., & Singh, V. (2018). Prediction of software reliability using bio inspired soft computing techniques. *Journal of Medical Systems*, 42, 1–16.

Fields, B. G. (2020). Regulatory, legal, and ethical considerations of telemedicine. *Sleep Medicine Clinics*, 15(3), 409–416.

Furlepa, K., Tenderenda, A., Kozłowski, R., Marczak, M., Wierzba, W., & Śliwczyński, A. (2022). Recommendations for the development of telemedicine in Poland based on the analysis of barriers and selected telemedicine solutions. *International Journal of Environmental Research and Public Health*, 19(3), 1221.

Galle, A., Semaan, A., Huysmans, E., Audet, C., Asefa, A., Delvaux, T., Afolabi, B. B., El Ayadi, A. M., & Benova, L. (2021). A double-edged sword-telemedicine for maternal care during COVID-19: findings from a global mixed-methods study of healthcare providers. *BMJ Global Health*, 6(2), e004575. doi: 10.1136/bmjgh-2020-004575. PMID: 33632772; PMCID: PMC7908054.

Gambhir, S., Malik, S. K., & Kumar, Y. (2016). Role of soft computing approaches in healthcare domain: a mini review. *Journal of Medical Systems*, 40, 1–20.

Gambhir, S., Malik, S. K., & Kumar, Y. (2016). Role of soft computing approaches in HealthCare domain: a mini review. *Journal of Medical Systems,* 40(12), 287. doi: 10.1007/s10916-016-0651-x. Epub 2016 Oct 29. PMID: 27796841.

Gogineni, J. (2012). An ingenious decision support system for remote health monitoring of pregnant women (Doctoral dissertation, Queensland University of Technology).

Gupta, U., & Gupta, D. (2022). Bipolar fuzzy based least squares twin bounded support vector machine. *Fuzzy Sets and Systems,* 449, 120–161.

Gupta, U., & Gupta, D. (2023). Least squares structural twin bounded support vector machine on class scatter. *Applied Intelligence,* 53(12), 15321–15351.

Gupta, D., Gupta, U., & Sarma, H. J. (2024). Functional iterative approach for Universum-based primal twin bounded support vector machine to EEG classification (FUPTBSVM). *Multimedia Tools and Applications,* 83(8), 22119–22151.

Gupta, M., Srivastava, D., Pantola, D., & Gupta, U. (2022). Brain tumor detection using improved Otsu's thresholding method and supervised learning techniques at early stage. In *Proceedings of Emerging Trends and Technologies on Intelligent Systems: ETTIS 2022* (pp. 271–281). Singapore: Springer Nature Singapore.

Heřman, H., Faridová, A., Tefr, O., Farid, S., Ayayee, N., Trojanová, K., ... & Feyereisl, J. (2022). Telemedicine in prenatal care. *Central European Journal of Public Health,* 30(2), 131–135.

Hossain, M. M., Kashem, M. A., Islam, M. M., Sahidullah, M., Mumu, S. H., Uddin, J., ... & Samad, M. A. (2023). Internet of things in pregnancy care coordination and management: a systematic review. *Sensors,* 23(23), 9367.

Kaur, K., Singh, C., & Kumar, Y. (2023). Diagnosis and detection of congenital diseases in newborns or fetuses using artificial intelligence techniques: a systematic review. *Archives of Computational Methods in Engineering,* 30(1–28).

Loane, M., & Wootton, R. (2002). A review of guidelines and standards for telemedicine. *Journal of Telemedicine and Telecare,* 8(2), 63–71.

Malik K, Sadawarti H, Sharma M, Gupta U, Tiwari P, editors. *Computational Techniques in Neuroscience.* CRC Press; 2023 Nov 14.

Morgan, A., Goodman, D., Vinagolu-Baur, J., & Cass, I. (2022). Prenatal telemedicine during COVID-19: patterns of use and barriers to access. *JAMIA Open,* 5(1), ooab116.

Nguyen, E., Engle, G., Subramanian, S., & Fryer, K. (2021). Telemedicine for prenatal care: a systematic review. *medRxiv,* 2021–05.

Nittari, G., Khuman, R., Baldoni, S., Pallotta, G., Battineni, G., Sirignano, A., ... & Ricci, G. (2020). Telemedicine practice: review of the current ethical and legal challenges. *Telemedicine and e-Health,* 26(12), 1427–1437.

Phillippi, J. C. (2009). Women's perceptions of access to prenatal care in the United States: a literature review. *Journal of Midwifery & Women's Health,* 54(3), 219–225.

Sahoo, A. J., & Kumar, Y. (2014). Seminal quality prediction using data mining methods. *Technology and Health Care,* 22(4), 531–545.

Singh, S., Bhardwaj, A., Budhiraja, I., Gupta, U., & Gupta, I. (2023). Cloud-Based Architecture for Effective Surveillance and Diagnosis of COVID-19. In: Danda B. Rawat, Lalit K. Awasthi (Eds.) *Convergence of Cloud with AI for Big Data Analytics: Foundations and Innovation,* (pp. 69–88). Springer.

Singh, D., Singh, D., Manju, & Gupta, U. (2023). Smart Healthcare: a Breakthrough in the Growth of Technologies. In: Manju, K., & Sardar (Eds.) *Artificial Intelligence-based Healthcare Systems* (pp. 73–85). Cham: Springer Nature Switzerland.

Sharma, M., Deswal, S., Gupta, U., Tabassum, M., & Lawal, I. (Eds.). (2023). *Soft Computing Techniques in Connected Healthcare Systems.* CRC Press.

Sharma, S., Singh, G., & Sharma, M. (2021). A comprehensive review and analysis of supervised-learning and soft computing techniques for stress diagnosis in humans. *Computers in Biology and Medicine*, 134, 104450.

Sharma, D., Singh Aujla, G., & Bajaj, R. (2021). Deep neuro-fuzzy approach for risk and severity prediction using recommendation systems in connected health care. *Transactions on Emerging Telecommunications Technologies*, 32(7), e4159.

Su, Z., Li, C., Fu, H., Wang, L., Wu, M., & Feng, X. (2022). *Review of the Development and Prospect of Telemedicine*. Intelligent Medicine.

Sword, W., Heaman, M. I., Brooks, S., Tough, S., Janssen, P. A., Young, D., ... & Hutton, E. (2012). Women's and care providers' perspectives of quality prenatal care: a qualitative descriptive study. *BMC Pregnancy and Childbirth*, 12(1), 1–18.

3 Detection of Abnormality in Heart Rhythm Using a Machine Learning Approach

Prabhudutta Ray, Raj Rawal, and Ahsan Z. Rizvi

3.1 INTRODUCTION

In today's competitive world it has become even harder to stay healthy due to the stress and strain put upon the mind and body. Stress and strain lead to long-term anxiety resulting in palpitations, fluctuation in blood pressure, and finally creating an adverse and negative effect on our health, especially our heart. Heart disease, in particular arrhythmia, is one of the most common diseases affecting humans today. Patients suffering from arrythmia have problems of insufficient blood flow within their body. Here the heart fails to properly distribute blood to various parts of the body. As a result, extra blood either gets accumulated in the walls of the heart, or in organs like lungs or feet, which can lead several health hazards as follows [1].

Congestive heart failure (see Figure 3.1): *Heart blockage occurs when the arteries are congested due to bad cholesterol or fat and unable to pass sufficient blood to the other parts of the body. Insufficient blood flow to the different parts of the body can lead to heart failure.* There are three degrees of heart block. First-degree heart block may cause minimal problems, but third degree heart block can be life-threatening. High output cardiac failure is both a rare and critical type of heart disease. Literature [1,2] suggests that this type of heart disease is becoming more prevalent today due to changes in lifestyle. According to researchers, six million people in the United States are suffering from this disease. Detecting high output cardiac failure is difficult. Patients suffering from this type of heart-related problem often have no symptoms, or mild symptoms, but at time it becomes so critical that immediate hospitalization becomes necessary. Symptoms like difficulty in breathing, chest pain, irregular heart bit, feeling uneasy, etc., are some common reasons behind heart blockage [2].

Several factors are responsible for congestive heart diseases. Most common among them are 1) blockage in the coronary artery, 2) presence of cardio myopathy (genetic or viral), 3) congenital heart diseases (presence from the time of birth), 4) irregular high BP, 5) obesity (if BMI is more than 30), 6) smoking, 7) consumption of alcohol,

DOI: 10.1201/9781003508403-3

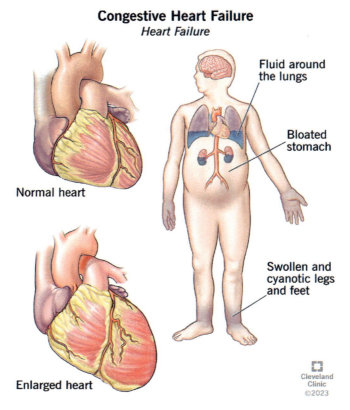

FIGURE 3.1 Congestive heart failure symptoms.

8) intake of medicine meant for cancer, 9) kidney diseases, and 10) most vital among them is the diseases of heart (arrythmia).

A person affected with anthemia suffers irregular heartbeats. It generally affects the lower chamber or the ventricles of the heart, which causes obstruction to the heart as it fails to pump enough blood to the body. A heart suffering with anthemia does not always show any sign. But some common symptoms are: 1) a very rapid or very slow heartbeat, 2) chest pain, 3) anxiety, 4) dizziness, 5) abnormal sweating, 6) feeling of tiredness, 7) fainting or semiconscious falling etc.

Arthemia (see Figure 3.2): A heart that pumps blood to the arteries from the veins and to different parts of our body. But when a person is suffering from arethemia, the body fails to receive proper circulation of blood to his body; as a result of insufficient flow of blood, the normal functioning of the heart and various other body parts gets affected. According to research there are five types of arthemia: 1) atrial fibrillation (AFib} this is a very crucial and common type of arthymia. Report suggest that majority of the people gets affected by this type of arthymia, throughout the world. In USA millions of peoples get affected by this type of arthymia. 2) atrial flutter: any problem caused by the electrical system of the heart that is short circuit in te heart

Detection of Abnormality in Heart Rhythm 51

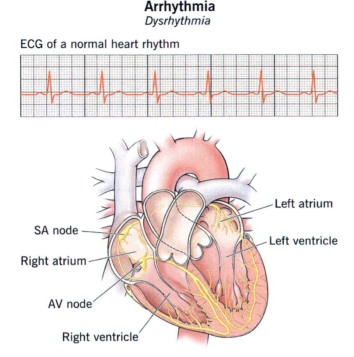

FIGURE 3.2 Common normal heart rhythm.

results in artial flatter [3]. It is a type of heart rhyme disorder of the heart. Severe type of artial flatter may results in stroke and may even cause death or permanent immobility of the body. 3) Super ventricular tachycardia (SVT): trachycardia results in rapid movement of heart beat (on an average 100 to 120 beats per minute). The main difference between tachycardia and SVT is that the former begins in the ventricles and is called trachucardia, while SVT begins from the above so is called super ventricular tachycardia (SVT).

4) Ventricular tachycardia: when the heart beat of a person reduces to less than 50 bpm it is called as ventricular tachycardia. Patients suffering with ischemic heart diseases are prone to this type of ventricular tachycardia. Symptoms include palpitations, shortness of breath, cardiac failure, syncope etc.

5) Ventricular fibrillation: – It is another type of arthemia. This results in irregular heart rhythms. Fibrillation is a muscular stretching which causes individual muscle fiber to function in an irregular manner -without proper co-ordination.

6) vrady cardia: – another type of irregular or abnormal heart beat where the beats are less than 60 bpm. The symptoms include filling of cold, tiredness, uneasy-filling etc. Treatment like change in life style, proper medication can improve the symptoms of vrady cardia. To find out the type of arthymia a person is suffering from, various tests are recommended. Most common among them are discussed below [4].

For detection of arrhythmia it is important to find the irregular heart beat during an examination by taking pulse and listening to rhythm of the heart. By listening the heart rhythm with a stethoscope and checking the pulse rate can give initial indication regarding the abnormality of the heart but for more deep investigation electrocardiogram test is more necessary.

- **Electrocardiogram (EKG or ECG):** it is the most common test for persons with heart problem. It helps to find out the electrical signal of the heart. It monitors the condition of the heart and finds out if any irregularities are occurring in the heart. This is a most common and painless test, primarily recommended by the doctors. The graph paper that is delivered by the machine, after detecting the condition of the heart, helps the doctor to have a vivid picture of the condition of the patient and the stability of his heart.
- **Echocardiogram (echo):** an echocardiogram is another method applied buy the medical **practioner** to detect the status of the organ. Through echo the polarization activity of the heart gets checked. Echo is also a pain less procedure and is helpful to detect the condition of both the patient and his heart.

Blood Tests: – sometimes blood tests are **also recommended by the doctors, to detect** the abnormalities of the heart. Blood test helps to find the level of sodium, potassium, thyroid, cholesterol level etc. present in the blood, as there also responsible of silently causing arthymia in the heart [5].

3.1.1 Detection of Abnormality Using ML Approach

As discuss earlier ECG is a device which detect arthymia through electrical signal that gets generated in the patient heart. These temporary electrodes help to monitor activities of the heart. The readings/ signals are then converted to the computer screen, whose information is then diagnosed by the doctors. This is the painless procedure and reports gets diagnosed and delivered on the same day or on the following day. There are three kinds of ECG tests like

a) Holter Monitor: A portable type of electrocardiogram, which can record electrical activities of the heart for 24 to 48 hours, is called a holter monitor. It is a type of machine which was continuously; even if one is away from the doctors chambers. It is a simplest and fastest method to detect any abnormalities is present in the heart. It helps to find out the reasons behind a) irregular heart palpitation, b) arthymia c) unexplained dizziness. It also helps to detect ischemic and heart blockages, it also provides accurate systolic and diastolic reading, along with heart rate, mean arterial pressure, BP and pulse pressure. But in spite of several advantages one major disadvantage of holter monitor is that patients might not experienced any symptoms of cardiac problems during the monitoring periods [6].

b) Event Monitor: another method used to record heart electrical activities is through a portable device called event monitor. Its activities are more or less like ECG but this device can record events for longer duration of time. It can be worn during normal activities, even while sleeping. Event monitor is a small monitoring device and only starts recording or functioning when if symptom arises.

c) Pacemaker: it is a device used for stimulating the heart chamber to regulate its contraction. It sense electrical pulses to understand the heart beat at normal rhythm and rate. It also helps the heart chamber beat in normal sync, for effective pumping of blood in our body. Persons who have vardy cardia, fainting symptom are at an urgent need of pacemaker. If pacemakers are not inserted properly one may have the chances of developing infections around the wires of the pacemaker, which at times can be fatal. Afeter inserting pacemaker's person can lead a normal life of 5 to 15 years at an easy [7]. An EKG reads that signal and monitor its impact on the heart as it contracts and relaxes during each heart beat. The ECG signal contains the P-wave, QRS and T-wave which uses onset, offset and peak points which are known as the fiducial points. Applying these and a set of known condition helps to generate data set, then apply several machine learning algorithms that are helps to identify various abnormality conditions. For example, a simplified ML algorithm can be used to detect abnormality condition of the heart from ECG report like

Algorithm Description: -
a. Identify the principal **fiducial points** on a generated ECG report.
b. Compute the statically calculated variance in the generated **R—R gap** by applying certain threshold value.
c. Identify abnormal activity – with the help of **P waves**
   - No regular atrial activity identified?
   - Any irregular abnormality activity reported.?
   - High frequency abnormality waves are present?

Using the above algorithmic approach the results are generated to detect the abnormality of the heart and reported accordingly [8,9].

3.2 RELETATED WORK

Different researchers apply distinct ML approach to solve heart disorder-related predictions using ECG signal using classification models. **Table 3.1** describes related work, different ML methods, classifier used, and inference taken from the paper in tabular form.

3.3 METHODOLOGY

The techniques of separation of a data set into classes by the use of classification techniques is highly used in the medical field. Actual separation of both different types of data are performed. The starting steps in the procedure is finding the class for available data points, several names that include target, output etc. Different mathematical theories such as LP, DT, and NN involved in categorization. Coronary diseases detection can be done through categorization steps because it has two parts,that is one has CVD or not.

TABLE 3.1
Related works, different ML methods, classifiers used, and inference

Related work	Method	Classifier used	Inference
ECG, M,L and time series analysis [10]	Time series and meta analysis of algorithms	ALL ML algorithms	Performance of ML for ECG classification
ML for ECG risk and treatment [11]	Patients with occlusion myocardial infarction (OMI)	ML with classifier	RF model shows the good results
Configuration of electrodes affects the result [12]	Developing algorithms to detect electrode misplacement	RF, NN, and DT	Better education regarding ECG acquisition
AI-enabled ECG [13]	Adoption of the AI-enabled ECG	AI methods	Better diagnosis
AI-Enabled Electrocardiogram Analysis for Disease Diagnosis [14]	AI using ECG data accompanied by modern wearable biosensors	AI with ML for classification	Health monitoring and early diagnosis
ECG reflects electrical activity and check condition of the cardiac cycle [15]	AI-enabled ECG analysis	AI methods	Adoption of AI-enabled ECG
ECG approach used for identifying CVD problems [16]	Detecting ECG anomalies using DL models.	DL, NN	ECG and arrhythmia classification.
ECG analysis and classification for CVD [17]	ECG analysis for classification	CNN, DL, ML, LSTM	Classification accuracy is 99.13%
Automated diagnosis of CVD diseases [18]	ECG data analysis	CNN, SVM	A system that helps for the patient's autonomy
Interpretable ML techniques for CVD [7]	Identify and characterization of ECG signals.	ML, DL	Progress for ECG signal identification
ML for detecting atrial fibrillation from ECGs [19]	Meta-analysis of diagnostic accuracy	ML, DL, CNN	ML is effective for detecting AF from ECGs
ML-based disease classification techniques. [20]	Optimized framework named as WbGAS for prediction	ML, CNN, DL	comparing the outcomes of different ML-based approach
Arrhythmia that can modify the heart's rhythms and its potentially impact [21]	MMPA to identify AF in brief ECG data.	SVM, AF	For ECG recordings, HRV is effective and Reliable for AF identification

Detection of Abnormality in Heart Rhythm 55

TABLE 3.1 (Continued)
Related works, different ML methods, classifiers used, and inference

Related work	Method	Classifier used	Inference
Classifying patients using ML to builds a DSS [22]	Building DSS for prediction	SVM, grid search	Proposed DSS same as standard ECG
An intelligent hybrid classification model for data classification [23]	Hybrid model for handles class imbalance	Adaboost, Bagging, RF, K-NN and SVM	building intelligent and accurate IoT-enabled healthcare systems

A) Support Vector Machines (SVM): SVM used for ramification techniques for data. A non linear mapping technique is used for converting the data into a higher dimension for training. To differentiate the points for te input variables of a hyper plane, for classes ranging from 0 to 1. A 2D plane heps to shows this as line and it is predicted that each point can be completely separated from their original line. The coordinating distance from the hyperplane through the adjacent data is called the margin. The line which has a lagest margin is helpful for distinguishing of two classes uing an optimal hyper plane. The points of this hyper plane is known as support vector as the name suggest, they help to define or support the structure of the hyper plane. In general, optimization techniques are used to calculate the value of the parameters which helps in the maximization of the margin level. Depending on the several kernels the hyperplane can be decided. Kernels are different types like linear, polynomial, radial, and sigmoid . The hyperplane is used to separate the locations in the available variable space that containing their class either 0 or 1. Margin denotes the distance between the hyperplane and adjacent data coordinates. Optimal hyperplane denotes the line that has the largest margin can distinguish between the two classes. These points are called support vectors, as they define or support the hyperplane. The SVM is widely considered due to its efficiency in pattern classification techniques. Kim et al. [24,2] proved that the SVM in the classification for prognostic prediction. The brief mathematical description based on of the SVM model is described below for the calculation. CVD with the convention of linear divisibility for training samples, we have

$$S = \{(x_1, y_1), (x_2, y_2), \ldots \ldots (x_n, y_n)\} \quad (3.1)$$

where $x_i \in (|R|)$, such that the design matrix X belongs to the d dimensional response space, and the response variable, CVD, is represented by y_i, which has a binary class in the vector Y with $y_i \in (0,1)$ in the study. The appropriate discriminating equation is given by

$$f(x) = sign\{(z, x) + \beta\}] \quad (3.2)$$

Similarly, Z represents the vector that determines the coordination of the hyperplane (discriminating plane), and so Z, X, and β are offsets. There are infinite numbers of possible hyperplanes that are efficiently classified by the training data that can be applied to the validation dataset. The optimal classifier shows that the similar optimal generalized hyper planes that are nearer or even away from each cluster of objects. The input set of coordinates is considered optimally separated by the hyperplane.

B) Random Forest (RF): This ML algorithms that uses concepts of Bagging or Bootstraping aggregation. To estimate a value from a data sample use the mean, bootstrap which are powerful statistical approach. Lots of samples of data are taken, respective mean is calculated, after that, all of the mean values are averaged to give a real mean value. In bagging, the sampling method is used, but instead of estimating the mean of every data sample, decision trees are generally used. Here, several samples of the training data are considered and models are generated for every data sample.

C) Simple Logistic Regression: In bimary classification method the values are identified in two classes. The aim of both LR and linear regresion is to correctly calculate the coeffcents values for every input variables. The logistic function acts as a non linear function, which helps to transform any range of value from 0 to 1. In logictic regression the prediction made is mainly used for the purpose of predicting the probabilty of a data instance which consists of either class 0 or class 1. It is necessary for solving problems where rationality is mostly preferred for any particular prediction. A better work from LR can be expected when attributes are not by any chance are related with output variables . It uses the sigmoid function for classification like

$$P = \frac{1}{1+e^{-x}} \quad (3.3)$$

In this case, the LR coefficients for each example are given as $x_1, x_2, x_3, x_4, \ldots\ldots\ldots\ldots x_n$ will be $b_0, b_1, b_2, b_3, \ldots\ldots\ldots\ldots b_n$ during the training phase . Here stochastic gradient used to calculate and update values like

$$y = b_1 x_1 + b_2 x_2 + b_3 x_3 + b_4 x_4 + \ldots\ldots\ldots b_n x_n \quad (3.4)$$

Again

$$b = b + 1*(y-p)*(1-p)*p*x \quad (3.5)$$

Here y is represented is output value for each training phase. LR depends on the actual representation of the data.

D) Decision Tree: In practical approaches, **DT** is the most important predictive modeling and classification method. DT algorithm can help to detect different ways by splitting the data sets based on numerous situations. A responsive point value is treated

Detection of Abnormality in Heart Rhythm 57

as a actual set of values for any classification tree for a tree method-based model. The purpose of the DT is to solve decesion making problems that can be helpful for making building models more challenging. The steps for decision tree are as follows

a) it divides the data set into two sub-data.
b) the total trainig data is considered as a root for an initial stage.
c) contineous values needs to get classified before any model building. But in case of *categorical values* are given preferences for detecting feature attributes.
d) in case of any established subset, each subset includes data which are useful for predicting future attributes.
e) At last repetion of steps-(a) to steps(d) continues unless we get a perfect leaves.

In case of DT classification get starts for recording from the root level where values gets compared with root features for succedding record characterization. In this comparison the equivalent values of the coming node gets sucessfully analyzed.

E) Approach of K-NN: KNN is another supervised ML approach used for both regression and classification. For categorization techniques, the use of k labels is alloted and for the regression, the returned value is the mean of k labels. KNN is the basic technique used for classification where earlier knowledge of data is missing. Manhattan distance used as distance metrics for distance units for calculating the nearest data points. Knn gives better results when data is large and noisy [25,2].

3.4 RESULTS AND DISSCUSSION

The proposed method uses collected ECG data set of patients suffers from arthymia diseases and using ML algorithms there is a need to classify the ECG signal for abnormality or normality. The collected input signals are analyzed by using different filter methods like low pass and high pass filter to check the level of noise present on the signal. Detecting the peak present on the QRS area and extracting the important characteristics for the ECG signal helps to detect the presence of arthymia diseases. Figures 3.3 and 3.4 show the difference between normal and abnormal heart rhythm from the ECG signal.

Model Training: this phase uses dataset consists of (4045, 10505) where 4045 are normal ECG record and there are 10505 records are for abnormal ECG signal. The split the data set into training and testing with random state is 1.

Here 70% and 30% splitting ratio are used. During training checked the corresponding loss also. Figure 3.5 illustrates the corresponding statistics of the model training loss and corresponding validation loss [26].

During the model construction for the input, reconstruction of signals and corresponding error level were checked. These are shown in Figures 3.6 and 3.7 with training loss details.

Next we show the graph for training loss and test loss in Figures 3.8 and 3.9.

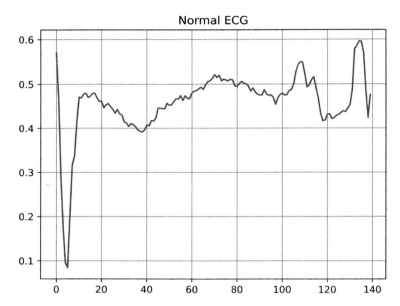

FIGURE 3.3 Normal ECG.

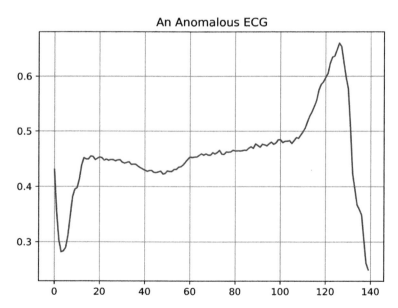

FIGURE 3.4 Anomalous ECG.

Detection of Abnormality in Heart Rhythm

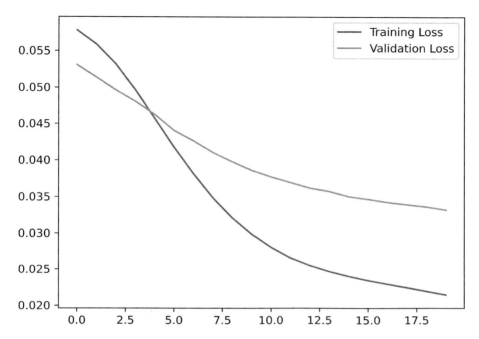

FIGURE 3.5 Model training.

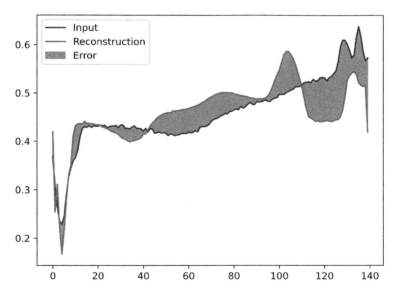

FIGURE 3.6 Model construction.

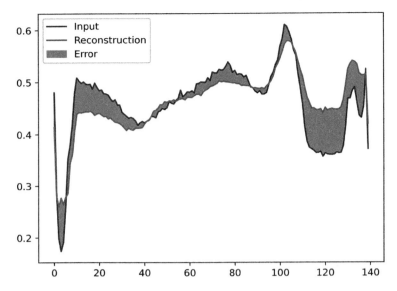

FIGURE 3.7 Model loss.

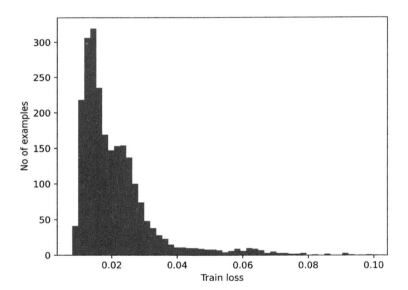

FIGURE 3.8 Training loss.

In the developed model shows the following statistics which are used for classification of abnormality in the heart rthyms which are as follows Accuracy = 0.94, Precision = 0.992243, Recall = 0.90892854

Several machine learning algorithms applied to classify abnormal and normal ECG signals are shown in Table 3.2.

Detection of Abnormality in Heart Rhythm

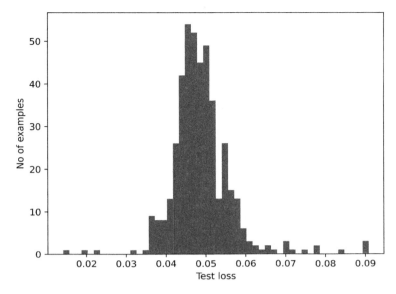

FIGURE 3.9 Test loss.

TABLE 3.2
Comparison matrix for ML algorithms

SL. No	Algorithm	Accuracy
1	Naive Bayes	0.6975690152451587
2	SVN	0.8030490317264112
3	Decision Tree	0.8908117016893284
4	**Random Forest**	0.9546765554182118
5	KNN	0.6464580152451587

From the above table it is observed that random forest algorithm gives good accuracy as compared to other machine learning algorithms.

Discussions: Acquisition of polarization working of the heart is capture by the use of ECG signal. The polarized signals are collected up by the associated electrodes of the corresponding 12 lead ECG machines that record the working of the heart from distinct angels over times. Polarization and depolarization of the of the heart is essential as it is an continuous processes which are responsible for contraction and de-contraction of the heart for pumping of blood among several parts of the body. Without polarization and depolarization it is never happened. In an one cycle of an ECG consists of patterns like P, Q, R and S signals or waves . Details of the waves are already mentioned in the above. Examining the chances of several AI techniques for interpreting an ECG signals regarding the diagnosis of the heart condition is a common activity. The difficulty of

ML algorithm model's interpretability has hindered doctors from having confidence in the diagnosis results of ML models. Each recording undergoes pre-processing stage in-order to extract the important feature vectors from the ECG signals. From 0 Hz range to 85 Hz the power spectrum density and HRV-based characteristic are in consideration. The pre-processing activity generates a feature vector of several dimensions. Purifying the signal by avoiding noise level is an important challenge as the noise level increases the performance decreases. The arterial fibrillation (AF) is an atypical electrical activity that helps to detect abnormal activity of the heart. Since chambers of the heart or atria cannot pump the blood normally. atrial tachycardia which is responsible for the characterization of the heart rate that is excesses of 100 beats per minute which is presence of an abnormality. Moreover, early detection is very essential for avoiding fatality. Predicting mortality demands analyzing the polarization and depolarization activity of the cardiovascular system is essential. The ECG signals help to show the artria and ventricular activity of the heart with the help of polarization and depolarization. The early detection of the symptoms related to AF and the prognosis of AF activity are both needs use of ML and AI techniques. Further identifying abnormality and correctly classifying the abnormality are the primary concerns related to capture ECG signal analysis. Below figure 8 shows ECG signal abnormality detection graph for better understanding the abnormality [27,2].

3.5 CONCLUSION

The above research helps to classify the abnormality of heart rhythm from the ECG signal. By reducing the noise level helps to improve the better classification accuracy. Further Random Forest algorithm approach can perform better as compared to other classification techniques. Diagnosis of heart diseases from tracing of ECG signal is complex for clinical physicians working at respective levels. These difficulties give opportunity for the involvement of the ML techniques to analyze the ECG signals more deeply for better prediction by extracting important features. The black box nature of these ML algorithms and their respective performances helps to detect abnormality of the heart rhythm more precisely. Random forest algorithm creates different decision trees that help us to find more clarification of heart rhythms. From the available heart ECG data set are preprocessed for removing noise level then application of several ML methods, their results and complexity can easily chose the best methods. In conclusion the respective results achieved help us better diagnosis of the patients with good physicians. Further new approach should be discovered and existing methods needs to be formalized to achieve physicians level innovation behind the use of ML models decisions approach.

REFERENCES

[1] Huiyi Wu, Kiran Haresh Kumar Patel, Xinyang Li, et al., "A fully-automated paper ECG digitisation algorithm using deep learning", Scientific Reports, Volume 12, 2022, Pages 20963. https://doi.org/10.1038/s41598-022-25284-1.

[2] Ananya Mantravadi, Siddharth Saini, Sai Chandra, Sparsh Mittal, Shrimay Shah, Sri Devi, and Rekha Singhal, "CLINet: A novel deep learning network for ECG signal

classification", *Journal of Electrocardiology*, Volume 83, March–April 2024, Pages 41–48. https://doi.org/10.1016/j.jelectrocard.2024.01.004

[3] K.A. Sharada, KSN Sushma, V. Muthukumaran, T.R. Mahesh, B. Swapna, and S. Roopashree, "High ECG diagnosis rate using novel machine learning techniques with Distributed Arithmetic (DA) based gated recurrent units", Microprocessors and Microsystems, Volume 98, April 2023, Pages 104796. https://doi.org/10.1016/j.micpro.2023.104796.

[4] Mohammed B. Abubaker, and Bilal Babayiğit, "Detection of Cardiovascular Diseases in ECG Images Using Machine Learning and Deep Learning Methods", IEEE Transactions on Artificial Intelligence, Volume 4, Issue 2, April 2023, Pages 373–382. ISSN: 2691-4581 doi: 10.1109/TAI.2022.3159505, Publisher: IEEE.

[5] Eugenio Vocaturo, and Ester Zumpano, "ECG Analysis via Machine Learning Techniques: News and Perspectives", Published in: 2021 IEEE International Conference on Bioinformatics and Biomedicine (BIBM), Date of Conference: 09-12 December 2021, Date Added to IEEE Xplore: 14 January 2022, ISBN Information: Electronic ISBN:978-1-6654-0126-5, Print on Demand(PoD) ISBN:978-1-6654-2982-5. doi: 10.1109/BIBM52615.2021.9669776, Publisher: IEEE, Conference Location: Houston, TX, USA.

[6] Arunashis Sau, and Fu Siong Ng, "The emerging role of artificial intelligence enabled electrocardiograms in healthcare", BMJ Medicine, Volume 2, 2023, Pages e000193. doi:10.1136/ bmjmed-2022-000193.

[7] Anupreet Kaur Singh, and Sridhar Krishnan, "ECG signal feature extraction trends in methods and applications", BioMedical Engineering OnLine, Volume 22, Article number: 22, 08 March 2023. https://doi.org/10.1186/s12938-023-01075-1..

[8] Pierre Elias, Timothy J. Poterucha, Vijay Rajaram, Luca Matos Moller, Victor Rodriguez, Shreyas Bhave, Rebecca T. Hahn, Geoffrey Tison, Sean A. Abreau, Joshua Barrios, Jessica Nicole Torres, J. Weston Hughes, Marco V. Perez, Joshua Finer, Susheel Kodali, Omar Khalique, Nadira Hamid, Allan Schwartz, Shunichi Homma, Deepa Kumaraiah, David J. Cohen, Mathew S. Maurer, Andrew J. Einstein, Tamim Nazif, Martin B. Leon, and Adler J. Perotte, "Deep learning electrocardiographic analysis for detection of left-sided Valvular Heart Disease", Journal of the American College of Cardiology, Volume 80, Issue 6, 9 August 2022, Pages 613–626. https://doi.org/10.1016/j.jacc.2022.05.029.

[9] Armin Shoughi, and Mohammad Bagher Dowlatshahi, "A practical system based on CNN-BLSTM network for accurate classification of ECG heartbeats of MIT-BIH imbalanced dataset", Published in: 2021 26th International Computer Conference, Computer Society of Iran (CSICC), Date of Conference: 03-04 March 2021 Date Added to IEEE Xplore: 07 May 2021. doi: 10.1109/CSICC52343.2021.9420620, Publisher: IEEE, Conference Location: Tehran, Iran.

[10] Salah S. Al-Zaiti, Christian Martin-Gill, Jessica K. Zègre-Hemsey, Zeineb Bouzid, Ziad Faramand, Mohammad O. Alrawashdeh, Richard E. Gregg, Stephanie Helman, Nathan T. Riek, Karina Kraevsky-Phillips, Gilles Clermont, Murat Akcakaya, Susan M. Sereika, Peter Van Dam, Stephen W. Smith, Yochai Birnbaum, Samir Saba, Ervin Sejdic, and Clifton W. Callaway, "Machine learning for ECG diagnosis and risk stratification of occlusion myocardial infarction", *Nature Medicine*, Volume 29, 2023, Pages 1804–1813.

[11] Khaled Rjoob, Raymond Bond, Dewar Finlay, Victoria McGilligan, Stephen, J. Leslie, Ali Rababah, Aleeha Iftikhar, Daniel Guldenring, Charles Knoery, Anne McShane, Aaron Peace, and Peter W. Macfarlane, "Machine learning and the electrocardiogram over two decades: Time series and meta-analysis of the algorithms, evaluation metrics

and applications", *Artificial Intelligence in Medicine*, Volume 132, October 2022, Pages 102381. https://doi.org/10.1016/j.artmed.2022.102381.
[12] Raymond R. Bond, D. D. Finlay, C. D. Nugent, Cathal Breen, D. Guldenring, and M. J. Daly, "The effects of electrode misplacement on clinicians' interpretation of the standard 12-lead electrocardiogram", European Journal of Internal Medicine, Volume 23, Issue 7, October 2012, Pages 610–615.
[13] Shaan Khurshid, "Clinical perspectives on the adoption of the artificial intelligence-enabled electrocardiogram", Journal of Electrocardiology, Volume 81, November–December 2023, Pages 142–145.
[14] Mohammad Mahbubur Rahman Khan Mamun, and Tarek Elfouly, "AI-Enabled electrocardiogram analysis for disease diagnosis", Applied System Innovation, Volume 6, Issue 5, 20 October 2023, Pages 95.https://doi.org/10.3390/asi6050095.
[15] Shaan Khurshid, "Clinical perspectives on the adoption of the artificial intelligence-enabled electrocardiogram", Journal of Electrocardiology, Volume 81, November–December 2023, Pages 142–145. https://doi.org/10.1016/j.jelectrocard.2023.08.014.
[16] Yaqoob Ansari, Omar Mourad, Khalid Qaraqe, and Erchin Serpedin, "Deep learning for ECG Arrhythmia detection and classification: an overview of progress for period 2017–2023", Frontiers in Physiology, Volume 14, 15 September 2023, Pages 1246746. https://doi.org/10.3389/fphys.2023.1246746
[17] Md Moklesur Rahman, Massimo Walter Rivolta, Fabio Badilini, and Roberto Sassi, "A systematic survey of data augmentation of ECG signals for AI applications", *Sensors (Basel)*, Volume 23, Issue 11, June 2023, Pages 5237. doi: 10.3390/s23115237.
[18] Yehualashet Megersa Ayano, Friedhelm Schwenker, Bisrat Derebssa Dufera, and Taye Girma Debelee, "Interpretable machine learning techniques in ECG-based heart disease classification: A systematic review", *Diagnostics*, Volume 13, Issue 1, 29 December 2022, Pages 111. https://doi.org/10.3390/diagnostics13010111.
[19] Zahra Ebrahimi, Mohammad Loni, Masoud Daneshtalab, and Arash Gharehbaghi, "A review on deep learning methods for ECG arrhythmia classification", *Expert Systems with Applications*, Volume 7, September 2020, Pages 100033. https://doi.org/10.1016/j.eswax.2020.100033.
[20] Giovanni Baj, Ilaria Gandin, Arjuna Scagnetto, Luca Bortolussi, Chiara Cappelletto, Andrea Di Lenarda, and Giulia Barbati, "Comparison of discrimination and calibration performance of ECG-based machine learning models for prediction of new-onset atrial fibrillation", *BMC Medical Research Methodology*, Volume 23, Article number: 169, 22 July 2023.
[21] Shradha Naik, Saswati Debnath, and Vijin Justin, "A Review of Arrhythmia Classification with Artificial Intelligence Techniques: Deep vs Machine Learning", in: *2021 2nd International Conference for Emerging Technology (INCET), 21-23 May 2021*, 22 June 2021. doi: 10.1109/INCET51464.2021.9456394.
[22] Panteleimon Pantelidis, Maria Bampa, Evangelos Oikonomou, and Panagiotis Papapetrou, "Machine learning models for automated interpretation of 12-lead electrocardiographic signals: a narrative review of techniques, challenges, achievements and clinical relevance", *Journal of Medical Artificial Intelligence*, Volume 6, 30 May 2023. doi: 10.21037/jmai-22-94.
[23] Fajr Ibrahem Alarsan, and Mamoon Younes, "Analysis and classification of heart diseases using heartbeat features and machine learning algorithms", *Journal of Big Data*, Volume 6, Article number: 81, 31 August 2019, Pages 1–5.
[24] John Irungu, Timothy Oladunni, Andrew C. Grizzle, Max Denis, Marzieh Savadkoohi, and Esther Ososanya, "ML-ECG-COVID: A machine learning-electrocardiogram

signal processing technique for COVID-19 predictive modeling" *IEEE Access*, 8 December 2023. doi: 10.1109/ACCESS.2023.3335384.

[25] Shi Su, Zhihong Zhu, Shu Wan, Fangqing Sheng, Tianyi Xion, Shanshan Shen, Yu Hou, Cuihong Liu, Yijin Li, Xiaolin Sun, and Jie Huang, "An ECG signal acquisition and analysis system based on machine learning with model fusion", *Sensors*, Volume 23, Issue 17, 3 September 2023, Pages 7643. https://doi.org/10.3390/s23177643.

[26] Hanna Vitaliyivna Denysyuk, Rui João Pinto, Pedro Miguel Silva, Rui Pedro Duarteb, Francisco Alexandre Marinho, Luís Pimenta, António Jorge Gouveia, Norberto Jorge Gonçalves, Paulo Jorge Coelho, Eftim Zdravevski, Petre Lameski, Valderi Leithardt, Nuno M. Garcia, and Ivan Miguel Pires, "Algorithms for automated diagnosis of cardiovascular diseases based on ECG data: A comprehensive systematic review", *Heliyon*, Volume 9, Issue 2, February 2023, Pages e13601. https://doi.org/10.1016/j.heliyon.2023.e13601.

[27] Yehualashet Megersa Ayano, Friedhelm Schwenker, Bisrat Derebssa Dufera, and Taye Girma Debelee, "Interpretable machine learning techniques in ECG-Based heart eisease classification: A systematic review", Diagnostics (Basel), Volume 13, Issue 1, January 2023, Pages 111. PMCID: PMC9818170 PMID: 36611403. doi: 10.3390/diagnostics13010111.

4 AI in Drug Discovery and Development

Tahreem Shahzad, Arooj Fatima Tul Zahra, Ayesha Naeem, and Mujahid Tabassum

4.1 INTRODUCTION

Artificial intelligence (AI) has proven to be a highly beneficial drug discovery and development technology. The application of AI in drug design is expected to increase the speed of drug discovery, reduce costs, and eventually reduce development time, increasing the success rate in clinical trials (Perumal et al., 2022). One of the main applications of AI in drug discovery is targeting potential drugs (Mak, Wong, and Pichika 2022). By analyzing vast amounts of biological data, AI can accurately identify drug targets, enabling the discovery of potential drugs. AI algorithms, such as machine learning models and deep learning techniques, have also helped predict the efficacy and safety profiles of potential drug candidates, allowing researchers to focus on the most promising options. In this way, AI could indirectly contribute to improved drug discovery by predicting a given drug candidate's optimum dosage, formulation, and delivery method. This predictive power indirectly helps prevent any probable risks of side effects and boosts drug efficacy. An emergent scenario is virtual screening, where many advances are being made in applying AI to drug discovery. AI algorithms help rapidly assess compound binding to a target protein by identifying its binding sites and, at the same time, help comb through large compound databases, speeding up the virtual identification of hits. Moreover, AI helps analyze massive clinical trial data, enabling researchers to identify the correct groups of patient demographics who are most likely to benefit from the treatment. This focused approach aims to increase drug development success rates in clinical trials and reduce the time and costs associated with getting a new drug to the market (Gupta et al. 2021).

AI is poised to disrupt the drug discovery and development space; it will hasten the drug discovery process by identifying new drug targets, streamline the drug discovery and development process, and contribute to the potential of personalized medicine. Thus, with the help of large datasets that include genetic information, protein structures, and disease pathways, AI also helps cut the time and costs of finding new drug targets (You et al. 2022).

4.2 OBJECTIVE

This chapter's main objective is to describe in detail the transformational role AI plays in revolutionizing the pharmaceutical industry, particularly drug discovery and development. It will also touch on how technology driving AI has pushed challenges in the sector. These challenges include high R&D costs, long timelines, and increasing disease complexity.

4.3 HISTORICAL CONTEXT OF AI IN PHARMACEUTICALS

4.3.1 Early Development

The early history of AI within the pharmaceutical industry must be considered against the backdrop of the developments in computer science and machine learning. The very first serious attempts to introduce AI into drug discovery could be noticed back in the 1960s and 1970s when computational chemistry and molecular modeling, for that matter, had nothing to do with routine practical applications. Scientists back in those days would develop algorithms to predict the structures of molecules or simulate chemical reactions. This began computer-aided drug design (Yang et al. 2019).

4.3.2 Rule-Based Systems and Expert Systems

Based on if-then logic, rule-based systems developed in the 1970s and 1980s aimed to capture the knowledge of experienced chemists and pharmacologists. Early expert systems like DENDRAL and MYCIN were designed to assist with chemical analysis and diagnosis. Their principles were later adapted to explore potential drug compounds (Alther and Reddy, 2015.).

4.3.3 Molecular Modeling and Docking

Computer-aided drug design methods evolved through the '80s and '90s, and molecular docking gained traction. This approach used algorithms to simulate the interaction between small molecules (drug candidates) and target proteins, predicting how well a molecule would bind to a receptor (Godwin, Melvin, and Salsbury 2016).

4.3.4 Quantitative Structure-Activity Relationship (QSAR)

Another significant development was the acceptance of QSAR models that applied statistical and machine learning methodologies to derive relationships between a molecule's chemical structure and biological activity. Such models provided means for prediction to researchers of the kind of compound that would be most effective against a specific type of disease, thus rationalizing the early part of drug discovery (Staszak et al. 2022).

4.3.5 Databases and High-Throughput Screening

With the setup of large chemical and biological databases in the late 1990s and early 2000s, high-throughput screening technologies enabled the production of a lot of data. This provided a fertile ground for running machine learning algorithms to analyze trends, find novel drug targets, and optimize chemical structures (Pyzer-Knapp et al. 2015).

4.3.6 Integration of Genomics Data

Its completion in 2003 was a watershed, opening the potential use of AI for a much more extensive reach in pharmaceuticals. Integration of genomic data was done, and algorithms were developed for a specific disease to find the genetic markers as well as predict the most effective drug for a patient based on their genetic set (Udegbe et al. 2024).

4.4 AI IN DRUG DISCOVERY

The vast chemical diversity, which encompasses over 10^{60} possible molecular structures, provides a rich and diverse landscape for discovering many new medicinal compounds. However, integrating AI could have immense potential in drug development. Doing so hastened the affirmative and the phase of design optimization of targets (Xue et al. 2018). The AI algorithms have been very successful in predicting blockbuster drug candidates and the optimization of molecular structures for such drugs. Another sign of their speed in this work is the optimization of molecular structures for such drugs, which would position AI to enable drug development at reduced cost with quicker timelines. The virtual representation of chemical space resembles an abstract map that illustrates the distribution of various molecular structures, helping researchers identify and prioritize promising candidates for further investigation in drug discovery. The illustration is quite helpful in identifying bioactive compounds, hence allowing one to select those with promise for further experiments with the help of virtual screening (Maia et al. 2020).

In silico techniques, such as structure-based and ligand-based methods, rapidly exclude the analysis of compound profiles and find potential drug molecules at low cost. Advanced drug design algorithms, including Coulomb matrices and molecular fingerprinting, evaluate various factors such as physical features, chemical properties, and toxicological profiles, aiding in the formulation of effective drug candidates. The combination of AI and drug development promises to overhaul the existing system, ushering in a new era of rapid and cost-effective medication development that benefits patients worldwide (Ahmadi 2023); see Figure 4.1. The integrated machine learning approach, combined with molecular docking, plays a crucial role in drug repurposing. First, drug and disease data are sourced and formatted in a feature matrix for training a predictive model. Next, COVID-19 host target data is fed into it to predict the repurposed drugs. Next, COVID-19 host target data is fed into it to predict the repurposed drugs. Literature mining and evidence collection are essential for validation, particularly in the context of drug repurposing for COVID-19, where researchers utilize existing studies and data to identify potential therapeutic candidates (Ahmed et al. 2022).

AI in Drug Discovery and Development

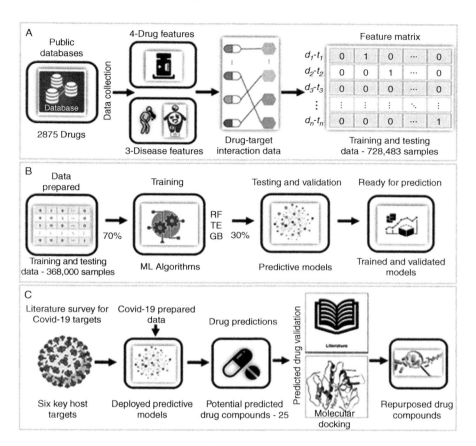

FIGURE 4.1 The AI used in natural product-inspired drug discovery, i) subsets comprising known and unknown drug-target interaction data, which are enriched with diverse drug and disease features (Figure 4.1A); ii) correspondingly, machine learning models can be trained (Figure 4.1B). The trained models would further need to iii) identify new DTI pairs for proteins associated with diseases of viral origin like COVID-19 (Figure 4.1C). Literature surveys on molecular docking to further validate the interactions through binding affinity against the predicted drug-target pairs, as shown in Figure 1C. We were able to predict 25 possible drug candidates against COVID-19. Literature validation identified 12 of these 25 drug compounds at 48% that were already against COVID-19. Molecular docking was then applied to the remainder of the 13 compounds. Four potential drugs with the best docking results and prediction scores were suggested further for validation at both preclinical and clinical levels to be used against COVID-19 (Ahmed et al. 2022).

4.5 AI IN DRUG SCREENING

In drug discovery, the way from the first screening to the market is an ordeal, usually spanning over a decade, with average massive financial inputs required to the tune of about US$2.8 billion. This is made even more difficult because nine out of ten drug candidates that show great promise at an earlier stage of development do not make it through the lengthy trial and error processes of clinical trials—never progressing to phase III, let alone getting approval. From that perspective, AI integration could provide transforming solutions toward speeding up and optimizing the drug screening mechanism (Tripathi et al. 2023). Many modern algorithms are leading the AI-powered drug discovery forefront, each embracing a new way to make headway within the vast chemical space to identify promising therapeutics. Some of these algorithms include Nearest-Neighbor Classifiers, Random Forests (RF), Extreme Learning Machines (ELM), Support Vector Machines (SVMs), and others the use of advanced AI algorithms is a significant player in the field of virtual screening (VS), offering high efficiency and precision that allows researchers to effectively traverse chemical compound libraries (N. Singh, Chaput, and Villoutreix 2021).

The biopharmaceutical industry is a part of transforming the potential value of AI in drug discovery (Murray 2022). Major players, including Bayer, Roche, and Pfizer, have realized the strategic imperative of leveraging AI-driven technologies to complement their drug discovery pipelines. Engaging with IT tech giants, these pharmaceutical behemoths are at the forefront of developing cutting-edge platforms engineered for therapeutic discovery across diverse areas, including but not limited to immuno-oncology and cardiovascular diseases, ultimately aiming to improve patient outcomes (Turner 2023). This represents a massive tectonic shift from the confluence of AI and drug discovery to the power it will bring to bear on innovation and efficiency. AI will enable researchers to speed up finding likely drug candidates and decrease the risks of developing these drugs so that life-saving medicines can reach patients faster. AI integration promises democratic drug discovery (i.e., flattening drug discovery and breaking traditional barriers to entry), letting researchers from all over the world engage in seeking new therapeutics. With AI as a force multiplier, the future of drug discovery will be agile, precise, and above all, remain steadfast in commitment to improving global healthcare outcomes (Salim 2020).

4.6 PREDICTION OF PHYSICOCHEMICAL PROPERTIES

The physicochemical properties predicted are one of the most critical aspects of the drug development process, as they govern solubility, the partition coefficient (logP), the degree of ionization, and intrinsic permeability that affects the pharmacokinetic profile and interaction at the site of action. These physicochemical properties must be considered seriously while designing any new drug. The appearance of AI in recent years is going to pave the way for the revolution in predicting physicochemical properties by offering a suite of powerful tools that could help in easing and improving this aspect of drug design (Han et al. 2023). This is achieved using machine learning approaches that deploy big data generated during the optimization of the drug formulation process to develop predictive models. All this data has to be further perused by the ML algorithms so that the patterns and relationships of molecular structures with physicochemical properties can be found to make precise predictions for the novel compounds (Goldsmith et al. 2018).

Among the AI-based drug designs, molecular descriptors such as SMILES strings, implicit energy measures, electron density distributions, and three-dimensional geometrical parameters play a core role. The feeding of these descriptions to deep neural networks (DNNs) enables high-accuracy predictions of the physicochemical properties of the drug molecule. Furthermore, since AI methods are scalable and efficient, the investigation of large chemical spaces can be accelerated with the respective identification of which of these compounds will be worth pursuing at an early stage as candidates for potential drugs, far earlier than conventional methods (H. Chen et al. 2018). The ALGOPS software was used to predict solubility and lipophilicity for various compounds. In this software, a neural network-based predictor was used to indicate the physicochemical properties mentioned earlier, such as solubility, lipophilicity, and other relevant characteristics (Bhattamisra et al. 2023). The Estimation Programme Interface (EPI) Suite was developed by Zang et al. in 2017 and utilizes a quantitative structure-property relationship (QSPR) approach to examine six environmental physicochemical properties of chemicals compiled by the Environmental Protection Agency (EPA) (Zang et al. 2017).

All these innovations will enhance drug discovery programs by leveraging AI's predictive power for more informed decision-making, thereby accelerating the development of novel therapeutic approaches. Beyond the realm of drug design, there seems to be some promise for AI-powered tools in physicochemical property prediction for applications in pharmacokinetics and pharmacodynamics modeling (Gangwal and Lavecchia 2024). Incorporating AI-powered predictive models into computational frameworks provides deep insights and optimization of dosing regimens, leading to improved therapeutic outcomes related to the drug's behavior inside the body The integration of AI into the drug discovery and development process will revolutionize the field, offering new opportunities to enhance the quality of drug design and optimization.

4.7 PREDICTION OF BIOACTIVITY

Predicting the bioactivity of drug molecules is essential in drug discovery, as bioactivity is defined by the effectiveness of a drug in binding to its target proteins or receptors. A molecule lacking adequate affinity for the target protein or receptor will not elicit the desired therapeutic response (Srinivasarao and Low 2017). Such drugs may bind randomly to non-desired proteins or receptors and, thus, are more likely to produce side effects or toxicity. Predicting this drug-target binding affinity (DTBA) is central to predicting the binding efficiency of the drug molecules in the respective targets and assessing the efficacy and safety of potential therapeutics (Thafar et al. 2019; Aleb 2022). In the last couple of years, methods based on AI to predict drug-target interactions have become popular as quite efficient tools to provide a clue about the chances of successful pharmacological outcomes. On the other hand, different AI-based methodologies approach the evaluation of the affinity binding between drugs and targets for a selection of promising drug candidates (S. Chen et al. 2022). One methodology is point-based modeling approaches that concentrate on chemical structures for the drug molecule and the target protein. The point-based models analyze the respective molecular features and properties of these entities, generating point vectors that represent the features. The compatibility of these vectors is checked, offering clues in predicting the likelihood of drug binding to the target (Giguère et al. 2015).

On the contrary, similarity-based modeling directly compares and estimates the closeness of the drug molecule to the ligand or compound known to specifically bind with the target protein. This principle acts on the postulate that if two molecules have analogous structures or similar properties, then the two molecules are likely to have analogous activity toward a common target. The models that predict drug bioactivity have a comparable basis in using the property of the drug concerning similarity to some known pharmacologically active compounds (Alberga et al. 2019). Using AI the researcher can tap into substantial chemical spaces and, at the same time, evaluate possible drug-target interactions in a computationally cost-effective way. The methodologies analyze vast sets of data quickly and provide very accurate predictions, thus considerably speeding up the process of drug discovery when the two are combined (Lavecchia 2015).

Such bioactivity prediction powered by AI offers the possibility to apply it in areas exceeding the scope of drug-target interactions, including toxicity assessment and drug repurposing. While minimizing the risk of off-target interactions and adverse effects, it becomes possible to identify safer and more effective drug candidates (Han et al. 2023). AI technologies implemented during drug development have changed the landscape of bioactivity prediction. The predictive ability of drug-target interaction with preciseness at a faster pace suggests that AI will help shape the future landscape of pharmaceutical research and innovation (Jiménez-Luna et al. 2021).

4.8 AI IN DESIGNING DRUG MOLECULES

4.8.1 Prediction of the Target Protein Structure

The success of treatment outcomes lies in identifying the exact target in drug solution development. In many diseases, several proteins are involved that play crucial roles in manifestation and development, and some of the proteins are overexpressed, which adds to the need for the targeting of interventions (Hare et al. 2017). This makes predicting the structure of the target protein crucial for enabling and fostering the design of tailor-made medicinal formulations targeted at specific pathologies. The current development in structure-guided drug discovery places AI as a powerful ally that offers sophisticated tools for predicting 3D protein structures. Predicting the structure of the target protein synchronizes the design process with the chemical landscape of that protein (Hameduh et al. 2020).

A significant stride in this direction is AlphaFold, which is the salient AI tool primarily based on the use of DNNs for the prediction and analysis of complex 3D structures of proteins. AlphaFold utilizes a method that calculates the distances between neighboring amino acids and the corresponding angles of peptide bonds, enabling it to accurately predict the three-dimensional structures of target proteins (Sanjeevi et al. 2022). AlphaFold is showing great promise in structure-based drug discovery for the prediction of target protein structures for attaining molecular interaction insight of potential drug candidates and their intended targets (Rehman et al. 2024). This will allow, therefore, a rational design of the corresponding therapeutic interventions. This predictive ability shortens the drug development process and enhances the likelihood of identifying a compound with the best efficacy and safety balance. Moreover, this capability to predict drug-target interactions with AI will enable the in-silico screening of many potential drug-target interactions, and it will save not just valuable resources but also avoid requiring expensive and time-consuming experimental assays (Zhou et al. 2016).

This computational approach, utilizing predictive modeling and structural analysis, empowers researchers to rapidly assess the viability of numerous drug candidates, ultimately accelerating the pace of drug discovery and development. The more successful AlphaFold's predictions are, the more accurately it can determine protein structures, highlighting AI's transformational potential to advance our understanding of complex biological systems (Niazi and Mariam 2024). Deciphering the structural arrangement of proteins to such fine detail, AI-driven approaches will open doors for novel therapeutics tailored to basic mechanisms of disease interventions. Understanding protein folding is essential for determining protein function and developing optimized biologics, which directly relates to the capabilities of AlphaFold in accurately predicting protein structures.

4.8.2 Modern Methods of Drug Discovery and Development

The search for new and improved drugs is as an essential tool for humanity when struggling with a broad spectrum of diseases, from new epidemics to old, persistent evils (Murray 2022). In this continuous search, classical pharmacology and reverse pharmacology are two fundamental approaches to drug discovery (Zhou et al. 2016). Classical pharmacology is a historical term that refers to systematic investigations of natural products and plant extracts to pursue their therapeutic properties. Such approaches are old school; these are very cautiously searched among some of the biological sources to find helpful compounds for medicinal purposes (David, Wolfender, and Dias 2015). At the same time, researchers will continue to tap such immense reservoirs of nature's diversity for novel bioactive small molecules to find new leads to fight diseases and human suffering. On the other hand, reverse pharmacology is a modern, upcoming drug discovery process describing the understanding of complex molecular mechanisms of pathological diseases. In reverse pharmacology, the characterizing and identification of molecular targets, such as genes or proteins, helps in understanding the onset and progression of certain diseases (Hampel et al. 2018). Such experiments not only contribute to a better elucidation of the role of such targets but also aid in identifying how they could be modulated pharmacologically, leaving the route open for targeting drugs designed to affect disease-related pathways. Therefore, some potential drug targets have been identified, both with classical and reverse pharmacology, forming a basis for further exploration and refinement. Target validation and its pre-clinical testing are some very crucial steps in the drug discovery pipeline after the identification of a target (H. Zhang et al. 2019). This includes the effectiveness and safety of lead compounds through extensive experimental candidate drug assays and animal studies to confirm that promising leads are ready for clinical development. The drug discovery process, encompassing target identification through clinical testing, is often lengthy and complex, despite progress in drug discovery technologies and methodologies.

Among the modern technologies that appeared over the last decade in the fight against cumbersome drug discovery processes were high-throughput screenings and AI. The use of "AI" allows you to work with large volumes of data quickly and get the most accurate predictions regarding the possible interaction of the drug with its targets. It is only with the advent of high-throughput screening technologies that we can efficiently assess even an extensive library of compounds in the shortest possible time and detect, if any, a hint of a likely drug candidate (Hessler and Baringhaus 2018).

AI-based small molecule drug discovery is gaining momentum at the intersection of computer science and life sciences. Fragment-based drug discovery has now been proposed as a new method for compound discovery. Chemical structures can be encoded in much the same way as the representation of natural language text, using linear representations. Much like sentences are tokenized in NLP, molecular fragments can be defined as words in a sentence. This idea can be incorporated in the case of fragment-based drug discovery, as demonstrated in Figure 4.2. Across a broader chemical space, molecular fragments that consist of atoms and chemical bonds can be used to construct more effective representation scenarios that offer flexibility in the molecular representation process (Jinsong et al. 2024). Equally important is the active collaboration of academia, industry, and regulatory agencies in drug discovery and development. With this interdisciplinary collaboration and shared expertise, these stakeholders can better handle challenges confronting them on their way to translating a new science discovery into new treatment types.

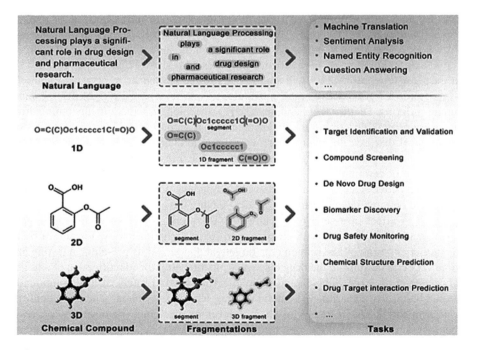

FIGURE 4.2 Development advantages of molecular fragmentation (Jinsong et al. 2024).

AI in Drug Discovery and Development

4.9 OVERVIEW OF AI TECHNOLOGIES IN DRUG DEVELOPMENT

4.9.1 Supervised Learning Models

Supervised learning is a type of machine is characterized by algorithms learning from training labeled data to make predictions or classifications from new, unseen data. It serves as the basis for prediction properties of the drug compound tasks and classification tasks based on the available outcomes (Southwest Jiaotong University, China et al. 2015).

4.9.2 Unsupervised Learning Models

The unsupervised learning models uncover the hidden structures and patterns in a data set that lacks labeled outcomes and thus have value in investigating high-dimensional biological and chemical data. Unsupervised learning includes (Mahmud et al. 2021) the following.

1.8.2.1 Clustering of Compounds: Clustering groups similar compounds, either structurally or functionally, to assist researchers in finding new classes of drugs or identifying compo

1.8.2.2 Anomaly Detection: Such deviations within compounds or in biological responses from the normal pattern may identify unique therapeutic effects or potential off-target interactions.

1.8.2.3 Dimensionality Reduction: One of the approaches that achieves this is the use of Principal Component Analysis (PCA) and t-Distributed Stochastic Neighbor Embedding (t-SNE), both of which are techniques for reducing the dimensionality of complex data sets while preserving their structure and relationships (Skublov, Gavrilchik, and Berezin, 2022).

4.9.3 Deep Learning Models

Deep learning is a branch of machine learning that uses multi-layered neural networks to model complex patterns in large data sets (Tabassum et al., 2021). It has great value in drug development. Some important applications include (Bikku 2020):

1.8.3.1 Convolutional Neural Networks (CNNs): CNNs have been in use for the determination of disease marker or cellular response to the drug compound by processing microscopy or histopathology images in many biomedical image analyses (Moen et al. 2019).

1.8.3.2 Recurrent Neural Networks (RNNs): Suitable for the analysis of sequential data like time series information, from which it detects trends or predicts outcomes in something like clinical trials or patient records (Yadav et al. 2018).

1.8.3.3 Graph Neural Networks (GNNs): These models structure and model the interactions of molecules to assist a researcher in predicting molecular properties, finding target interactions, and developing new structures that will have some targeted activity. Deep learning models offer breakthrough techniques to derive insights from high-dimensional data and are poised to drive drug discovery and development. It is one of the main tools utilized in drug development, as AI technologies enable machine learning models to provide researchers with a more efficient way to analyze large data sets. It comes with a lot of benefits in the prediction of drug efficacy, shortening development timelines and increasing the power for benefit prediction by and large (Z. Zhang et al. 2022).

4.10 APPLICATIONS OF AI IN DRUG DISCOVERY

4.10.1 Target Identification and Validation

Target validation is a pivotal process in the discovery of effective treatment against disease. The identification and validation of a therapeutic target have historically been rather laborious and time-consuming processes, involving haphazard searches followed by empirical experimentation (Koscielny et al. 2017). Today, with the upsurge of AI, researchers have an extremely viable tool that is helping them accelerate and improve their target identification and validation. Machine learning and deep learning techniques of AI revolutionize the ways targets are identified through the analysis of big biological datasets. This is achieved through the exploration of genomics, proteomics, and clinical data sources of origin from where small variations that reveal potential targets linked to diseases emanate (Gupta et al. 2021; Nayarisseri et al. 2021). AI also authenticates these targets through predictive modeling and simulation. The AI algorithms synthesize data from structural biology, bioinformatics, and pharmacology in making predictions of functional consequences when targeting specific proteins or pathways. This allows researchers to find the most promising targets for therapeutic success. Such AI-driven methods speed up the process of identification and validation of the targets, hence bringing more focus to the attention of the researchers on the targets that are clinically relevant to hoped for outcomes (Quazi 2021). For the pharmaceutical companies, accessing AI's predictive capability makes drug discovery quite efficient, cost-friendly, and more successful in providing new treatment to patients within a shorter time frame (Alizadehsani et al. 2024).

4.10.2 Compound Screening

With the solid identification of drug targets, the next critical step during the drug discovery process is compound screening. In this process, the screening of hundreds of thousands of chemical compounds takes place to establish those that have the desired biological activity against a selected target (Fu et al. 2019). The process is normally costly and lengthy. AI offers a virtual screening technique to solve this. Virtual screening (VS) predicts whether a given compound binds to the target protein, computed through various mathematical models and algorithms by structural and chemical properties. With the availability of virtual compound libraries, many promising drug candidates can undergo screening at very early stages of development using powerful AI algorithms (Maia et al. 2020). VS is divided into two broad categories: structure-based virtual screening, typically denoted as SBVS, and ligand-based virtual screening, typically denoted as LBVS. SBVS results from three-dimensional models of compounds and targets that are mainly validated by experimental methods such as X-ray diffraction and nuclear magnetic resonance or NMR. The primary step of virtual screening—molecular docking—has two phases: the binding of ligands to the binding site of the receptor is done virtually using steric, physical, and chemical properties at the time of screening with the database platform. Thereafter, a mathematical scoring function is employed to count the energetic binding affinity. Significant tools for docking are provided by Auto Dock, Glide, and DOCK. The docking scoring functions have the potential to be improved through the use of methods such as random forest (RF) and support vector machines (SVM) in the machine learning approach (Wu et al. 2024). AI can also open large chemical space to find new chemical scaffolds and leading compounds that otherwise might remain unnoticed with current methods. These are AI-powered approaches for VS that subsequently save time and resources to find a suitable drug candidate (Gryniukova et al. 2023).

4.10.3 Lead Optimization

Lead optimization further refines the potency and selectivity of candidate compounds to a maximum efficacy and minimal toxicity value. Further optimization of a lead involves development of properties like potency and selectivity to candidate compounds with the aim of maximal efficacy and minimal toxicity (D. B. Singh 2018). Promising drug candidates have been identified. This is where AI plays a key role in optimizing leads through predictive models of compound activity and safety profiles. The machine learning techniques set the base for structure-activity relationships (SAR) that are modeled computationally. These techniques are used to predict how changes in the chemical structure of a particular compound affect its biological activity against the target protein. AI also predicts compound absorption, distribution, metabolism, excretion, and toxicity (ADMET). AI also predicts compound absorption, distribution, metabolism, excretion, and toxicity (ADMET). AI algorithms can predict the likelihood of a compound presenting the expected ADMET properties by combining data originating from in vitro assays, animal models, and clinical trials (Sucharitha et al. 2022). They allow the combination of chemistry and exploration of chemical space to develop and synthesize newer libraries of compounds that should have better activity and selectivity profiles. In this way, relatively much less time and resources are required for lead optimization by AI-driven approaches. AI-based lead optimization could change the drug discovery game, allowing researchers to more rapidly, more efficiently, and more effectively identify and improve viable drug candidates. This will ensure that AI-driven drug discovery is cost-effective, with a focused approach and eventually more successful in getting new treatments to patients (Yang et al. 2019).

4.11 PRECLINICAL AND CLINICAL DEVELOPMENT OF AI IN MEDICINE

4.11.1 Preclinical Testing

Preclinical testing is a stage in drug development in which the potential drug candidate's safety, efficacy, and pharmacokinetic properties are fully tested before the compound is allowed to advance into clinical trials. This has been typically driven by the required in vitro and in vivo experiments, most times laborious and time-consuming, thus delaying further progress of potentially valuable drugs (Sinha and Vohora 2018). AI in this regard has revolutionized preclinical testing so that it is undertaken faster and more efficiently than ever before. In that respect, it may thus be used to come up with optimal experimental designs in a bid to mine key biomarkers and predict drug responses within complex biological systems. Predictions in multiple cellular contexts are made by in vitro models out of the high dimensional data generated by the interaction between drugs and cells. In the same vein, in vivo animal-based models are critical tools to provide estimations of safety and efficacy of a drug candidate. Study designs and derived predictive biomarkers from AI-driven analytics unlock data streams very complex in nature from animal experiments (A. V. Singh et al. 2023). This has accelerated the process of drug development through preclinical testing, has reduced costs, and has lowered the industry's dependency on an animal model approach through strategies driven by AI. AI promises to empower researchers with the most rational decisions about which candidates to further develop into clinical trials and that can reach patients safely and efficiently (Xu et al. 2021).

4.11.2 Clinical Trials

Clinical trials are usually expensive, complex, and resource-intensive, and are characterized by high financial costs, long times, and unpredictable results. AI is a game changer, making clinical trials tick in terms of improved design, resulting in efficient execution and decision support of the researchers, and all other processes and sub-processes that will lead to the development of a new drug. AI and big data analytics tools can mine datasets that are replete with patient demographics, clinical outcomes, and genetic information to identify relevant biomarkers and stratify patient populations into appropriate treatment guidelines. This contributes to patient recruitment and retention, identifying eligible candidates by predefined criteria, even able to predict adherence and dropout rates of patients (Sahu et al. 2022). This will boost the rate of recruitment, reduce costs, and in general, make clinical trials more efficient. AI is also very likely to contribute to the revolution of real-time monitoring and data analysis in clinical trials, covering monitoring data from wearable sensors and electronic health records (EHRs) longitudinally and in real time, enabling the early detection of adverse effects and improvement of the prediction of the response to treatments. AI potential goes further to the analysis of real-world clinical data, whereby EHRs form a treasure of insights into disease progression and drug responses, quite ideal for post-market research. Van Laar et al. compared the accuracy of data mining in EHRs through automated versus manual extractions and focused on the outcomes of treatments for renal cell carcinoma. The evaluation showed an F1 score of 100% for variables like sex and mortality, and more than 90% for structured data as laboratory measurements, showing robust precision and recall. However, unstructured data such as adverse drug reactions and comorbidities, often embedded in free text within the EHRs, presented challenges, with F1 scores between 53% and 90%, demonstrating an uphill task in extracting such information well (van Laar et al. 2020). AI-enabled clinical trials offer a unique opportunity to increase efficiency, accuracy, and cost-effectiveness in the process of drug development. They allow researchers to develop and deliver new treatments more rapidly to patients. This would increase both patient outcomes and the entire field of precision medicine (Chopra et al. 2023).

4.11.3 Personalized Medicine

Personalized medicine is a new paradigm in health care, offering patients custom treatment strategies according to their individual genomic, phenotypic, and environmental features. However, the traditional one-size-fits-all treatments often lead to suboptimal care and frequent side effects as compared to such personalized treatment strategies. AI is allowing the analysis of vast amounts of patient data to design better, more accurate, and optimal options for treatment (Goetz and Schork 2018). An example of this: the ability to design treatment regimens wherein human responses are taken into consideration will be one of those ways that genomic medicine is transformed by AI. It will also be powering algorithms through the analysis of genetic variations and clinical data to define how treatment regimens can be designed according to patient response (Sharma et al., 2023). AI algorithms driving genomic medicine will furthermore be used to help detect disease predisposition and susceptibility, drug response, and metabolism in genetic variations. The use of genomic information in clinical decisions is enabled by this, aiding in the customization of treatment strategies to optimize therapeutic benefits while minimizing adverse effects (Cesario et al. 2021). This is further coupled with the consolidation of EHRs, imaging data, and data from wearable sensors in relation to the health status of a patient and its response to treatment. AI algorithms are used in the detection of patterns in data streams from these different sources, in predicting the progress of a disease, and in optimizing treatment plans in real time (Mousavi, Bakar, and Vakilian 2015).

The AI-based processes are a significant improvement over conventional techniques, reducing both time and cost. More particularly, in integration with machine learning applications, AI improves each step of the drug discovery pipeline to be faster and more cost-effective. The strength of AI lies in its ability to provide researchers with recommendations for promising drug candidates through predictive analyses, which rely heavily on the quality and nature of the input data (Nguyen and Vo 2024). Table 4.1 gives a comprehensive comparison between conventional and AI-based protocols, defining the time and cost reduction and success rate improvement at each step of drug discovery.

TABLE 4.1
Comprehensive comparison of Conventional vs. AI-Driven Drug Discovery Processes

Process Stage	Traditional Approach	AI-Driven Approach	Time Reduction	Cost Reduction	Success Rate Increase
Target Identification	It is the resource-expensive screenings of thousands of molecules using biochemical assays.	Uses machine learning algorithms to search through genetic and biomedical data, predicting possible attack sites.	AI cuts the time required to find possible target sites by searching through vast amounts of data very rapidly.	Reduce costs by reducing the consumption of physical reagents and manpower when conducting initial screenings.	Allows for improved accuracy and relevancy of target sites found, means a higher success rate of follow-on processes.
Lead Generation	Built on random screening and serendipity, compound screening in high throughput.	Uses deep learning to simulate and model molecular interactions, to design targeted molecules.	Reduces drug discovery time by directly screening predictive leads, as opposed to trial-and-error testing.	Saves the cost of large-scale screening and chemical synthesis of numerous compounds.	Leads more likely to succeed in later pipeline stages because of more advanced preliminary screening.
Lead Optimization	Requires synthesis and testing cycle to optimize drug candidates to utmost levels of safety and effectiveness.	Uses reinforcement learning to optimize molecules based on desired properties and minimized toxicity to accelerate optimization.	Saves on synthesis bottleneck by predicting results and optimizing the synthesis path.	Reduces cost of redundant synthesis and testing of modified compounds.	Increases chances of clinical successes by more effectively predicting and optimizing drug properties.

AI in Drug Discovery and Development

Preclinical Experiments	Carries out a huge number of in vivo and in vitro experiments for safety and efficacy testing.	Predicts the output of virtual simulations using previous data using neural networks, thereby minimizing the number of animal tests.	Reduces time to low reliance on in vivo experiments that are time-consuming through rapid in silico predictions.	Protects they are cost-focused through a reduction in the number of physical experiments.	Increases accuracy of prediction of pharmacokinetics and toxicology, resulting in overall more realistic development.
Clinical Trials	Uses standard trial designs, which tend to be time-consuming and highly variable in the outputs.	Uses approaches like natural language processing to optimize the design of the trial and the choice of patients, making the trial more efficient.	Saves trial time by more closely matching patients as well as simulating outcomes.	Saves the cost of the trial by improving the efficiency of recruitment so that fewer patients drop out.	Success of trials enhanced, provided the design and patient selection are optimal.
Regulatory Review	Manual submission and review processes that typically involve in-depth documentation and time-delaying, iterative feedback leading to lags.	Includes automated systems for preparation of dossiers and compliance models, enabling streamlined submissions.	Minimizes regulatory review timelines with automated documentation and predictive compliance models.	Reduces costs in terms of resources involved in manual documentation preparation and iterative submissions.	Does not have a direct impact on success rate, but significantly improves efficiency and compliance.

4.12 CHALLENGES AND ETHICAL CONSIDERATIONS

4.12.1 Data Quality and Bias

The key issue in AI for drug discovery lies in the quality and integrity of data. Inadequate or biased data weaken the performance and reliability of AI models through wrong inferences and flawed outcomes. Proactive measures need to be undertaken to reduce possible sources of bias and to ensure representativeness (Kolluri et al. 2022). This approach encompasses strong protocols for data collection, integrates diverse data sources to provide a comprehensive perspective of the target population, and includes quality control mechanisms to identify and rectify bias. Transparency and accountability are major drives in ensuring data quality and lessening bias. In the meantime, researchers must make it explicit which data sources have been used in the development of an AI model and which actions have been taken to decrease the bias and to keep the data clean. Continuous monitoring of AI models is required for uncovering and correcting biases and inconsistencies that creep in over time (Fraisl et al. 2022).

4.12.2 Interpretability and Transparency

Yet another issue in the development of AI drug design is the interpretability and transparency of AI models: complex AI algorithms, especially deep learning models, are normally turned into "black boxes," and their predictions or recommendations end up being hard to understand. Researchers are working in the direction of the explainability of AI models. These could be all manner of techniques, from feature importance analysis and model visualization to explanation methods used to throw light on how these AI models operate and the driving factors of their predictions. In highly regulated industries, such as the pharmaceutical one, full transparency in decision-making is required to be able to assure compliance with regulatory requirements (Hassija et al. 2024).

4.12.3 Perspectives

Regulatory agencies, such as the FDA, ensure that drugs are safe, effective, and of good quality. However, this task is becoming increasingly challenging due to the proliferation of AI-based technologies. Regulatory agencies' major concerns in AI-based drug development are model validation and model verification. As witnessed in the recently released guidance documents by the FDA on AI and machine learning-based medical devices, key points underscore transparency, reliability, and reproducibility in AI-driven technologies. Regulators also consider how best to adapt existing frameworks to enable them to respond appropriately to the intrinsic properties of AI technologies. This could be in the form of new regulatory guidelines and/or standards and a review process designed and meant to review AI models in a drug development setting. Effective collaboration among the agencies, stakeholders, and researchers is needed to help solve these pressing challenges (Thakkar et al. 2023).

4.12.4 ETHICAL IMPACT

This massive uptake of AI in drug development results in huge ethical issues regarding privacy, autonomy, and decision-making. The first relates to concerns about data safety and security, alongside several potential unwanted consequences of algorithmic bias and discrimination. Therefore, the careful protection of individual rights and reasonable access to care require a balance between the benefits that AI can offer in this sector and the respective ethical issues. It is in this respect that the need for strong methodologies of data protection through anonymization and encryption of sensitive patient information, under which transparency and accountability are promoted in the use of AI-driven technologies, is justified (Morley et al. 2020). More broadly, consideration of the societal impact AI has on drug development is attributed to the contribution of disparities in health, differences in access to treatment, and the changing relationship between doctor and patient. These AI-driven technologies at work should be in the interest of the public, ensuring equitable outcomes in health. Although AI is set to revolutionize drug development because much better patient outcomes are expected, it at the same time raises great challenges and ethical concerns. Resolving issues relating to the quality of data and bias, interpretability, transparency, compliance with regulations, and ethical impact will thus render real the potential of AI in health in such a way that individual rights and fair access by researchers and policymakers are guaranteed (Ahmadi and Ganji 2023).

4.13 FUTURE PERSPECTIVES AND INVESTIGATION TOPICS

4.13.1 INNOVATIVE TECHNOLOGIES

Recent technological advances in the application of AI for rapid research and development in the pharmaceutical industry have been recognized as significant breakthroughs. One of these recent breakthroughs is the utilization of generative AI to enable fast design of new molecular structures with predicted properties. In so doing, the former can utilize powerful, generative AI models, like generative adversarial networks and variational autoencoders, in designing new drug candidates with higher efficacy, better safety, and an improved pharmacokinetic profile (Romanelli, Cerchia, and Lavecchia 2024). It is in such areas that generative AI holds great transformational potential. It has opened the promising opportunities of optimum design in workflow and decision processes that can be created in the case of drug development using reinforcement learning techniques (Tabassum et al., 2021). The reinforcement learning algorithms are derived based on experience and behavior and adapt to getting the maximum long-term return in each environment. They are perfect for such applications as compound screening, lead optimization, or design of clinical trials in the framework of hard scientific activity. The use of RL means Reinforcement Learning allows increasing the efficiency of discovering better treatments for a broad spectrum of diseases AI-driven approaches in systems biology and network pharmacology also provide new insights into the disease and drug action mechanisms. AI-driven approaches in systems biology and network pharmacology also provide new insights into the disease and drug action mechanisms. Integration of multi-omics data with computational modeling techniques further assists in specifying the detailed molecular make-up of complex biological networks, hence making it possible for precise molecular targets to be identified for therapeutic intervention. This puts the search for a cure in a holistic approach that allows for tailoring accurate treatments to a particular patient (Erlanson et al. 2016).

4.13.2 Research Gaps

Much progress has now been made at the time of the AI revolution with AI-driven drug design; however, many gaps remain to be filled by further research. This allows one to note the scarce robustness of the developed methodologies for AI model validation and interpretation within the drug development process. Although the predictions given by AI algorithms are often fairly accurate, one of the challenges is interpreting how the predictions have been arrived at. Building confidence in AI-driven drug discovery would involve developing techniques to interpret and validate. Another challenge in the research is the availability of large, representative, and diverse datasets for training AI models. Most of the current datasets display biases and limitations that, in turn, may affect not only AI algorithm performance but also generalizability. It will then be the task of researchers to develop high-quality, representative datasets capturing the whole scope of biological diversity and disease heterogeneity to facilitate the making of correct predictions over the varied populations by AI models.

4.13.3 Changing Landscape

With this in perspective, AI is going to continue playing a major disruptive role in redefining the landscape of pharmaceutical research and bringing dynamism into the trend of shortening the time to market for drugs. Personalized medicine will be taken further with this, as both AI and genomic technologies march ahead. In such a scenario, using genomic and clinical data analysis can help identify patient-specific biomarkers and thereby develop effective treatment strategies. In fact, the idea of adopting this approach is to bring revolutionary changes in patient management so that treatments are more efficient and produce fewer side effects (Zielinski 2021). AI-led collaboration and innovation between academia, industry, and the regulatory environment will be improved. Enhanced sharing of data, tools, and insights between the three domains of academia, industry, and regulation will improve the speed of research collaboration, improving knowledge sharing for drug discovery and better translation into clinical applications. The future of AI in pharmaceutical research and development is nothing short of promising. Groundbreaking technologies will change terrains and business approaches. Addressing research gaps, enabling emerging technologies, and fostering collaboration and innovation with a human touch will help us realize the potential of AI in full force to harness AI for an accelerated drug discovery process, improved patient outcomes, and bettering medicine (Mak and Pichika 2019).

4.14 CONCLUSION

As discussed, the increased disruptive power in the pharmaceutical industry is coming with new opportunities fueled by AI for drug discovery and development. This chapter peered into the myriad ways AI is reshaping the landscape of pharmaceutical research and development, from target identification to compound screening, clinical trials, and even personalized medicine. Its key findings underline the massive potential for AI to shorten the process of drug discovery and assist in the reduction of clinical trials, while, at the same time, ensuring that the strategies of treatment to a patient are being done individually, in a bid towards their personalization. Researchers now humanly invade once-unthinkable insights into the mechanisms of disease development for unprecedentedly precise, personalized therapies with generative AI, reinforcement learning, and systems biology at hand. Yet, some of the existing constraints and deficits need to be acknowledged and rectified. These relate to data quality and bias, interpretability and transparency, regulatory compliance, and ethical considerations—largely meaning key barriers and challenges to the full realization of the promise of AI in pharma. Thus, overcoming challenges, promoting collaborations, and upholding the highest ethical standards will help maximize the potential of AI in accelerating drug discovery, improving patient outcomes, and contributing to further developments in medical science for the greater good of the patient and society.

REFERENCES

Ahmadi, Ali. 2023. "Machine Learning for Drug Discovery: Bridging Computational Science and Medicine." *International Journal of BioLife Sciences* 2 (2): 211–220.

Ahmadi, Ali, and Nima Rabie Nezhad Ganji. 2023. "AI-Driven Medical Innovations: Transforming Healthcare through Data Intelligence." *International Journal of BioLife Sciences* 2 (2): 132–142.

Ahmed, Faheem, Jae Wook Lee, Anupama Samantasinghar, Young Su Kim, Kyung Hwan Kim, In Suk Kang, Fida Hussain Memon, Jong Hwan Lim, and Kyung Hyun Choi. 2022. "SperoPredictor: An Integrated Machine Learning and Molecular Docking-Based Drug Repurposing Framework With Use Case of COVID-19." *Frontiers in Public Health* 10 (June). Frontiers. doi:10.3389/fpubh.2022.902123.

Alberga, Domenico, Daniela Trisciuzzi, Michele Montaruli, Francesco Leonetti, Giuseppe Felice Mangiatordi, and Orazio Nicolotti. 2019. "A New Approach for Drug Target and Bioactivity Prediction: The Multifingerprint Similarity Search Algorithm (MuSSeL)." *Journal of Chemical Information and Modeling* 59 (1). American Chemical Society: 586–96. doi:10.1021/acs.jcim.8b00698.

Aleb, Nassima. 2022. "A Mutual Attention Model for Drug Target Binding Affinity Prediction." *IEEE/ACM Transactions on Computational Biology and Bioinformatics* 19 (6): 3224–32. doi:10.1109/TCBB.2021.3121275.

Alizadehsani, Roohallah, Solomon Sunday Oyelere, Sadiq Hussain, Senthil Kumar Jagatheesaperumal, Rene Ripardo Calixto, Mohamed Rahouti, Mohamad Roshanzamir, and Victor Hugo C. De Albuquerque. 2024. "Explainable Artificial Intelligence for Drug Discovery and Development: A Comprehensive Survey." *IEEE Access* 12: 35796–812. doi:10.1109/ACCESS.2024.3373195.

Alther, Martin, and Chandan K Reddy. (2015). "Clinical Decision Support Systems." In *Healthcare Data Analytics*, 27–50. Chapman and Hall/CRC.

Ben-Menahem, Shiko M., Georg von Krogh, Zeynep Erden, and Andreas Schneider. 2016. "Coordinating Knowledge Creation in Multidisciplinary Teams: Evidence from Early-Stage Drug Discovery." *Academy of Management Journal* 59 (4). Academy of Management: 1308–38. doi:10.5465/amj.2013.1214.

Bhattamisra, Subrat Kumar, Priyanka Banerjee, Pratibha Gupta, Jayashree Mayuren, Susmita Patra, and Mayuren Candasamy. 2023. "Artificial Intelligence in Pharmaceutical and Healthcare Research." *Big Data and Cognitive Computing* 7 (1). Multidisciplinary Digital Publishing Institute: 10. doi:10.3390/bdcc7010010.

Bikku, Thulasi. 2020. "Multi-Layered Deep Learning Perceptron Approach for Health Risk Prediction." *Journal of Big Data* 7 (1): 50. doi:10.1186/s40537-020-00316-7.

Cesario, Alfredo, Marika D'Oria, Francesco Bove, Giuseppe Privitera, Ivo Boškoski, Daniela Pedicino, Luca Boldrini, et al. 2021. "Personalized Clinical Phenotyping through Systems Medicine and Artificial Intelligence." *Journal of Personalized Medicine* 11 (4). Multidisciplinary Digital Publishing Institute: 265. doi:10.3390/jpm11040265.

Chen, Hongming, Ola Engkvist, Yinhai Wang, Marcus Olivecrona, and Thomas Blaschke. 2018. "The Rise of Deep Learning in Drug Discovery." *Drug Discovery Today* 23 (6): 1241–50. doi:10.1016/j.drudis.2018.01.039.

Chen, Siqi, Tiancheng Li, Luna Yang, Fei Zhai, Xiwei Jiang, Rongwu Xiang, and Guixia Ling. 2022. "Artificial Intelligence-Driven Prediction of Multiple Drug Interactions." *Briefings in Bioinformatics* 23 (6): bbac427. doi:10.1093/bib/bbac427.

Chen, Wei, Xuesong Liu, Sanyin Zhang, and Shilin Chen. 2023. "Artificial Intelligence for Drug Discovery: Resources, Methods, and Applications." *Molecular Therapy – Nucleic Acids* 31 (March). Elsevier: 691–702. doi:10.1016/j.omtn.2023.02.019.

Chopra, Hitesh, Dong K. Shin, Kavita Munjal, Kuldeep Dhama, and Talha B. Emran. 2023. "Revolutionizing Clinical Trials: The Role of AI in Accelerating Medical Breakthroughs." *International Journal of Surgery* 109 (12): 4211. doi:10.1097/JS9.0000000000000705.

David, Bruno, Jean-Luc Wolfender, and Daniel A. Dias. 2015. "The Pharmaceutical Industry and Natural Products: Historical Status and New Trends." *Phytochemistry Reviews* 14 (2): 299–315. doi:10.1007/s11101-014-9367-z.

Erlanson, Daniel A., Stephen W. Fesik, Roderick E. Hubbard, Wolfgang Jahnke, and Harren Jhoti. 2016. "Twenty Years on: The Impact of Fragments on Drug Discovery." *Nature Reviews Drug Discovery* 15 (9). Nature Publishing Group: 605–19. doi:10.1038/nrd.2016.109.

Fraisl, Dilek, Gerid Hager, Baptiste Bedessem, Margaret Gold, Pen-Yuan Hsing, Finn Danielsen, Colleen B. Hitchcock, et al. 2022. "Citizen Science in Environmental and Ecological Sciences." *Nature Reviews Methods Primers* 2 (1). Nature Publishing Group: 1–20. doi:10.1038/s43586-022-00144-4.

Fu, Yanwei, Jiaoyang Luo, Jiaan Qin, and Meihua Yang. 2019. "Screening Techniques for the Identification of Bioactive Compounds in Natural Products." *Journal of Pharmaceutical and Biomedical Analysis* 168 (May): 189–200. doi:10.1016/j.jpba.2019.02.027.

Gangwal, Amit, and Antonio Lavecchia. 2024. "Unleashing the Power of Generative AI in Drug Discovery." *Drug Discovery Today* 29 (6): 103992. doi:10.1016/j.drudis.2024.103992.

Giguère, Sébastien, François Laviolette, Mario Marchand, Denise Tremblay, Sylvain Moineau, Xinxia Liang, Éric Biron, and Jacques Corbeil. 2015. "Machine Learning Assisted Design of Highly Active Peptides for Drug Discovery." *PLOS Computational Biology* 11 (4). Public Library of Science: e1004074. doi:10.1371/journal.pcbi.1004074.

Godwin, Ryan C., Ryan Melvin, and Freddie R. Salsbury. 2016. "Molecular Dynamics Simulations and Computer-Aided Drug Discovery." In *Computer-Aided Drug Discovery*, edited by Wei Zhang, 1–30. New York, NY: Springer. doi:10.1007/7653_2015_41.

Goetz, Laura H., and Nicholas J. Schork. 2018. "Personalized Medicine: Motivation, Challenges, and Progress." *Fertility and Sterility* 109 (6): 952–63. doi:10.1016/j.fertnstert.2018.05.006.

Goldsmith, Bryan R., Jacques Esterhuizen, Jin-Xun Liu, Christopher J. Bartel, and Christopher Sutton. 2018. "Machine Learning for Heterogeneous Catalyst Design and Discovery." *AIChE Journal* 64 (7): 2311–23. doi:10.1002/aic.16198.

Gryniukova, Anastasiia, Florian Kaiser, Iryna Myziuk, Diana Alieksieieva, Christoph Leberecht, Peter P. Heym, Olga O. Tarkhanova, Yurii S. Moroz, Petro Borysko, and V. Joachim Haupt. 2023. "AI-Powered Virtual Screening of Large Compound Libraries Leads to the Discovery of Novel Inhibitors of Sirtuin-1." *Journal of Medicinal Chemistry* 66 (15). American Chemical Society: 10241–51. doi:10.1021/acs.jmedchem.3c00128.

Gupta, Rohan, Devesh Srivastava, Mehar Sahu, Swati Tiwari, Rashmi K. Ambasta, and Pravir Kumar. 2021. "Artificial Intelligence to Deep Learning: Machine Intelligence Approach for Drug Discovery." *Molecular Diversity* 25 (3): 1315–60. doi:10.1007/s11030-021-10217-3.

Haleem, Abid, Mohd Javaid, Ravi Pratap Singh, and Rajiv Suman. 2022. "Medical 4.0 Technologies for Healthcare: Features, Capabilities, and Applications." *Internet of Things and Cyber-Physical Systems* 2 (January): 12–30. doi:10.1016/j.iotcps.2022.04.001.

Hameduh, Tareq, Yazan Haddad, Vojtech Adam, and Zbynek Heger. 2020. "Homology Modeling in the Time of Collective and Artificial Intelligence." *Computational and Structural Biotechnology Journal* 18 (January): 3494–3506. doi:10.1016/j.csbj.2020.11.007.

Hampel, Harald, Andrea Vergallo, Lisi Flores Aguilar, Norbert Benda, Karl Broich, A. Claudio Cuello, Jeffrey Cummings, et al. 2018. "Precision Pharmacology for Alzheimer's Disease." *Pharmacological Research* 130 (April): 331–65. doi:10.1016/j.phrs.2018.02.014.

Han, Ri, Hongryul Yoon, Gahee Kim, Hyundo Lee, and Yoonji Lee. 2023. "Revolutionizing Medicinal Chemistry: The Application of Artificial Intelligence (AI) in Early Drug Discovery." *Pharmaceuticals* 16 (9). Multidisciplinary Digital Publishing Institute: 1259. doi:10.3390/ph16091259.

Hare, Jennifer I., Twan Lammers, Marianne B. Ashford, Sanyogitta Puri, Gert Storm, and Simon T. Barry. 2017. "Challenges and Strategies in Anti-Cancer Nanomedicine Development: An Industry Perspective." *Advanced Drug Delivery Reviews*, Editor's Collection 2016, 108 (January): 25–38. doi:10.1016/j.addr.2016.04.025.

Hassija, Vikas, Vinay Chamola, Atmesh Mahapatra, Abhinandan Singal, Divyansh Goel, Kaizhu Huang, Simone Scardapane, Indro Spinelli, Mufti Mahmud, and Amir Hussain. 2024. "Interpreting Black-Box Models: A Review on Explainable Artificial Intelligence." *Cognitive Computation* 16 (1): 45–74. doi:10.1007/s12559-023-10179-8.

Hessler, Gerhard, and Karl-Heinz Baringhaus. 2018. "Artificial Intelligence in Drug Design." *Molecules* 23 (10). Multidisciplinary Digital Publishing Institute: 2520. doi:10.3390/molecules23102520.

Jiménez-Luna, José, Francesca Grisoni, Nils Weskamp, and Gisbert Schneider. 2021. "Artificial Intelligence in Drug Discovery: Recent Advances and Future Perspectives." *Expert Opinion on Drug Discovery* 16 (9). Taylor & Francis: 949–59. doi:10.1080/17460441.2021.1909567.

Jinsong, Shao, Jia Qifeng, Chen Xing, Yajie Hao, and Li Wang. 2024. "Molecular Fragmentation as a Crucial Step in the AI-Based Drug Development Pathway." *Communications Chemistry* 7 (1). Nature Publishing Group: 1–9. doi:10.1038/s42004-024-01109-2.

Kolluri, Sheela, Jianchang Lin, Rachael Liu, Yanwei Zhang, and Wenwen Zhang. 2022. "Machine Learning and Artificial Intelligence in Pharmaceutical Research and Development: A Review." *AAPS Journal* 24 (1): 19. doi:10.1208/s12248-021-00644-3.

Koscielny, Gautier, Peter An, Denise Carvalho-Silva, Jennifer A. Cham, Luca Fumis, Rippa Gasparyan, Samiul Hasan, et al. 2017. "Open Targets: A Platform for Therapeutic Target Identification and Validation." *Nucleic Acids Research* 45 (D1): D985–94. doi:10.1093/nar/gkw1055.

Lavecchia, Antonio. 2015. "Machine-Learning Approaches in Drug Discovery: Methods and Applications." *Drug Discovery Today* 20 (3): 318–31. doi:10.1016/j.drudis.2014.10.012.

Mahmud, Mufti, M. Shamim Kaiser, T. Martin McGinnity, and Amir Hussain. 2021. "Deep Learning in Mining Biological Data." *Cognitive Computation* 13 (1): 1–33. doi:10.1007/s12559-020-09773-x.

Maia, Eduardo Habib Bechelane, Letícia Cristina Assis, Tiago Alves de Oliveira, Alisson Marques da Silva, and Alex Gutterres Taranto. 2020. "Structure-Based Virtual Screening: From Classical to Artificial Intelligence." *Frontiers in Chemistry* 8: 343 (April). Frontiers. doi:10.3389/fchem.2020.00343.

Mak, Kit-Kay, and Mallikarjuna Rao Pichika. 2019. "Artificial Intelligence in Drug Development: Present Status and Future Prospects." *Drug Discovery Today* 24 (3): 773–80. doi:10.1016/j.drudis.2018.11.014.

Mak, Kit-Kay, Yi-Hang Wong, and Mallikarjuna Rao Pichika. 2022. "Artificial Intelligence in Drug Discovery and Development." In *Drug Discovery and Evaluation: Safety and Pharmacokinetic Assays*, edited by Franz J. Hock and Michael K. Pugsley, 1–38. Cham: Springer International Publishing. doi:10.1007/978-3-030-73317-9_92-1.

Moen, Erick, Dylan Bannon, Takamasa Kudo, William Graf, Markus Covert, and David Van Valen. 2019. "Deep Learning for Cellular Image Analysis." *Nature Methods* 16 (12). Nature Publishing Group: 1233–46. doi:10.1038/s41592-019-0403-1.

Morley, Jessica, Caio C. V. Machado, Christopher Burr, Josh Cowls, Indra Joshi, Mariarosaria Taddeo, and Luciano Floridi. 2020. "The Ethics of AI in Health Care: A Mapping Review." *Social Science & Medicine* 260 (September): 113172. doi:10.1016/j.socscimed.2020.113172.

Mousavi, Maryam, Azuraliza Abu Bakar, and Mohammadmahdi Vakilian. 2015. "Data Stream Clustering Algorithms: A Review." *International Journal of Advances in Soft Computing & Its Applications* 7 (3): 1–15.

Murray, Scott. 2022. "Management Perceptions of the Value of Artificial Intelligence in Drug Discovery and Development." Dissertation, Georgia State University. doi:10.57709/29080507.

Nayarisseri, Anuraj, Ravina Khandelwal, Poonam Tanwar, Maddala Madhavi, Diksha Sharma, Garima Thakur, Alejandro Speck-Planche, and Sanjeev K. Singh. 2021. "Artificial Intelligence, Big Data and Machine Learning Approaches in Precision Medicine & Drug Discovery." *Current Drug Targets* 22 (6): 631–55. doi:10.2174/1389450122999210104205732.

Nguyen, Tien V., and Nhut T. Vo. 2024. *Using Traditional Design Methods to Enhance AI-Driven Decision Making*. IGI Global.

Niazi, Sarfaraz K., and Zamara Mariam. 2024. "Computer-Aided Drug Design and Drug Discovery: A Prospective Analysis." *Pharmaceuticals* 17 (1). Multidisciplinary Digital Publishing Institute: 22. doi:10.3390/ph17010022.

Perumal, Sundresan, Mujahid Tabassum, Moolchand Sharma, and Sanju Mohanan. 2022. "Next Generation Communication Networks for Industrial Internet of Things Systems." In *Next Generation Communication Networks for Industrial Internet of Things Systems*, edited by Sundresan Perumal and Saju Mohanan. CRC Press. https://doi.org/10.1201/9781003355946

Pyzer-Knapp, Edward O., Changwon Suh, Rafael Gómez-Bombarelli, Jorge Aguilera-Iparraguirre, and Alán Aspuru-Guzik. 2015. "What is High-Throughput Virtual Screening? A Perspective from Organic Materials Discovery." *Annual Review of Materials Research* 45. Annual Reviews: 195–216. doi:10.1146/annurev-matsci-070214-020823.

Quazi, Sameer. 2021. "Role of Artificial Intelligence and Machine Learning in Bioinformatics: Drug Discovery and Drug Repurposing." Preprints. doi:10.20944/preprints202105.0346.v1.

Rehman, Ashfaq Ur, Mingyu Li, Binjian Wu, Yasir Ali, Salman Rasheed, Sana Shaheen, Xinyi Liu, Ray Luo, and Jian Zhang. 2024. "Role of Artificial Intelligence in Revolutionizing Drug Discovery." *Fundamental Research*, May. doi:10.1016/j.fmre.2024.04.021.

Romanelli, Virgilio, Carmen Cerchia, and Antonio Lavecchia. 2024. "Unlocking the Potential of Generative Artificial Intelligence in Drug Discovery." In *Applications of Generative AI*, edited by Zhihan Lyu, 37–63. Cham: Springer International Publishing. doi:10.1007/978-3-031-46238-2_3.

Sahu, Mehar, Rohan Gupta, Rashmi K. Ambasta, and Pravir Kumar. 2022. "Chapter Three – Artificial Intelligence and Machine Learning in Precision Medicine: A Paradigm Shift in Big Data Analysis." In *Progress in Molecular Biology and Translational Science*, edited by David B. Teplow, 190: 57–100. Precision Medicine. Academic Press. doi:10.1016/bs.pmbts.2022.03.002.

Salim, Sheikh. 2020. *Understanding the Role of Artificial Intelligence and Its Future Social Impact*. IGI Global.

Sanjeevi, Madhumathi, Prajna N. Hebbar, Natarajan Aiswarya, S. Rashmi, Chandrashekar Narayanan Rahul, Ajitha Mohan, Jeyaraman Jeyakanthan, and Kanagaraj Sekar. 2022. "Chapter 25 – Methods and Applications of Machine Learning in Structure-Based Drug Discovery." In *Advances in Protein Molecular and Structural Biology Methods*, edited by Timir Tripathi and Vikash Kumar Dubey, 405–37. Academic Press. doi:10.1016/B978-0-323-90264-9.00025-8.

Sharma, Moolchand, Suman Deswal, Umesh Gupta, Mujahid Tabassum, and Isah Lawal. (2023). "Soft Computing Techniques in Connected Healthcare Systems." In *Soft Computing Techniques in Connected Healthcare Systems*, edited by Moolchand Sharma. CRC Press. https://doi.org/10.1201/9781003405368

Singh, Ajay Vikram, Vaisali Chandrasekar, Namuna Paudel, Peter Laux, Andreas Luch, Donato Gemmati, Veronica Tisato, Kirti S. Prabhu, Shahab Uddin, and Sarada Prasad Dakua. 2023. "Integrative Toxicogenomics: Advancing Precision Medicine and Toxicology through Artificial Intelligence and OMICs Technology." *Biomedicine & Pharmacotherapy* 163 (July): 114784. doi:10.1016/j.biopha.2023.114784.

Singh, Dev Bukhsh. 2018. "Natural Lead Compounds and Strategies for Optimization." *Frontiers in Computational Chemistry*, 4: 1–47. doi:10.2174/9781681084411118040003.

Singh, Natesh, Ludovic Chaput, and Bruno O Villoutreix. 2021. "Virtual Screening Web Servers: Designing Chemical Probes and Drug Candidates in the Cyberspace." *Briefings in Bioinformatics* 22 (2): 1790–1818. doi:10.1093/bib/bbaa034.

Sinha, Sandeep, and Divya Vohora. 2018. "Chapter 2 – Drug Discovery and Development: An Overview." In *Pharmaceutical Medicine and Translational Clinical Research*, edited by Divya Vohora and Gursharan Singh, 19–32. Boston: Academic Press. doi:10.1016/B978-0-12-802103-3.00002-X.

Skublov, Sergey G., Aleksandra K. Gavrilchik, and Aleksey V. Berezin. 2022. "Geochemistry of Beryl Varieties: Comparative Analysis and Visualization of Analytical Data by Principal Component Analysis (PCA) and t-Distributed Stochastic Neighbor Embedding (t-SNE)." *Записки Горного Института* 255 (eng). Россия, Санкт-Петербург: Федеральное государственное бюджетное образовательное учреждение высшего образования «Санкт-Петербургский горный университет»: 455–69.

Southwest Jiaotong University, China, Iqbal Muhammad, Zhu Yan, and Southwest Jiaotong University, China. 2015. "Supervised Machine Learning Approaches: A Survey." *ICTACT Journal on Soft Computing* 05 (03): 946–52. doi:10.21917/ijsc.2015.0133.

Srinivasarao, Madduri, and Philip S. Low. 2017. "Ligand-Targeted Drug Delivery." *Chemical Reviews* 117 (19). American Chemical Society: 12133–64. doi:10.1021/acs.chemrev.7b00013.

Staszak, Maciej, Katarzyna Staszak, Karolina Wieszczycka, Anna Bajek, Krzysztof Roszkowski, and Bartosz Tylkowski. 2022. "Machine Learning in Drug Design: Use of Artificial Intelligence to Explore the Chemical Structure–Biological Activity Relationship." *WIREs Computational Molecular Science* 12 (2): e1568. doi:10.1002/wcms.1568.

Sucharitha, P., K. Ramesh Reddy, S. V. Satyanarayana, and Tripta Garg. 2022. "Chapter 15 – Absorption, Distribution, Metabolism, Excretion, and Toxicity Assessment of Drugs Using Computational Tools." In *Computational Approaches for Novel Therapeutic and Diagnostic Designing to Mitigate SARS-CoV-2 Infection*, edited by Arpana Parihar, Raju Khan, Ashok Kumar, Ajeet Kumar Kaushik, and Hardik Gohel, 335–55. Academic Press. doi:10.1016/B978-0-323-91172-6.00012-1.

Tabassum, M., Perumal, S., Afrouzi, H. N., Abdul Kashem, S. Bin, & Hassan, W. (2021). Review on Using Artificial Intelligence Related Deep Learning Techniques in Gaming and Recent Networks. In *Deep Learning in Gaming and Animations*, edited by Vikas Chaudhary, 65–90. CRC Press. https://doi.org/10.1201/9781003231530-4

Tabassum, M., Perumal, S., Mohanan, S., Suresh, P., Cheriyan, S., & Hassan, W. (2021). IoT, IR 4.0, and AI Technology Usability and Future Trend Demands: Multi-Criteria Decision-Making for Technology Evaluation. In *Design Methodologies and Tools for 5G Network Development and Application*, 109–144). IGI Global. https://doi.org/10.4018/978-1-7998-4610-9.ch006

Thafar, Maha, Arwa Bin Raies, Somayah Albaradei, Magbubah Essack, and Vladimir B. Bajic. 2019. "Comparison Study of Computational Prediction Tools for Drug-Target Binding Affinities." *Frontiers in Chemistry* 7 (November): 782. doi:10.3389/fchem.2019.00782.

Thakkar, Shraddha, William Slikker, Frank Yiannas, Primal Silva, Burton Blais, Kern Rei Chng, Zhichao Liu, et al. 2023. "Artificial Intelligence and Real-World Data for Drug and Food Safety – A Regulatory Science Perspective." *Regulatory Toxicology and Pharmacology* 140 (May): 105388. doi:10.1016/j.yrtph.2023.105388.

Tripathi, Anushree, Krishna Misra, Richa Dhanuka, and Jyoti P. Singh. 2023. "Artificial Intelligence in Accelerating Drug Discovery and Development." *Recent Patents on Biotechnology* 17 (1): 9–23. doi:10.2174/1872208316666220802151129.

Turner, Zak. 2023. "Edison to AI: Intellectual Property in AI-Driven Drug R&D." May. https://hdl.handle.net/2152/120316.

Udegbe, Francisca Chibugo, Ogochukwu Roseline Ebulue, Charles Chukwudalu Ebulue, and Chukwunonso Sylvester Ekesiobi. 2024. "Precision Medicine and Genomics: A Comprehensive Review of IT-Enabled Approaches." *International Medical Science Research Journal* 4 (4): 509–20. doi:10.51594/imsrj.v4i4.1053.

Van Laar, Sylvia A., Kim B. Gombert-Handoko, Henk-Jan Guchelaar, and Juliëtte Zwaveling. 2020. "An Electronic Health Record Text Mining Tool to Collect Real-World Drug Treatment Outcomes: A Validation Study in Patients With Metastatic Renal Cell Carcinoma." *Clinical Pharmacology & Therapeutics* 108 (3): 644–52. doi:10.1002/cpt.1966.

Vora, Lalitkumar K., Amol D. Gholap, Keshava Jetha, Raghu Raj Singh Thakur, Hetvi K. Solanki, and Vivek P. Chavda. 2023. "Artificial Intelligence in Pharmaceutical Technology and Drug Delivery Design." *Pharmaceutics* 15 (7). Multidisciplinary Digital Publishing Institute: 1916. doi:10.3390/pharmaceutics15071916.

Wu, Hongjie, Junkai Liu, Runhua Zhang, Yaoyao Lu, Guozeng Cui, Zhiming Cui, and Yijie Ding. 2024. "A Review of Deep Learning Methods for Ligand Based Drug Virtual Screening." *Fundamental Research* 52 (4, March): 867–1078. doi:10.1016/j.fmre.2024.02.011.

Xu, Yongjun, Xin Liu, Xin Cao, Changping Huang, Enke Liu, Sen Qian, Xingchen Liu, et al. 2021. "Artificial Intelligence: A Powerful Paradigm for Scientific Research." *The Innovation* 2 (4): 100179. doi:10.1016/j.xinn.2021.100179.

Xue, Hanqing, Jie Li, Haozhe Xie, and Yadong Wang. 2018. "Review of Drug Repositioning Approaches and Resources." *International Journal of Biological Sciences* 14 (10): 1232–44. doi:10.7150/ijbs.24612.

Yadav, Pranjul, Michael Steinbach, Vipin Kumar, and Gyorgy Simon. 2018. "Mining Electronic Health Records (EHRs): A Survey." *ACM Computing Surveys* 50 (6): 85:1–85:40. doi:10.1145/3127881.

Yang, Xin, Yifei Wang, Ryan Byrne, Gisbert Schneider, and Shengyong Yang. 2019. "Concepts of Artificial Intelligence for Computer-Assisted Drug Discovery." *Chemical Reviews* 119 (18). American Chemical Society: 10520–94. doi:10.1021/acs.chemrev.8b00728.

You, Yujie, Xin Lai, Yi Pan, Huiru Zheng, Julio Vera, Suran Liu, Senyi Deng, and Le Zhang. 2022. "Artificial Intelligence in Cancer Target Identification and Drug Discovery." *Signal Transduction and Targeted Therapy* 7 (1). Nature Publishing Group: 1–24. doi:10.1038/s41392-022-00994-0.

Zang, Qingda, Kamel Mansouri, Antony J. Williams, Richard S. Judson, David G. Allen, Warren M. Casey, and Nicole C. Kleinstreuer. 2017. "In Silico Prediction of Physicochemical Properties of Environmental Chemicals Using Molecular Fingerprints and Machine Learning." *Journal of Chemical Information and Modeling* 57 (1). American Chemical Society: 36–49. doi:10.1021/acs.jcim.6b00625.

Zhang, Haiping, Jianbo Pan, Xuli Wu, Ai-Ren Zuo, Yanjie Wei, and Zhi-Liang Ji. 2019. "Large-Scale Target Identification of Herbal Medicine Using a Reverse Docking Approach." *ACS Omega* 4 (6). American Chemical Society: 9710–19. doi:10.1021/acsomega.9b00020.

Zhang, Zehong, Lifan Chen, Feisheng Zhong, Dingyan Wang, Jiaxin Jiang, Sulin Zhang, Hualiang Jiang, Mingyue Zheng, and Xutong Li. 2022. "Graph Neural Network Approaches for Drug-Target Interactions." *Current Opinion in Structural Biology* 73 (April): 102327. doi:10.1016/j.sbi.2021.102327.

Zhou, Wei, Yonghua Wang, Aiping Lu, and Ge Zhang. 2016. "Systems Pharmacology in Small Molecular Drug Discovery." *International Journal of Molecular Sciences* 17 (2). Multidisciplinary Digital Publishing Institute: 246. doi:10.3390/ijms17020246.

Zielinski, Adam. 2021. "AI and the Future of Pharmaceutical Research." *arXiv*. doi:10.48550/arXiv.2107.03896.

5 Transforming Healthcare
Leveraging Machine Learning Algorithms for Diagnosis, Treatment, and Management

Indu Joseph Thoppil, K. Ashtalakshmi, and Ramesh Chundi

5.1 INTRODUCTION

Healthcare's journey is a fascinating tale, evolving from ancient herbal remedies documented on scrolls to today's data-driven approach. Early civilizations relied on spiritual beliefs and herbal concoctions, with limited knowledge passed down orally. The scientific revolution ushered in microscopes, dissections, and vaccinations, while governments began collecting basic public health data. The 20th century saw a boom in medical advancements with X-rays, antibiotics, and the rise of paper-based medical records. The healthcare system functions as a multifaceted web designed to ensure our well-being. Its primary focus lies in preventing, diagnosing, and treating illnesses and injuries. This complex system relies on three key components: healthcare professionals (doctors, nurses, therapists, etc.); healthcare facilities (clinics, hospitals, and diagnostic centers); and a financing mechanism (insurance, government programs, or individual payments) to support the others. The severity of the need determines the level of care given. Primary care serves as the first point of contact for routine checkups and minor issues. Secondary care involves specialists who address urgent situations requiring advanced skills. For complex illnesses and treatments, hospitals with advanced technology provide tertiary care.

Finally, quaternary care refers to highly specialized procedures performed by top experts at select centers. Throughout these levels, healthcare professionals safeguard a wealth of confidential patient data, encompassing medical history, test results, and personal details. Traditionally, this information was stored in paper form using handwritten notes or typed reports. The landscape of healthcare has undergone a dramatic transformation in the information age. The patient files in paper formats of the past have been replaced by Electronic Health Records (EHRs), creating a vast trove of digital data on patients, diagnoses, and treatment histories. Today, the information age has brought EHRs and big data analysis, leveraging massive datasets to improve diagnostics, personalize medicine, and manage population health. Data's impact is undeniable, but challenges remain. While the potential benefits are vast, significant challenges remain in harnessing the true power of this information.

This vast trove of digital data serves as the fuel for the powerful engines of Artificial Intelligence (AI) and Machine Learning (ML). By analyzing these EHRs and other healthcare data, AI and ML algorithms can revolutionize this field by groundbreaking advancements. The information age has not only brought EHRs but also the transformative power of AI and ML. These technologies can analyze vast datasets, uncovering hidden patterns and insights that traditional methods might miss. While EHRs offer a wealth of information, unlocking their true potential requires advanced analytics. AI and ML provide the key to extracting valuable insights from this data, leading to advancements in diagnostics, personalized medicine, and population health management. But how do we make sense of it all? Enter Deep Learning (DL), a powerful technique within ML. DL uses special algorithms called deep neural networks, which became popular around 2012. Imagine these networks as intricate webs that can sift through big data and for hidden insights and patterns. The more data they process, the better they become at recognizing these patterns, leading to highly accurate results.

From medical professionals like doctors and nurses to service providers like insurance companies and clinics, the system encompasses a diverse range of participants. Hospitals with their equipment, mental health specialists, government regulators, pharmacies, drug manufacturers with research teams, and even medical device companies all contribute to this intricate web. This very diversity creates a challenge in managing the system effectively. Furthermore, the healthcare sector relies heavily on other critical infrastructure sectors like energy, water, communication networks, information technology, and transportation. If any of these supporting sectors experience issues or uncertainties, it can have a cascading effect, impacting the stability and reliability of healthcare delivery. Healthcare's complexity becomes particularly evident when considering infectious diseases, especially disease outbreaks. Predicting where and how diseases emerge and spread is a complex task. Imagine sensors failing to detect the initial source of an infection, or a model predicting infection rates relying on faulty data. These seemingly isolated issues can have a cascading impact. Inaccurate data or malfunctioning equipment within one area, like disease surveillance, can disrupt the entire healthcare system. This highlights the interconnectedness of the system – a single point of failure can trigger a chain reaction, jeopardizing overall healthcare effectiveness. Another challenge is the correctness of data. Imagine, training a model to recommend treatments. If the data contains errors, like including the wrong medication, the model could suggest that very medication – potentially with disastrous consequences. Pre-processing safeguards against such errors by ensuring data accurately reflects clinical realities.

Conventional approaches struggle to identify outliers within large datasets and can lead to inaccurate data analysis and skewed results. Also, analysis techniques for data collected through surveys and questionnaires, often lack the necessary sophistication to fully capture the complex relationships within data. Fortunately, ML is able to cover all of these limitations. Data helps in the creation of features that are fed into ML models. Good features can significantly improve model performance. Ensuring that the data is representative of the real-world distribution is crucial to avoid biases that can lead to poor model performance. To ensure a model's effectiveness on unseen data, we employ a three-way data split: training, validation, and testing. This comprehensive approach allows for model training, hyperparameter tuning, and final performance evaluation. (Hyperparameters are settings that control the training process of a model.). Balancing bias and variance is the key to developing robust ML models.

5.2 MACHINE LEARNING IN HEALTHCARE SYSTEMS

ML algorithms, a subset of AI, play a crucial role in realizing these advancements in the medical field. By leveraging techniques such as DL, support vector machines, and decision trees, ML models can uncover hidden patterns within medical records, images, and genetic information (Mishra, 2024). Additionally, ML algorithms enable continuous patient monitoring, providing real-time analysis of vital signs and symptoms, which facilitates timely medical interventions. In drug discovery (Gupta U., 2024) and personalized medicine, these algorithms help identify potential therapeutic compounds and tailor treatments to individual patients, thereby improving the efficacy and specificity of medical care. Through these applications, ML algorithms significantly contribute to the transformative impact of AI in healthcare. Data is the cornerstone of ML; without data, ML models (Gupta, 2024) cannot learn. The quality, quantity, and relevance of the data (Adane, 2019; Dash, 2019) significantly influence the performance and accuracy of Fuzzy ML models (Gupta, 2022). The large set of data in healthcare is accumulated from different types of sources such as hospital admission notes, discharge summaries, pharmacies, insurance companies (Ho, 2020), imaging sets, pathological laboratory reports, sensor-based devices (Kadhim, 2020), genomics, social media as well as articles in medical journals.

While ML offers significant potential for healthcare advancement, it is crucial to acknowledge its limitations. Predictive models, for instance, may identify disease progression (e.g., stage 3 to 4 chronic renal failure) without offering actionable treatment options beyond existing modalities (dialysis, transplant). Similarly, models predicting treatment ineffectiveness based solely on genetic data (e.g., lung cancer) raise ethical concerns regarding treatment withholding. However, these extreme scenarios do not negate the value of ML in healthcare. A significant middle ground exists where predictive models can empower proactive interventions, optimize resource allocation, and ultimately improve patient care while reducing costs.

5.2.1 APPLICATIONS OF MACHINE LEARNING IN HEALTHCARE

ML is weaving its magic throughout the healthcare tapestry, offering a diverse set of tools that empower healthcare professionals, personalize patient care, and ultimately pave the way for a healthier future. This surge in ML adoption stems from a fundamental shift in healthcare. Traditionally, medicine relied heavily on experience and intuition. By ingesting vast quantities of information – from EHRs and medical images to genomic data and wearable sensor readings (Harerimana, 2018) – ML models can identify subtle patterns and uncover hidden relationships within the data. This newfound knowledge empowers healthcare professionals mainly in four ways:

Descriptive Analytics: At the core lies descriptive analytics, the bedrock upon which all further analysis rests (Khalifa, 2016; Batko, 2022; Ahmed, 2021). ML algorithms can analyze mountains of data to paint a clear picture of what has transpired. Imagine a hospital grappling with high patient readmission rates. ML can delve into EHRs, revealing which patient demographics or diagnoses correlate most strongly with readmissions. Through data visualization tools, healthcare providers can create dashboards and reports that highlight trends, patterns, and anomalies in patient data (Singh, 2023). For instance, visualizing trends in-patient admissions can help hospitals prepare for seasonal surges in demand. Additionally, summary statistics provide insights into average treatment times, patient demographics, and common diagnoses, allowing for a better understanding of the patient population and resource allocation. This newfound clarity allows for targeted interventions to address the root causes. (Ball, 2015.)

Diagnostic Analytics: Descriptive analytics provides a "what," but diagnostic analytics delves into the "why." It goes a step further by delving into the data to understand the reasons behind observed outcomes. This type of analytics employs techniques like anomaly detection to identify irregularities in patient data, which could indicate potential health issues (Hauskrecht, 2013). Association rule mining can uncover relationships between different health conditions, such as common comorbidities in patients with chronic diseases. Root cause analysis helps healthcare providers pinpoint the underlying causes of medical events, such as high readmission rates, enabling them to implement targeted interventions to address these issues. Building upon the readmission example, diagnostic analytics might analyze medication adherence patterns (Razzak, 2020; Xu Y. X., 2023), follow-up care details, or even social determinants of health that could be influencing readmissions. This deeper understanding empowers healthcare providers to tailor strategies that address the specific reasons behind the trend.

Predictive Analytics: Predictive analytics leverages past medical data and identifies patterns to forecast future events. It is perhaps the most well-known application of ML in healthcare (Razzak, 2020; Passos, 2019; Obermeyer, 2016). By analyzing historical data, predictive models can forecast future events, such as disease outbreaks, patient readmissions, and treatment responses. These models leverage a variety of algorithms, from linear regression to neural networks. For example, predictive analytics can help identify patients who are likely to develop diabetes based on their medical history and lifestyle factors (Tuppad, 2022). This allows for early intervention and personalized treatment plans, ultimately improving patient outcomes and reducing healthcare costs. This also allows for proactive interventions, such as medication adjustments, targeted education programs, or even additional home healthcare support (Philip, 2021).

Prescriptive Analytics: Prescriptive analytics goes beyond prediction. It takes predictive analytics a step further by recommending specific actions based on the predictions. It utilizes optimization algorithms (Miner, 2023; Nti, 2022) and decision support systems to suggest the best course of action for healthcare providers.

For example, prescriptive analytics can optimize staff schedules in hospitals based on predicted patient inflow, ensuring that there are enough healthcare professionals available during peak times. Recommendation systems can provide personalized treatment suggestions for patients, considering their past medical records and preferences. By offering actionable insights, prescriptive analytics (Galli, 2021) empowers healthcare field to make judicious decisions that enhance patient care and operational efficiency. Imagine the ML model predicting a high readmission risk for a specific patient. Prescriptive analytics would use that information to recommend targeted interventions – from medication adjustments (Mosavi, 2020) and follow-up appointments to additional home healthcare support – all aimed at preventing the predicted readmission.

These four types of analytics, working in concert, offer a comprehensive view of patient health, allowing for proactive interventions and personalized care plans. However, the story does not end there. As we delve deeper, we will explore how ML extends its reach beyond analytics, transforming various aspects of healthcare delivery through:

- **Drug Discovery and Development** by accelerating the identification of potential life-saving medications.
- **Robot-Assisted Surgery** by enhancing precision and minimizing human error in the operating room.
- **Medical Imaging Analysis** by aiding radiologists in detecting abnormalities with exceptional accuracy.
- **Personalized Medicine** by tailoring treatment plans to each patient's unique needs and genetic makeup.
- **Virtual Nursing Assistants** by providing patients with around-the-clock monitoring and support.
- **Administrative Workflow Optimization** by automating routine tasks, freeing up valuable time for healthcare professionals.

5.2.1.1 Drug Discovery and Development

The drug discovery and development (Jiménez-Luna, 2021; Boniolo, 2021) designed scientific process aimed at identifying and developing novel therapeutic agents targeting validated biological targets is intrinsically associated with a specific disease state. AI and ML techniques are finding extensive application within the domain of early drug development. This encompasses various crucial stages, including compound screening, drug design, target identification, target selection, and target prioritization. Drug discovery has traditionally been a long and arduous process. Identifying the right targets – the molecules responsible for disease (Tanoli, 2021) – was a complex process relying on various techniques. First, we will explore the pre-AI/ML era methods and then delve into the transformative power of AI and ML in this crucial stage. Scientists employed a diverse toolbox to identify potential drug targets. Before the age of AI, discovering good targets required scientists to consider a range of possibilities, a bit like finding those one-in-a-hundred needles. By employing methodologies (e.g., genomics and proteomics) focused on identifying the disease-associated genes and the proteins they encode, follow-up biochemical and cell-based studies

Transforming Healthcare 97

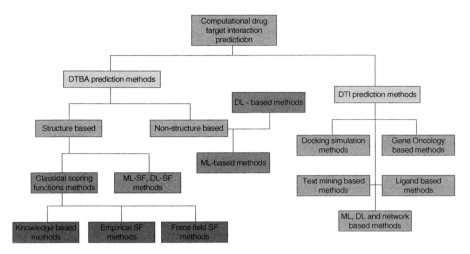

FIGURE 5.1 Deep Learning Techniques for Drug Target Prediction and Identification (Nag, 2022)

were used to further unravel the roles of these putative targets and determine how they could be modulated.

Additionally, animal models were used by researchers to establish disease progression, and to identify proteins, as well as pathways of relevance. High-throughput screening enabled the screening of thousands and millions of compounds against these targets very quickly and structure-based drug design involved utilizing the 3D target protein structure to design new drugs that could reach and interact with them. Phenotypic screening, a broader approach, allowed for the identification of compounds with specific effects beyond their direct cellular targets. This discovery revealed that these compounds could have additional, unexpected effects. Finally, researchers mined the wealth of scientific literature and patient data that join the dots between disease mechanisms and potential drug targets. Through collaboration with academia, fundamental biological research also played a vital role in uncovering promising targets for drug development. These techniques, while instrumental, had limitations. They were time-consuming, labor-intensive, and often relied on educated guesses. AI/ML brings a powerful set of tools using various DL techniques to predict and identify drug targets. These techniques are depicted in Figure 5.1.

Enhanced Predictive Modeling: By analyzing vast datasets of existing drug-target interactions, biological pathways, and disease mechanisms, these algorithms can learn to identify promising targets with a high likelihood of success. This significantly reduces the time and resources needed for experimental validation. DL algorithms being neuromorphic, excel at pattern recognition in complex datasets. In drug discovery, they can be used to analyze vast amounts of biological data. A DL model trained on protein-protein interaction data can predict how a new drug candidate might interact with a specific target protein, even if the 3D structure is unknown.

This helps prioritize promising candidates for further investigation. Support Vector Machines (SVM) algorithms excel at finding hyperplanes that best separate data points into distinct categories. In drug discovery, SVMs can be used to analyze data on known drug targets (Xu J. C., 2022) and their properties. Identifying potential targets (Klionsky, 2021) for cancer therapy by analyzing gene expression data from tumor samples and classifying genes likely to be involved in cancer cell growth or survival. Network Analysis Algorithms analyze biological networks, where nodes represent genes, proteins, or other molecules, and edges represent interactions between them. A network analysis algorithm can identify key nodes (proteins) within a disease-associated network. These key nodes might be good drug targets because disrupting their function could have a significant impact on the disease process. Identifying potential drug targets for neurodegenerative diseases involves analyzing protein-protein interaction networks in the brain. By prioritizing proteins with high connectivity and central roles in disease progression, researchers can focus on therapeutic interventions that may have a significant impact on these conditions. The key advantage of these algorithms lies in their ability to analyze massive datasets and identify subtle patterns that might be missed by traditional methods. This significantly accelerates the drug discovery process and holds immense promise for the development of new and effective therapies.

In conclusion, AI/ML has transformed drug target identification from a laborious endeavor into a data-driven and efficient process. This revolution holds immense promise for accelerating more effective drug discovery and personalized treatments for a wide range of diseases.

High-Throughput Screening (HTS): This is essentially a large-scale sifting process to identify potential drug candidates from a vast library of possibilities (Singh, 2024). However, this process can be akin to sifting through grains of sand to find a single pearl – time-consuming and laborious. Imagine being able to shortlist promising drug candidates before they even touch a test tube. This is the magic of AI in Virtual Screening (VS) (Murugan, 2022). AI algorithms, particularly DL models, are trained on vast datasets of known drugs, their targets, and their properties. These models then analyze libraries containing millions of potential drug molecules, predicting how they might interact with the target protein based on their structure and other characteristics (Gupta, 2021). This in silico (computer-based) approach allows researchers to predict a compound's binding affinity and potential efficacy. Traditional HTS methods might be limited to pre-defined libraries of compounds. AI-powered VS can explore a much vaster chemical space, identifying novel drug candidates with unique structures and mechanisms of action that might have been missed by traditional approaches. The HTS process begins with the identification of a promising drug target protein (Sadybekov, 2023). Researchers then meticulously prepare a vast library of compounds, often numbering in the millions. These chemical contenders can originate from diverse sources, including natural product libraries derived from plants or microorganisms, synthetic compound libraries created in the lab, or even virtual libraries generated computationally. The heart of HTS lies in automated testing (AI is a viable alternative to HTS: a 318-target study, 2024).

Specialized equipment allows researchers to rapidly assess how each compound in the library interacts with the target molecule. This might involve measuring changes in the target's activity or its binding affinity with each compound. Through this high-throughput analysis, researchers can identify those elusive "hits" – the compounds that hold promise for further development into potential drugs. Scenarios like identifying novel targets and potential therapies for complex cancers are a major challenge. Neural Network models analyze vast datasets of cancer genomics and protein interaction data to identify potential drug targets. VS can then be used to screen millions of compounds against these targets, prioritizing those with the potential to selectively target cancer cells while minimizing harm to healthy tissues. he emergence of antibiotic-resistant bacteria is a growing threat. VS powered by AI can be used to identify novel targets within resistant bacteria. For instance, the model might identify a unique enzyme involved in a resistance mechanism. VS can then be used to identify potential drugs that inhibit this enzyme, offering a new weapon in the fight against antibiotic resistance. Diseases like Alzheimer's and Parkinson's lack effective treatments. VS facilitates the identification of compounds targeting disease-linked proteins (dos Santos Maia, 2020). For example, AI algorithms might analyze the 3D structure of a protein implicated in Alzheimer's and use VS to identify compounds that prevent its aggregation (Monteiro, 2023), a hallmark of the disease. VS can significantly reduce the number of compounds that need to be tested in animals, leading to a more humane and efficient drug discovery process. The synergy between AI and VS is revolutionizing drug discovery for the development of new and effective therapies for a wide range of diseases.

Structure-Based Drug Design (SBDD): The human body is a complex network of interacting molecules, and diseases often arise from disruptions in these intricate pathways. SBDD emerges as a powerful tool in this fight against disease, offering a rational approach to drug discovery. This approach leverages the power of structural biology and computational techniques to design drugs that specifically target the 3D structure of a disease-causing protein. Traditionally, SBDD (Guruprasad, 2023) relies on techniques like X-ray crystallography or cryo-electron microscopy to determine the 3D structure of the target protein, often referred to as its "active site." This active site is the region on the protein where other molecules bind and exert their effects. Once the structure is known, researchers can use computational modeling to virtually dock (Adelusi, 2022) potential drug candidates into the active site, analyzing how well they fit and interact with the surrounding amino acid residues. This helps identify compounds with the best fit and optimal interaction, leading to the selection of more potent and specific drug candidates. Once a promising lead compound is identified, AI can be used to virtually modify its structure and predict its impact on binding affinity and other properties. This iterative process allows researchers to optimize the lead compound and develop even more potent and effective drugs. Advanced docking simulations can account for the flexibility of both the protein and the drug candidate, providing a more realistic picture of how they might interact with each other. This allows for the identification of compounds that might undergo subtle conformational changes upon binding, potentially leading to a tighter fit and higher binding affinity.

Docking simulations (Neves, 2021) are a virtual ballet between a potential drug candidate (ligand) and its protein target. The stage is set by preparing the protein structure, ensuring all the necessary elements for binding are present and accounted for. The ligand, too, is readied, with its structure defined and its flexibility taken into account. The search algorithm then orchestrates the dance, exploring different poses where the ligand could bind to the protein's active site. For each pose, a scoring function acts as the judge, evaluating the favorability of the interaction based on various factors like electrostatic attractions and hydrogen bonding. The pose with the highest score is considered the most promising, offering a glimpse into how the drug candidate might bind to the protein. This information is then used to analyze the interaction in detail and potentially refine the structure of the ligand to create an even more potent and specific drug. While limitations exist, docking simulations are a powerful tool in drug discovery, and with the help of AI, they are poised to become even more accurate and play an even greater role in bringing new therapies to patients. Several software tools are available for docking simulations, each with its specific features and capabilities. Popular examples include AutoDock, Glide, and Dock (Maia, 2020). These tools provide user-friendly interfaces for protein and ligand preparation, search algorithms, scoring functions, and visualization tools. While docking simulations are a powerful tool in SBDD, they do have limitations. The accuracy of the predicted binding modes depends on the quality of the protein structure and the chosen scoring function. Additionally, docking simulations often assume static protein structures, while proteins are dynamic in nature. As computational power increases and algorithms become more sophisticated, docking simulations will become even more accurate and will play a key role in the discovery and development of novel drugs.

5.2.1.2 Disease Diagnosis

AI plays a transformative role in disease diagnosis by leveraging advanced algorithms and ML techniques to analyze complex medical data. AI algorithms, such as Convolutional Neural Networks (CNNs) and Recurrent Neural Networks (RNNs), are extensively used for image recognition in radiology, pathology, and dermatology. Google's DeepMind (Ting, 2019) developed an AI system that can distinguish over 50 eye diseases from retinal scans with high accuracy. Similarly, IBM Watson uses NLP and ML to analyze medical literature and patient data to assist in diagnosing cancer (Malik, 2023). Software tools like TensorFlow and PyTorch are widely used for developing and training these AI models, while specialized platforms like Zebra Medical Vision (Tariq, 2020) and Aidoc provide end-to-end AI solutions for radiology diagnostics. These technologies not only enhance the accuracy and speed of diagnoses but also help in early detection and personalized treatment planning, significantly improving patient outcomes. The research shows that ML models are widely used in Chronic Disease (CD) diagnosis (Battineni, 2020). SVMs are the most common model, used in about 45% of studies. Other popular models include K-Nearest Neighbors (KNN), Naive Bayes (NB), Logistic Regression (LR), and Random Forests (RF). These models are applied to various tasks, such as predicting liver disease, gas chromatography analysis, and pathological changes. For liver fibrosis diagnosis specifically, studies show that RF (Feng, 2021) outperforms other methods in identifying the severity of the condition. Additionally, decision trees (Chen, 2017) and multilayer perceptron NN have shown promise in liver fibrosis prediction. Studies have used Bayesian networks (Quan, 2021) to analyze Chronic Obstructive Pulmonary Disease (COPD) patient data. Additionally, SVM, LR, and KNN models have been used to predict flare-ups in COPD patients. Notably, SVM models (Koshta, 2024) show the best accuracy in these tasks. ML's applications extend to other conditions as well. ANN and LR models can be used to predict diabetes. Ensemble models have also been explored for estimating kidney function.

5.3 RADIOLOGY

In radiology, AI primarily uses image recognition algorithms, such as CNNs, to analyze medical imaging data like X-rays, CT scans, and MRIs. These algorithms are trained on vast datasets to identify patterns and anomalies that may indicate diseases such as tumors, fractures, or lung diseases. AI platforms like Zebra Medical Vision and AIDOC can identify abnormalities in imaging studies, aiding radiologists in improving diagnostic accuracy, efficiency, and reducing the risk of human error. This can be particularly beneficial in identifying conditions like pneumonia and brain tumors. AI is rapidly transforming the field of medical imaging. One exciting application involves algorithms utilizing CNNs to analyze chest and abdomen CT scans (Nicolaes, 2024). These algorithms can segment the spinal column and identify vertebral fractures, potentially aiding radiologists in diagnosis. Similarly, AI can detect calcium deposits in coronary arteries from CT scans, potentially offering insights into cardiovascular risk (Booz, 2017). This technology even holds promise as an alternative to traditional DEXA scans for bone mineral density estimation.

It is important to remember that AI in medicine is intended to augment, not replace, radiologists' expertise (Recht, 2020). AI excels at identifying subtle features in medical images that might escape the human eye, and can analyze vast amounts of data to provide valuable insights in a shorter timeframe. The effectiveness of DL in medical diagnosis is demonstrably successful, even achieving human-level or better performance in some areas. For instance, AI algorithms have shown promise in detecting lymph node metastasis and classifying breast cancer malignancy (Chen C. Y., 2021). Another area of success is the detection of meningiomas in brain MRIs (Laukamp, 2019). Laukamp et al. trained a DL model to identify and segment meningiomas with high accuracy, exceeding the performance of manual segmentation by radiologists in some cases (Laukamp, 2019).

The benefits of AI extend beyond medical imaging and into electronic medical records (EMRs) as well. ML algorithms incorporated into EMRs can improve diagnoses for conditions that are often missed. A case in point is hyperparathyroidism, where only half of patients receive the necessary surgery. A study by Somnay et al. employed a ML model trained on labeled data from EMRs, including demographics and blood test results, to achieve a 97% accuracy rate in identifying hyperparathyroidism (Somnay, 2017). This success story highlights the potential of AI in EMRs to significantly improve patient diagnosis and ultimately patient outcomes (Bukowski, 2020). This assistance helps in early detection and reduces the chances of human error.

5.4 PATHOLOGY

In pathology, AI aids in the analysis of histopathological images, which involve examining tissues and cells under a microscope. AI algorithms can classify and quantify cell types, detect abnormal cells, and identify markers of diseases such as cancer. For instance, Paige.AI utilizes DL models (Jiang, 2020) to analyze pathology slides (Shafi, 2023), helping pathologists detect prostate cancer and other malignancies more effectively. These AI systems can process large volumes of slides rapidly, providing pathologists with crucial diagnostic information and enabling more precise and timely diagnoses.

AI and digital pathology are a powerful duo with the potential to revolutionize disease diagnosis. While the full scope of their applications is still being explored, some clear benefits are already taking shape. AI can significantly improve the accuracy and consistency of diagnoses, leading to more reliable patient outcomes. Additionally, AI-powered workflows can streamline the work of pathologists who face ever-increasing workloads. Notably, many AI algorithms can work with standard H&E stained slides (Go, 2022), eliminating the need for additional and potentially time-consuming procedures.

Recent advancements like DL, more powerful computing resources, and improved NNs have significantly boosted the effectiveness of AI-powered image analysis. Digital pathology offers well-documented advantages in terms of quality, efficiency, and service resilience. AI image analysis is totally transforming the way pathologists work by giving them a handy system to sort through cases (Gupta, 2023). This technology can spot cases that are probably harmless or flag ones that look suspicious and could be cancerous. This helps pathologists prioritize their workload more effectively (Hu, 2021; Wetstein, 2022). Studies on prostate cancer, basal cell carcinoma, and breast cancer metastases in lymph nodes have shown that DL whole slide image analysis systems have huge potential for actual use in clinics. But AI's role could go beyond just sorting cases. Imagine AI confirming diagnoses that pathologists are already confident about, but where double reporting is currently required or preferred (Wetstein, 2022). Basically, the pathologist's workstation could become their virtual second opinion, which means less reliance on human reviewers. Prostate cancer's Gleason grading system is a good example of this challenge, as it often shows poor reproducibility due to its subjective nature (Wulczyn, 2021). Research has shown that DL systems trained by expert urological pathologists can actually outperform general pathologists when it comes to Gleason grading tasks (Mun, 2021). By combining the power of these AI systems with the insights of pathologists, we can create a winning combination that improves the text while staying true to the original intent and facts. AI is handy in the field of pathology, especially when it comes to analyzing images. It can potentially help with automated diagnosis, organizing cases for a smoother workflow and uncovering new insights. But of course there are still some challenges to tackle. We need to make sure it is validated across multiple centers, it takes time to train the AI networks, and integrating it into digital pathology practice can be costly.

5.4.1 DERMATOLOGY

AI's role in dermatology revolves around the analysis of skin images to diagnose conditions like melanoma, psoriasis, and acne. AI systems use ML algorithms, particularly CNNs, to examine photographs of skin lesions (Goceri, 2020) and differentiate between benign and malignant conditions. An example is the use of algorithms developed by companies like SkinVision and DermTech (Joseph, 2024), which can assess skin cancer risk from smartphone photos. These tools empower dermatologists and general practitioners by providing a second opinion and ensuring that suspicious lesions are investigated further, improving early detection rates and patient outcomes. For example, there is the DANAOS expert systems (Hoffmann, 2003) DBDermo-Mips, and MoleAnalyser expert systems (Strzelecki, 2024) to help with diagnosing pigmented skin lesions. Smartphone apps like SkinVision, DermaAId, Skin 10, and MoleScope can be used to screen for skin cancers and also keep track of the changes in moles on your skin and identifying skin lesions. Automated detection of skin lesion using images has extended beyond melanoma to encompass pigmentary skin lesions, non-melanocytic skin cancers, psoriasis, skin rash, and onychomycosis, among other skin diseases. AI has become an important tool in diagnosing psoriasis (Kalbande, 2020). Applications include computer-aided diagnostic systems for image classification and risk stratification. ML models (Gupta, 2019) have also been developed to predict treatment response to biologics and differentiate psoriasis from psoriatic arthritis using genetic markers (Mulder, 2021). Correa da Rosa et al. (2017) (da Rosa, 2017) employed ML to analyze gene expression profiles in psoriatic skin lesions. Their study suggests that analyzing these profiles within the first 4 weeks of biological agent treatment can accurately predict a patient's clinical response at 12 weeks. This approach has the potential to reduce the assessment gap by 2 months.

5.4.2 CARDIOLOGY

Echocardiography remains the mainstay imaging technique for cardiac assessment due to its advantages in timeliness and cost-effectiveness. AI presents a promising avenue to augment and standardize echocardiogram analysis, potentially improving diagnostic accuracy and workflow efficiency. AI-powered analysis offers a transformative approach, enabling automated quantification of measurements, identification of pathological features like valvular disease or regional wall motion abnormalities, and the seamless integration of outcome data directly into the point-of-care analysis workflow. Moreover, AI's ability to detect subtle or previously unrecognized features in echocardiogram images (Alabdaljabar, 2023) presents a promising avenue for identifying subclinical disease or predicting patient prognosis, enabling earlier interventions and significantly improves the patient outcome. However, challenges persist in harnessing the full potential of AI for echocardiography. Current research often suffers from limitations in data size, potentially leading to the development of models that overfit the data and lack generalizability. Additionally, training AI on interpretations by human experts carries the risk of perpetuating existing human variability in interpreting and measuring echocardiogram data. Single-Photon Emission Computed Tomography (SPECT) and Myocardial Perfusion Imaging (MPI) (Dietze,

TABLE 5.1
Different types of disease diagnosis techniques

Author	Methodology	Datasets Used	Outcome	Evaluation Metrics
(Feng, 2021)	Random Forest	Not disclosed	predicting significant fibrosis (stage F ≥ 2)in patients with biopsy-confirmed NAFLD.	Area Under the Receiver Operating Characteristic Curve (AUROC)
(Quan, 2021)	Bayesian networks	National Surgical Quality Improvement database	Pulmonary evaluation reduces their hospital stay	MLM
(Koshta, 2024)	Fourier Decomposition Method (FDM) +Discrete Cosine Transform (DCT) +Discrete Fourier Transform (DFT)	https://data.mendeley.com/dataset/jwyy9np4gv/3	detect asthma and COPD patients and distinguish normal lung sounds from adventitious lung sounds.	Accuracy,F1 Score. Sensitivity
(Jiang, 2020)	two-step attention-based DL	Specimens of 400 patients who underwent endoscopic treatment	reduced unnecessary additional surgery by 15.1%	AUC
(Goceri, 2020)	CapsNet	Total sample of 170 non dermoscopic images from Groningen Medical Center University	Classification of skinlesions with high accuracy	Accuracy, Precision, Recall
(Wetstein, 2022)	Neural Networks	Total Sample of 706 young(,40 years) with different tumour grade and mitotic rate	Tumour Identification	Cohen's Kappa
(Mun Y. P., 2021)	Dl based automated Gleason grading system	Prostate needle biopsy single core slides (689 and 99) from two instituitions	Prostate cancer	ROC AUC PR AUC

2022) scans (Mao, 2023), such as automated motion correction, reconstruction, and analysis, demonstrates the exciting possibilities of AI in this domain. Table 7.1 summarizes the different types of disease diagnosis techniques for different types of diseases.

5.4.2.1 Personalized Medicine

The healthcare industry is witnessing a paradigm shift towards personalized medicine, fueled by the integration of powerful AI tools. These AI technologies allow us to analyze massive amounts of complex data, paving the way for more accurate diagnoses and tailored treatment strategies. High-throughput technologies like DNA sequencing, advanced imaging techniques, and wireless health monitoring devices are generating vast quantities of data. AI can handle the sheer volume and complexity of biomedical data, providing valuable insights that could previously go unnoticed. But the benefits extend beyond data analysis. These data-driven technologies are uncovering a remarkable truth: humans are incredibly diverse at a genetic, biochemical, physiological, and behavioral level. This diversity applies to how individuals experience disease and respond to treatment thus tailored to each patient's unique characteristics.

AI facilitates a paradigm shift by leveraging individual-level data, encompassing genetic predispositions, lifestyle choices, and environmental factors. Through analysis of this data, AI algorithms can identify individuals at heightened risk for specific diseases. Early detection, facilitated by AI, equips healthcare professionals to implement preventive measures and interventions. This proactive approach has the potential to mitigate disease burden and improve overall population health. This iterative feedback loop allows for ongoing refinement and optimization of treatment plans, leading to better patient outcomes. Genomic profiles offer invaluable insights into an individual's unique genetic makeup (Cuomo, 2023), revealing potential genetic variations that might influence susceptibility to disease or response to treatment (Cecchin, 2020). This comprehensive understanding of an individual's genetic landscape paves the way for the development of truly tailored treatment plans. Thus, AI unlocks a new era of precision medicine by enabling early disease detection, continuous treatment optimization, and the powerful analysis of big data, particularly within the realm of genomics.

The growing popularity of wearable devices in healthcare (Aziz, 2024) presents a significant opportunity for personalized medicine. These devices, such as fitness trackers and smartwatches, continuously collect physiological data on heart rate, sleep patterns, and physical activity. Integrating this rich data with EMRs and genomic profile allows researchers and clinicians to leverage the power of AI algorithms to uncover previously hidden correlations within this vast dataset. Another area where big data and AI intersect in personalized medicine is the analysis of patient-reported outcomes (Silveira, 2022). The patient-reported outcomes can analyzed to get valuable information for personalized treatment.

Precision medicine, as described by Chintala (2023) represents a shift towards individualized healthcare. This approach leverages vast datasets to tailor diagnoses, treatments, and preventive measures for each patient. In contrast to a "one-size-fits-all" strategy, precision medicine considers a person's unique genetic makeup, environment, and lifestyle choices. This empowers researchers and doctors to predict how effectively specific interventions might work for different patient groups. Ultimately, precision medicine aims to prevent, diagnose, and treat illness with a more individualized approach.

Personalized medicine hinges on collaboration between multidisciplinary healthcare teams and integrated technologies, such as clinical decision support systems. This approach leverages our evolving understanding of disease at the molecular level to optimize preventive healthcare strategies. The availability of human genome information empowers providers to create tailored care plans throughout a disease's course, transitioning healthcare from reactive to preventive. One of the most exciting prospects of AI in personalized medicine is its potential to democratize access to these customized healthcare solutions (Bhatt, 2022; Bastaki, 2023). As AI technologies become more affordable and widespread, the scalability of personalized medicine is poised to see significant improvements. Cloud-based AI solutions could enable large-scale analysis of genetic data and health records, reducing the cost and technical barriers associated with genomic sequencing and data analysis. This shift would make personalized treatment plans more accessible to a broader population, extending personalized medicine beyond specialized centers and into primary care settings. By utilizing AI to streamline the development and delivery of personalized treatments, healthcare systems can improve efficiency and reduce costs, further expanding access to this level of care (Sharma, 2023).

5.4.2.2 Streamlining Administrative Tasks

Over the past several decades, the administrative landscape of the healthcare industry has evolved to a considerable level of complexity. This Nexus of Forces is driven by the increase in healthcare service availability, the transformation of payment structures, legislative flux, and unceasing technological evolution. The result is that healthcare executives face an unparalleled set of problems in trying to pull off operational efficiency, financial health, regulatory compliance, and the hand-in-glove ideal coordination of patient care. The rise of EHRs, telemedicine, and digital health apps, while improving healthcare delivery, also creates a burden of administrative tasks for healthcare organizations. These organizations must carefully integrate these new platforms with existing infrastructure while prioritizing data security, ensuring compatibility between systems, and maintaining compliance with ever-changing regulations.

Modern healthcare is grappling with a surge in administrative tasks stemming from factors like complex regulations, mountains of paperwork, convoluted billing, etc. As a result, healthcare workers spend more time on paperwork than interacting with patients, leading to burnout and reduced effectiveness. The financial toll is also significant, with compliance and intricate billing processes driving up administrative costs for healthcare organizations. The different types of payment models have increased the challenges for healthcare organizations while adapting to changing reimbursement systems (Adler-Milstein, 2017). While EHRs offer advantages, they also introduce administrative burdens for healthcare professionals (Chambers, 2019). EHR systems often necessitate customization and upkeep, further straining resources.

Additionally, incompatibility between various health IT systems hinders information sharing and care coordination, potentially compromising patient care quality and staff satisfaction (West, 2016). However, Natural Language Processing (NLP), a form of AI, offers promise. NLP enables computers to understand human language and can be applied in healthcare administration for tasks like automated coding and billing, improved clinical documentation analysis, and patient feedback sentiment analysis (Kreimeyer, 2017). By leveraging NLP and other AI technologies, healthcare administration has the potential to become more efficient, reduce staff burdens, and ultimately contribute to improved patient care.

Robotic Process Automation (RPA) is a technology transforming healthcare administration. RPA "robots" automate repetitive tasks like data entry, claims processing, and appointment scheduling, freeing up staff for more critical work. This translates to increased efficiency, fewer errors, faster processes, and ultimately, cost savings and higher productivity in healthcare administration. (Fridsma, 2019). The automation in healthcare extends far beyond the traditional tasks of appointment scheduling and billing. Automated eligibility verification systems can confirm patient insurance coverage in real-time, significantly reducing delays and denials associated with coverage issues. Automated systems for insurance authorization can expedite the process of obtaining prior approval for procedures and treatments, enabling patients' timely access to necessary healthcare services (Zhang, 2018). By automating these routine administrative tasks healthcare staff can dedicate more time to direct patient care, leading to a more efficient healthcare systems.

Apart from scheduling appointments and handling billing, automation can also be utilized for other regular administrative duties in the healthcare field, including verifying eligibility, obtaining insurance authorization, and managing documentation. One example is when automated eligibility verification systems verify patient insurance coverage in real-time, which helps in reducing delays and denials associated with insurance problems. Likewise, automated systems for insurance authorization can make the process of securing prior authorization for medical procedures and treatments more efficient, guaranteeing that patients receive timely access to healthcare (Zhang, 2018). Automating these tasks in healthcare organizations can enhance operational efficiency, decrease administrative burden, and improve the overall patient experience.

5.5 ETHICAL CONSIDERATIONS OF MACHINE LEARNING IN HEALTHCARE

There is ongoing debate about whether AI fits neatly into existing legal frameworks, or if its unique features require entirely new legal categories. While AI offers tremendous potential to improve healthcare in clinical settings, it also raises significant ethical concerns that need to be addressed. To fully unlock the potential of AI in healthcare, four key ethical issues must be tackled: informed consent for data use, safety and transparency, algorithmic fairness and bias, and data privacy. The legal implications of AI systems extend beyond just legal debates, becoming a politically charged topic as well.

The increasing use of AI in healthcare settings necessitates proactive consideration of the ethical challenges it presents (Morley, 2020). A key concern is the lack of transparency in algorithmic decision-making processes. This "black box" phenomenon hinders accountability and raises questions about fairness, particularly as AI influences clinical decision-making (Smith, 2021). The potential for bias within AI algorithms further exacerbates these concerns, jeopardizing the principle of justice in healthcare (Jeyaraman, 2023). Transparency is crucial not only for ethical reasons but also for patient autonomy. Patients have the right to understand how AI systems arrive at conclusions, including the limitations of the models, their underlying algorithms, and the data used for training (Allen J. W., 2024). This knowledge empowers both patients and healthcare professionals to make informed decisions regarding the appropriateness and reliability of AI-generated recommendations (Jeyaraman, 2023). Informed consent becomes paramount when LLMs (Large Language Models) are involved in patient care. Data security and patient privacy are additional ethical considerations intertwined with the principle of autonomy (Mirbabaie M. L., 2022) Robust data protection measures are essential to prevent breaches that could compromise patient privacy and lead to identity theft or other forms of misuse (Li, 2023; Weidman, 2023). Existing regulations like HIPAA (Health Insurance Portability and Accountability Act) provide a framework for safeguarding patient confidentiality in the United States (Varas, 2023). However, ensuring the transparency and interpretability of AI systems remains a critical challenge in upholding ethical principles within healthcare. Proactive policy development and ongoing dialogue among stakeholders are crucial to ensure that AI is implemented responsibly and ethically within the healthcare domain.

5.6 FUTURE OF MACHINE LEARNING IN HEALTHCARE

The future of healthcare is inextricably linked with the continued advancement of ML and its powerful synergy with bioinformatics. Bioinformatics (Majhi, 2019), the marriage of biology and information technology, stands as an unsung hero in the future of healthcare. It empowers researchers to analyze and interpret vast troves of biological data, encompassing DNA sequences (Posada, 2009), protein structures, gene expression patterns (Hossain, 2019), and intricate biological pathways. This analytical power translates into tangible benefits for healthcare professionals. Beyond the current benefits of improved patient management, this combined force holds the potential to revolutionize diagnostics, treatment planning, and the very fabric of the healthcare system. Enhanced imaging analysis by ML algorithms, informed by vast bioinformatics datasets, will lead to earlier and more precise diagnoses. Predictive models, fueled by bioinformatics data on genetic markers and environmental factors, will identify individuals at high risk of developing certain diseases, allowing for preventative measures. Drug discovery can be significantly accelerated with ML's ability to analyze vast datasets of molecules and bioinformatic data on protein structures and interaction pathways, pinpointing promising drug candidates with higher efficacy and fewer side effects. On the treatment side, ML can pave the way for precision medicine by analyzing a patient's genetic makeup (leveraging bioinformatics) to predict their response to specific therapies (Khan, 2023). AI-powered chatbots and virtual assistants can empower patients with information and basic health management tools. Remote patient monitoring with wearable devices and sensors (Boikanyo, 2023), coupled with ML analysis of the collected bioinformatics data, allows for closer monitoring of chronic conditions and earlier detection of complications. The impact of this powerful duo extends beyond individual patients. ML and bioinformatics can streamline administrative tasks, improve population health management through analysis of public health data and bioinformatic trends, and even address accessibility issues by bringing quality healthcare to underserved communities. This future powered by ML and bioinformatics holds immense promise for a more effective, efficient, and equitable healthcare system for all.

5.7 CONCLUSION

By harnessing the power of vast and complex datasets, ML algorithms have the potential to revolutionize disease detection, treatment personalization, and healthcare system optimization. This chapter has delineated the multifaceted applications of ML in healthcare, highlighting its efficacy in domains ranging from early disease diagnosis to targeted drug discovery and personalization treatment for patient and streamlining administrative task. Paramount considerations include ensuring robust data privacy safeguards, developing techniques to minimize algorithmic bias, and fostering the interpretability of complex models. Furthermore, bioinformatics tools are instrumental in managing, analyzing, and interpreting the ever-growing volume of biological data, such as genomic sequences and protein structures. This synergy between ML and bioinformatics will be crucial for advancing our understanding of disease mechanisms, developing personalized medicine approaches, and ultimately, improving patient outcomes. By addressing the aforementioned challenges and prioritizing responsible and ethical development, we can usher in an era of healthcare characterized by not only enhanced efficiency and accuracy but also a focus on individualized and equitable treatment for all.

REFERENCES

Adane, K. M. (2019). "The role of medical data in efficient patient care delivery: a review." *Risk Management and Healthcare Policy 12*, 67–73.

Adelusi, T. I.-Q. (2022). "Molecular modeling in drug discovery." *Informatics in Medicine Unlocked 29*, 100880.

Adler-Milstein, J. H. (2017). "Electronic health record adoption in US hospitals: the emergence of a digital "advanced use" divide." *Journal of the American Medical Informatics Association 24*, 1142–1148.

Ahmed, I. M. (2021). "A framework for pandemic prediction using big data analytics." *Big Data Research 25*, 100190.

Alabdaljabar, M. S. (2023). "Machine learning in cardiology: a potential real-world solution in low-and middle-income countries." *Journal of Multidisciplinary Healthcare*, 285–295.

Aljindan, F. K. (2023). "Utilization of ChatGPT-4 in plastic and reconstructive surgery: A narrative review." *Plastic and Reconstructive Surgery–Global Open 11, no. 10*, e5305.

Allen, J. W. (2024). "Consent-GPT: is it ethical to delegate procedural consent to conversational AI?" *Journal of Medical Ethics 50, no. 2 ():*, 77–83.

Aziz, R. F. (2024.). "Wearable IoT Devices in Rehabilitation: Enabling Personalized Precision Medicine." In *Medical Robotics and AI-Assisted Diagnostics for a High-Tech Healthcare Industry, IGI Global*, pp. 281–308.

Baid, U. S. (2021). "The rsna-asnr-miccai brats 2021 benchmark on brain tumor segmentation and radiogenomic classification." *arXiv preprint arXiv:2107.02314.*

Ball, J. R. (2015). *Improving Diagnosis in Health Care.* National Academies Press.

Bastaki, K. U. (2023). Personalized Medicine. In *Metabolomics,* National Academic Press., pp. 1–32.

Batko, K. A. (2022). "The use of Big Data Analytics in healthcare." *Journal of Big Data 9, no. 1.*

Battineni, G. G. (2020). "Applications of machine learning predictive models in the chronic disease diagnosis." *Journal of Personalized Medicine 10, no. 2.*

Beauchamp, T. L. (2001). *Principles of Biomedical Ethics.* Oxford University Press, USA, 2001.

Bhatt, P. L. (2022). "Emerging artificial intelligence–empowered mhealth: scoping review." *JMIR mHealth and uHealth 10, no. 6,* e35053.

Boikanyo, K. A. (2023). "Remote patient monitoring systems: Applications, architecture, and challenges." *Scientific African 20,* e01638.

Boniolo, F. E. (2021). "Artificial intelligence in early drug discovery enabling precision medicine." *Expert Opinion on Drug Discovery 16, no. 9,* 991–1007.

Booz, C. P. (2017). "Evaluation of bone mineral density of the lumbar spine using a novel phantomless dual-energy CT post-processing algorithm in comparison with dual-energy X-ray absorptiometry." *European Radiology Experimental 1,* 1–6.

Bukowski, M. R.-R. (2020). "Implementation of eHealth and AI integrated diagnostics with multidisciplinary digitized data: are we ready from an international perspective?" *European Radiology 30,* 5510–5524.

Cecchin, E. A. (2020). "Pharmacogenomics and personalized medicine.". *Genes 11, no. 6,* 679.

Chambers, E. G. (2019). "Patient and career involvement in palliative care research: an integrative qualitative evidence synthesis review." *Palliative Medicine 33,* 969–984.

Char, D. S. (2020). "Identifying ethical considerations for machine learning healthcare applications." . *The American Journal of Bioethics 20, no. 11,* 7–17.

Chen, C. Y. (2021). "A meta-analysis of the diagnostic performance of machine learning-based MRI in the prediction of axillary lymph node metastasis in breast cancer patients." *Insights into Imaging 12,* 1–12.

Chen, Y. (2017). "Machine-learning-based classification of real-time tissue elastography for hepatic fibrosis in patients with chronic hepatitis B." *Computers in Biology and Medicine 89,* 18–23.

Cheng, C.-T. (2024). "Applications of deep learning in trauma radiology: A narrative review." *Biomedical Journal,* 100743.

Chintala, S. (2023). "AI-Driven personalised treatment plans: The future of precision medicine. *Machine Intelligence Research 17, no. 02,* 9718–9728.

Cuomo, A. S. (2023). "Single-cell genomics meets human genetics." *Nature Reviews Genetics 24, no. 8,* 535–549.

da Rosa, J. C.-F. (2017). "Shrinking the psoriasis assessment gap: early gene-expression profiling accurately predicts response to long-term treatment." *Journal of Investigative Dermatology 137, no. 2,* 305–312.

Dash, S. S. (2019). "Big data in healthcare: management, analysis and future prospects." *Journal of Big Data 6,* 1–25.

Dietze, M. M. (2022). "Progress in large field-of-view interventional planar scintigraphy and SPECT imaging." *Expert Review of Medical Devices 19, no. 5,* 393–403.

dos Santos Maia, M. G.-J. (2020). "Identification of new targets and the virtual screening of lignans against Alzheimer's." *Oxidative Medicine and Cellular Longevity 2020, no. 1,* 3098673.

Feng, G. (2021). "Machine learning algorithm outperforms fibrosis markers in predicting significant fibrosis in biopsy-confirmed NAFLD." *Journal of Hepato-Biliary-Pancreatic Sciences 28, no. 7*, 593–603.

Fridsma, D. B. (2019). "The role of robotic process automation in healthcare administration." *Journal of the American Medical Informatics Association*, 440–441.

Galli, L. T. (2021). "Prescriptive analytics for inventory management in health care.". *Journal of the Operational Research Society 72, no. 10*, 2211–2224.

Go, H. (2022). "Digital pathology and artificial intelligence applications in pathology.". *Brain Tumor Research and Treatment 10, no. 2*, 76.

Goceri, E. A. (2020). "Comparative Evaluations of CNN Based Networks for Skin Lesion Classification.". *In 14th International Conference on Computer Graphics. Visualization, Computer Vision and Image Processing (CGVCVIP)*. Zagreb, Croatia,.

Gupta, R. D. (2021). "Artificial intelligence to deep learning: machine intelligence approach for drug discovery." *Molecular Diversity 25*, 1315–1360.

Gupta, U. (2019). "An improved regularization based Lagrangian asymmetric v-twin support vector regression using pinball loss function." *Applied Intelligence 49, no. 10*, 3606–3627.

Gupta, U. (2022). "Bipolar fuzzy based least squares twin bounded support vector machine." *Fuzzy Sets and Systems 449*, 120–161.

Gupta, U. (2023). "Least squares structural twin bounded support vector machine on class scatter." *Applied Intelligence, 53, no. 12*, 15321–15351.

Gupta, U. (2024). "The Contribution of Artificial Intelligence to Drug Discovery: Current Progress and Prospects for the Future." *In Microbial Data Intelligence and Computational Techniques for Sustainable Computing*, pp. 1-23.

Guruprasad, L. P. (2023.). "Structure-Based Methods in Drug Design." *In Cheminformatics, QSAR and Machine Learning Applications for Novel Drug Development,* Academic Press, pp. 205–237.

Harerimana, G. B. (2018). "Health big data analytics: A technology survey." *IEEE Access 6*, 65661–65678.

Hauskrecht, M. I. (2013). "Outlier detection for patient monitoring and alerting." *Journal of Biomedical Informatics 46, no. 1*, 47–55.

Ho, C. W. (2020). "Ensuring trustworthy use of artificial intelligence and big data analytics in health insurance." *Bulletin of the World Health Organization 98*, 263.

Hoffmann, K. T. (2003). "Diagnostic and neural analysis of skin cancer (DANAOS). A multicentre study for collection and computer-aided analysis of data from pigmented skin lesions using digital dermoscopy." *British Journal of Dermatology 149, no. 4*, 801–809.

Hossain, M. A. (2019). "Machine learning and bioinformatics models to identify gene expression patterns of ovarian cancer associated with disease progression and mortality." *Journal of biomedical informatics 100*, 103313.

Hu, Y. F. (2021). "Deep learning system for lymph node quantification and metastatic cancer identification from whole-slide pathology images." *Gastric Cancer 24*, 8.

Jeyaraman, M. S. (2023). "ChatGPT in action: Harnessing artificial intelligence potential and addressing ethical challenges in medicine, education, and scientific research." *World Journal of Methodology 13*, 170.

Jiang, Y. M. (2020). "Emerging role of deep learning-based artificial intelligence in tumor pathology." *Cancer Communications 40, no. 4*, 154–166.

Jiménez-Luna, J. F. (2021). "Artificial intelligence in drug discovery: recent advances and future perspectives." *Expert Opinion on Drug Discovery 16, no. 9*, 949–959.

Joseph, J. A. (2024). "The synergy of skin and science–A comprehensive review of artificial intelligence's impact on dermatology." *Cosmoderma 4*.

Kadhim, K. T. (2020). "An overview of patient's health status monitoring system based on internet of things (IoT). *Wireless Personal Communications 114, no. 3*, 2235–2262.

Kalbande, D. R. (2020). "Early stage detection of psoriasis using artificial intelligence and image processing." *In Soft Computing: Theories and Applications: Proceedings of SoCTA 2018*.

Khalifa, M. A. (2016). "Utilizing health analytics in improving the performance of healthcare services: A case study on a tertiary care hospital." *Journal of Infection and Public Health 9, no. 6*, 757–765.

Koshta, V., Singh, B. K., Behera, A. K., & Ranganath, T. G. (2024). Fourier Decomposition Based Automated Classification of Healthy, COPD and Asthma Using Single Channel Lung Sounds. *IEEE Transactions on Medical Robotics and Bionics*.

Mao, Y. Y. (2023). "Echocardiographic evaluation of the effect on left ventricular function between left bundle branch pacing and right ventricular pacing." *International Journal of General Medicine*.

McCradden, M. D. (2020). "Clinical research underlies ethical integration of healthcare artificial intelligence." *Nature Medicine 26, no. 9*, 1325–1326.

Menze, B. H.-C. (2014). "The multimodal brain tumor image segmentation benchmark (BRATS)." *IEEE Transactions on Medical Imaging 34, no. 10*, 1993–2024.

Miner, G. D. (2023). *Practical Data Analytics for Innovation in Medicine: Building Real Predictive and Prescriptive Models in Personalized Healthcare and Medical Research Using AI, ML, and Related Technologies*. Academic Press.

Mirbabaie, M. L. (2022). "Artificial intelligence in hospitals: providing a status quo of ethical considerations in academia to guide future research." *AI & Society 37, no. 4*, 1361–1382.

Mishra, S., Ahmed, T., Sayeed, M. A., & Gupta, U. Artificial Neural Network Model for Automated Medical Diagnosis. In *Soft Computing Techniques in Connected Healthcare Systems* (pp. 34-54). CRC Press.

Monteiro, A. R. (2023). "Alzheimer's disease: Insights and new prospects in disease pathophysiology, biomarkers and disease-modifying drugs." *Biochemical Pharmacology 211*, 115522.

Morley, J. C. (2020). "The ethics of AI in health care: a mapping review." *Social Science & Medicine 260*, 113172.

Mosavi, N. S. (2020). "How prescriptive analytics influences decision making in precision medicine." *Procedia Computer Science 177*, 528–533.

Mulder, M. L. (2021). "Clinical, laboratory, and genetic markers for the development or presence of psoriatic arthritis in psoriasis patients: a systematic review." *Arthritis Research & Therapy 23, no. 1*, 168.

Mun, Y. (2021). "Yet another automated Gleason grading system (YAAGGS) by weakly supervised deep learning." *NPJ Digital Medicine 4, no. 1*, 99.

Murugan, N. A. (2022). "Artificial intelligence in virtual screening: Models versus experiments." *Drug Discovery Today 27, no. 7*, 1913–1923.

Nag, S. A. (2022). "Deep learning tools for advancing drug discovery and development." *Biotech 12, no. 5*, 110.

Neves, B. J.-F. (2021.). "Best Practices for Docking-Based Virtual Screening." *In Molecular Docking for Computer-Aided Drug Design*. Academic Press, 75–98.

Nicolaes, J. Y. (2024). *External Validation of a Convolutional Neural Network Algorithm for Opportunistically Detecting Vertebral Fractures in Routine*. Springer.

Nti, I. K. (2022). "A mini-review of machine learning in big data analytics: Applications, challenges, and prospects." *Big Data Mining and Analytics 5, no. 2*, 81–97.

Obermeyer, Z. (2016). "Predicting the future—Big data, machine learning and clinical medicine." *New England Journal of Medicine 375*, 1216–1219.

Olaronke, I. A. (2016). "Big data in healthcare: Prospects, challenges and resolutions." In *Future Technologies Conference (FTC)*, pp. 1152–1157. IEEE.

Passos, I. C. (2019). "Big Data and Machine Learning Meet the Health Sciences". In *Personalized Psychiatry: Big Data Analytics in Mental Health*, pp. 1–13.

Philip, N. Y. (2021). "Internet of Things for in-home health monitoring systems: Current advances, challenges and future directions." *IEEE Journal on Selected Areas in Communications 39, no. 2, pp. 300-310.*

Posada, D. B. (2009). *Ioinformatics for DNA Sequence Analysis*. Humana Press.

Quan, D. J. (2021). "Exploring influencing factors of chronic obstructive pulmonary disease based on elastic net and Bayesian network. *"Scientific Reports 12*, 7563.

Rajkomar, A. O. (2018). "Scalable and accurate deep learning with electronic health records." *NPJ Digital Medicine 1*, 1–10.

Razzak, M. I. (2020). "Big data analytics for preventive medicine." *Neural Computing and Applications 32, no. 9*, 4417–4451.

Recht, M. P. (2020). "Integrating artificial intelligence into the clinical practice of radiology: challenges and recommendations." *European Radiology 30*, 3576–3584.

Sadybekov, A. V. (2023). "Computational approaches streamlining drug discovery." *Nature 616, no. 7958*, 673–685.

Shafi, S. A. (2023). "Artificial intelligence in diagnostic pathology." *Diagnostic Pathology 18, no. 1*, 109.

Sharma, M. (Eds.). (2023). *Soft Computing Techniques in Connected Healthcare Systems*. CRC Press.

Silveira, A. T. (2022). "Patient reported outcomes in oncology: changing perspectives—a systematic review." *Health and Quality of Life Outcomes 20, no. 1*, 82.

Singh, S. H. (2024). "Advances in Artificial Intelligence (AI)-assisted approaches in drug screening." *Artificial Intelligence Chemistry 2, no. 1*, 100039.

Singh, D., Singh, D., Manju, & Gupta, U. (2023). Smart Healthcare: A Breakthrough in the Growth of Technologies. In *Artificial Intelligence-based Healthcare Systems* (pp. 73-85). Cham: Springer Nature Switzerland.

Smith, H. (2021). "Clinical AI: opacity, accountability, responsibility and liability." *AI & Society 36, no. 2*, 535–545.

Somnay, Y. R. (2017). "Improving diagnostic recognition of primary hyperparathyroidism with machine learning." *Surgery 161, no. 4*, 1113–1121.

Sousa, A. M. D., Moreira, P. O. L., & Monte Neto, R. L. D. (2024). AI is a viable alternative to high throughput screening: a 318-target study.

Stephenson, J. (2021). "Who offers guidance on use of Artificial Intelligence in medicine." *JAMA Health Forum, 2, no. 7*, American Medical Association, e212467–e212467.

Strzelecki, M. M. (2024). "Artificial intelligence in the detection of skin cancer: state of the art." *Clinics in Dermatology*.

Tanoli, Z. M.-K. (2021). "Artificial intelligence, machine learning, and drug repurposing in cancer." *Expert Opinion on Drug Discovery 16, no. 9*, 977–989.

Tariq, A. S. (2020). "Current clinical applications of artificial intelligence in radiology and their best supporting evidence." *Journal of the American College of Radiology 17, no. 11*, 1371–1381.

Tigard, D. W. (2021). "There is no techno-responsibility gap." *Philosophy & Technology 34, no. 3*, 589–607.

Ting, D. S. (2019). "Artificial intelligence and deep learning in ophthalmology." *British Journal of Ophthalmology 103, no. 2*, 167–175.

Tuppad, A. A. (2022). "Machine learning for diabetes clinical decision support: a review." *Advances in Computational Intelligence 2, no. 2*, 22.

Varas, J. B. (2023). "Innovations in surgical training: exploring the role of artificial intelligence and large language models (LLM)." *Revista do Colégio Brasileiro de Cirurgiões 50*, e20233605.

Weidman, A. A. (2023). "OpenAI's ChatGPT and its role in plastic surgery research." *Plastic and Reconstructive Surgery 151, no. 5*, 1111–1113.

West, C. P. (2016). "Interventions to prevent and reduce physician burnout: a systematic review and meta-analysis." *Lancet 388*, 2272–2281.

Wetstein, S. C. (2022). "Deep learning-based breast cancer grading and survival analysis on whole-slide histopathology images." *Scientific Reports 12, no. 1*, 15102.

Wulczyn, E. K. (2021). "Predicting prostate cancer specific-mortality with artificial intelligence-based Gleason grading." *Communications Medicine 1, no. 1*, 10.

Xu, J. C. (2022). "Interpretable deep learning translation of GWAS and multi-omics findings to identify pathobiology and drug repurposing in Alzheimer's disease." *Cell Reports 41, no. 9*, 111717.

Xu, Y. X. (2023). "Exploring patient medication adherence and data mining methods in clinical big data: A contemporary review." *Journal of Evidence-Based Medicine 16, no. 3*, 342–375.

Zhang, P., Schmidt, D. C., White, J., & Lenz, G. (2018). Blockchain technology use cases in healthcare. In *Advances in computers* (Vol. 111, pp. 1–41). Elsevier.

6 Gestational Diabetes Prediction Using Hybrid Probabilistic Machine Learning Models

Lakshmi K., Umme Salma M., and Sangeetha Shathish

6.1 INTRODUCTION

Early detection of Gestational diabetes mellitus (GDM) is essential due to the risk factors affecting both pregnant women and their child. In order to support early detection and intervention to reduce related complications, this study presents a hybrid predictive modeling framework for detecting GDM. The model is built by combining the ensemble models like LightGBM, XGBoost, and Random Forest algorithms with the probabilistic model Naive Bayes.

K-Nearest Neighbors (KNN) was used for missing value imputation and Synthetic Minority Over-sampling Technique (SMOTE) for class balancing to address issues with imbalanced class distribution and missing feature values.

Initially, classification using ensemble models like Random Forest, XGBoost, and Light-GBM classifiers gave overfitted results. The issue of overfitting was tackled by integrating the robustness and predictive performance of ensemble methods with the simplicity and assumption-based learning of Naive Bayes. The proposed model gives better results than the ensemble models alone for GDM detection.

The overarching objective of this research is to develop accurate and reliable predictive models for GDM detection, leveraging the synergistic benefits of Naive Bayes and ensemble learning algorithms alongside advanced data preprocessing techniques. Through comprehensive testing and evaluation, we aim to demonstrate the efficacy of our hybrid model in improving maternal and fetal health outcomes and advancing predictive modeling in healthcare.

6.2 RELATED WORK

Over the past few years, there have been significant advancements in the field of GDM prediction using various Machine Learning (ML) and Deep Learning (DL) techniques.

Earlier models used traditional ML techniques such as Logistic Regression, SVM, etc., to classify gestational diabetes. Later, with the introduction of DL techniques, classifications were done using ANN, which gave better accuracy. Another recent trend in GDM prediction is combining two or more ML models to get more efficient predictions.

The work by Xiong et al. titled "Prediction of Gestational Diabetes Mellitus in the First 19 Weeks of Pregnancy Using ML techniques" [1] investigates the use of ML algorithms such as SVM and LGBM to predict the onset of GDM during the first 19 weeks of pregnancy using clinical and demographic data. It draws attention to the possibility of early GDM prediction with these techniques, implying consequences for interventions and preventive measures in maternal-fetal healthcare. The goal of the paper titled "Early Prediction of Gestational Diabetes Mellitus in the Chinese Population via Advanced Machine Learning" by Yan-ting Wu et al. [2] was to use advanced ML techniques to predict GDM in the Chinese population. Based on clinical, biochemical, and demographic factors, a predictive model was created to identify GDM early in pregnancy. This could result in timely interventions and better outcomes for the health of the mother and fetus.

The paper "Machine Learning Prediction Models for Gestational Diabetes Mellitus: Meta-analysis" investigated ML prediction models for GDM [3]. It evaluated the efficacy of these models and their possible influence on clinical management by comparing and contrasting sensitivity, specificity, and area under the curve (AUC) measurements. The paper "Comparison of Machine Learning Methods and Conventional Logistic Regressions for Predicting Gestational Diabetes Using Routine Clinical Data: A Retrospective Cohort Study" [4] found that several ML methods did not outperform logistic regression in predicting GDM.

Ref. [5] examined the possibility of predicting the GDM treatment mode from clinical data collected at different stages of pregnancy. In order to predict risks for pharmacologic treatment beyond Medical Nutrition Therapy (MNT), researchers compared transparent and ensemble ML prediction methods, such as least absolute shrinkage and selection operator (LASSO) regression and super learner, which contained classification and regression tree, LASSO regression, Random Forest, and extreme gradient boosting algorithms. Ref. [6] compared various ML models and got accuracies less than 80%.

In Ref. [7] the authors introduced a ML model using logistic regression and XGBoost to give better accuracy. In Ref. [8], the authors compared various ML models and Gradient Boosting Machine (GBM) gave the highest accuracy of 93%.

Ref. [9] uses various ML techniques to classify GDM and found Random Forest to have the highest accuracy as the dataset was large. Ref. [10] compares three models such as Random Forest, GBM, and XGBoost, and it was observed that XGBoost outperformed the others. Using the DCA dataset (gathered from Assam, India) and the PIMA Indian diabetic dataset, Ref. [14] analyzes many machine learning methods for diabetes diagnosis; logistic regression performs the best. It highlights the need of early detection and makes suggestions for possible advancements using hybrid models.

6.3 MATERIALS AND METHODS

6.3.1 GESTATIONAL DIABETES MELLITUS

GDM is characterized by hyperglycemia (elevated blood sugar) developing during the second or third trimester of pregnancy and resolving after childbirth. It arises from a combination of insulin resistance and insufficient insulin production. Hormonal changes and weight gain during pregnancy exacerbate insulin resistance. Several factors influence GDM susceptibility, including advanced maternal age, prepregnancy overweight/obesity, family history of diabetes, and prior occurrences of macrosomia or GDM. Additionally, certain ethnicities (Hispanic, African American, Native American, South Asian) and conditions like polycystic ovary syndrome (PCOS) are associated with increased GDM risk. A history of adverse pregnancy outcomes (stillbirth, unexplained miscarriage, congenital malformations) is another potential risk factor. Elucidating the interplay between these variables can facilitate early detection and management of GDM, thereby improving maternal and fetal health outcomes [11].

6.3.2 MACHINE LEARNING ALGORITHMS

Naive Bayes Classifiers: These classifiers, a family of probabilistic ML algorithms based on Bayes' theorem, leverage the assumption of feature independence to achieve efficient classification. This assumption simplifies calculations by presuming that individual features are conditionally independent given the class label. Despite this simplification, Naive Bayes models often exhibit surprisingly strong performance, particularly when dealing with high-dimensional data with a limited number of samples. Their efficacy stems from their ability to calculate the posterior probability of a class membership for a given feature vector. However, the performance of Naive Bayes classifiers can deteriorate when the underlying assumption of feature independence is significantly violated or when features exhibit strong correlations. Additionally, encountering unseen feature value combinations during prediction can lead to zero probability estimates, potentially impacting the model's accuracy. Nevertheless, their computational efficiency, effectiveness, and often competitive performance make Naive Bayes classifiers a popular choice for various classification tasks.

Random Forest: An ensemble learning technique, Random Forests address the limitations of individual decision trees, particularly their high variance and susceptibility to overfitting. This approach is especially valuable when dealing with noisy or complex datasets. Random forests leverage the power of ensemble learning by aggregating predictions from multiple decision trees. To introduce randomness and enhance generalization performance, two key aspects are randomized during forest construction. First, each tree is trained on a bootstrapped subset of the data. Second, only a random subset of features is considered at each split point within the trees. This process of randomization effectively reduces overfitting, increases robustness to noise and outliers, and ultimately leads to significantly improved generalization performance for the final model.

Compared to isolated decision trees, Random Forests exhibit superior prediction accuracy and robustness across diverse domains. Their resilience to noise, ease of use, and adaptability makes them well-suited for a wide range of applications in finance, bioinformatics, remote sensing, and beyond. Extensive adoption within the ML field has solidified the position of Random Forests as a crucial tool for developing reliable and accurate predictive models [12].

XGBoost (eXtreme Gradient Boosting): This is a powerful and scalable implementation of gradient boosting machines. A revolutionary technique called XGBoost handles large-scale datasets with millions of samples and features by utilizing distributed and parallel computing. It uses regularization strategies to reduce overfitting and regulate model complexity, such as L1 and L2 penalties. For supervised learning tasks like regression and classification, it also provides loss function customization, approximate learning algorithms, and sparse-aware approaches. Because of its efficiency and scalability, XGBoost is a well-liked option for both commercial and academic research. Its practical uses include bioinformatics, ad placement, and web search ranking [13].

LightGBM (Light Gradient Boosting Machine): This is a scalable and effective gradient boosting system that employs cutting-edge techniques to solve computing difficulties. In order to choose useful data points for tree construction, it employs Gradient-based One-Side Sampling (GOSS), which lowers computational costs without sacrificing prediction accuracy. Additionally, it discretizes feature values into bins using histogram-based techniques, which cuts down on training time and memory usage.

LightGBM consistently achieves higher prediction accuracy compared to traditional gradient boosting methods while delivering significantly faster training speeds and lower memory requirements. This efficiency translates to benefits beyond model creation. It expands the applicability of LightGBM, particularly in real-world scenarios where scalability and speed are paramount, such as in image classification and web search ranking [15–17].

6.4 PROPOSED WORK

The project uses CRISP methodology for the implementation of the objective. The flowchart for "Gestational Diabetes Mellitus Prediction using Probabilistic Machine Learning Models" is given in Figure 6.1.

6.4.1 Data Collection

The project involves data collection for a model to detect GDM in pregnant individuals. The dataset includes factors like age, pregnancies, gestation, BMI, HDL levels, family history, prenatal loss, large child or birth default, PCOS, blood pressure, OGTT results, hemoglobin levels, sedentary lifestyle, and prediabetes. The goal is to develop a more accurate diagnosis. The data collection was done with Kaggle, which encompassed clinical records and patient surveys. With a total of 3525 entries,

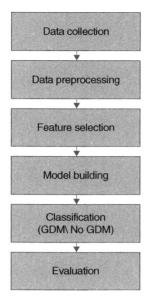

FIGURE 6.1 Flowchart of the proposed work.

each represents a unique individual. Noteworthy is the inclusion of the target variable labeled "Class Label," denoting whether an individual has been diagnosed with GDM. This comprehensive dataset forms the cornerstone for both training and evaluating the efficacy of the GDM detection model.

6.4.2 DATA PREPROCESSING

Data preprocessing is important to transform the raw data into a format that can be used efficiently. It is always important to clean the data before passing it to the model.

The dataset contained NaN values in features like BMI, HDL, Sys BP, and OGTT. KNN Imputer technique was used to fill these null values of these features because it gave more accurate results than replacing the NaN values with mean or median of the feature. The class label was unbalanced with more non-GDM labels than GDM labels. SMOTE was used to oversample the GDM labels because unbalanced class labels gave less accuracy (see Figure 6.2).

6.4.3 FEATURE SELECTION

Correlation of the columns in the dataset was calculated and a heatmap was created to visualize the correlation. The heatmap obtained is given in Figure 6.3.

From the heatmap, all features that are highly or moderately correlated to the target were chosen (i.e., any feature whose correlation was > 0.4 was chosen for model building). From the dataset, which had 15 features, 8 features were selected. Features selected for model building were PCOS, prediabetes, Dia BP, OGTT, age, gestation in previous pregnancy, BMI, and hemoglobin.

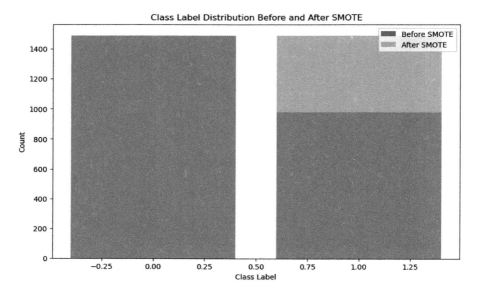

FIGURE 6.2 Class Label Distribution before and after applying SMOTE.

6.5 DATA MODELING

After preprocessing the model, ML algorithms like Random Forest, XGBoost, and LightGBM gave 100% accuracy for the training set, which meant that oversampling the class labels using SMOTE overfitted the model. To curb this issue, a probabilistic approach was implemented on these models by combining Naive Bayes classifier to these models to get a hybrid model, which gave good accuracy and solved the problem of overfitting [18–20].

ML uses two main dataset subsets: the training set and the testing set. The training set contains input features and target labels, while the testing set evaluates the model's performance. The training set uses these features for predictions, while the testing set assesses the model's generalization ability. The dataset is randomly divided into 70% for training and 30% for testing.

6.5.1 Mathematical Description

The proposed hybrid model aggregates probabilities generated by Naive Bayes with those from models prone to overfitting, such as Random Forest, XGBoost, and LightGBM, to enhance prediction accuracy for test instances. By combining the probabilistic output from Naive Bayes with these more complex classifiers, the model balances simplicity with robustness, aiming to achieve better generalization on unseen data.

To get the likelihood of each class label for each test case T_i, the probabilities from Naive Bayes and Random Forest are averaged in the first phase. By balancing the

Gestational Diabetes Prediction

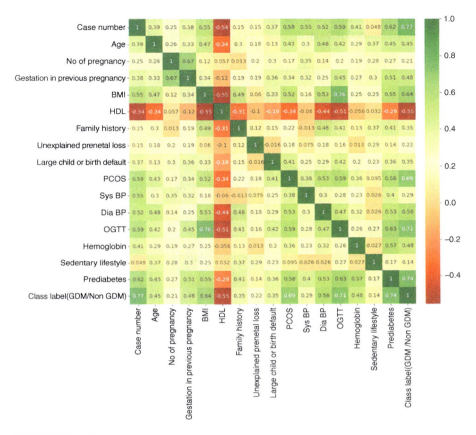

FIGURE 6.3 Heatmap.

predictions from two classifiers, this approach minimizes biases and inconsistencies. To ensure that the final prediction represents the most certain classification, each test case is assigned the predicted class label with the highest aggregated probability in Step 2. Choosing the class label with the highest aggregated probability yields the anticipated class for the test case [16–20].

Naive Bayes and Random Forest: The probability of class C_i for a given test instance T_i is calculated as the average of the probabilities obtained from Naive Bayes and Random Forest:

$$Probability(C_i) = \frac{\left(Probability(C_i \text{ from } NB) + Probability(C_i \text{ from } RF)\right)}{2} \qquad (6.1)$$

The class label assigned to the test instance T_i is the one with the highest probability value obtained in Step 1:

$$Predicted\,Class(T_i) = \arg\max_{C_i} \left[\frac{\left(Probability(C_i\,from\,NB) + Probability(C_i\,from\,RF)\right)}{2} \right]$$

(6.2)

Naive Bayes and XGBoost: The probability of class C_i for a given test instance T_i is calculated as the average of the probabilities obtained from Naive Bayes and XGBoost:

$$Probability(C_i) = \frac{\left(Probability(C_i\,from\,NB) + Probability(C_i\,from\,XGBoost)\right)}{2}$$

(6.3)

The class label assigned to the test instance T_i is the one with the highest probability value obtained in Step 1:

$$Predicted\,Class(T_i) = \arg\max_{C_i} \left[\frac{\left(Probability(C_i\,from\,NB) + Probability(C_i\,from\,XGBoost)\right)}{2} \right]$$

(6.4)

Naive Bayes and LightGBM: The probability of class C_i for a given test instance T_i is calculated as the average of the probabilities obtained from Naive Bayes and LightGBM:

$$Probability(C_i) = \frac{\left(Probability(C_i\,from\,NB) + Probability(C_i\,from\,LightGBM)\right)}{2}$$

(6.5)

The class label assigned to the test instance T_i is the one with the highest probability value obtained in Step 1:

$$Predicted\,Class(T_i) = \arg\max_{C_i} \left[\frac{\left(Probability(C_i\,from\,NB) + Probability(C_i\,from\,LightGBM)\right)}{2} \right]$$

(6.6)

6.5.2 Model Building

The study aims to improve classification task accuracy and robustness by merging predictions from two classifiers: Random Forest and Naïve Bayes. Naïve Bayes is a probabilistic classifier based on Bayes' theorem, while Random Forest is an ensemble learning technique. Both classifiers are trained using resampled data, and their combined predictive strength is calculated by averaging test instance probabilities. The technique was also applied to XGBoost and LightGBM, replacing Random Forest with the respective model.

6.6 RESULTS AND DISCUSSIONS

6.6.1 Training Accuracy

Training accuracy is a statistical measure of a ML model's performance on training data, calculating the percentage of cases the model correctly classifies out of all instances in the dataset. It is calculated by dividing the total number of examples by the number of correctly predicted instances. Mathematically, it can be expressed as:

$$Training\ Accuracy = \frac{(Number\ of\ Correct\ Predictions\ in\ Training\ Set)}{(Total\ Number\ of\ Instances\ in\ Training\ Set)} \times 100 \quad (6.7)$$

Training accuracy obtained by various models in this project is given in Table 6.1.

Inference

- From the accuracy score it is clear that the **ensemble models Random Forest, XGBoost, and LightGBM gave 100% accuracy,** which shows the possibility of **overfitting.**
- **The accuracy scores of the hybrid models shows that the overfitting is managed** and all the three models classify the GDM dataset well.
- Among the three models, the **hybrid model, which combines Naive Bayes classifier and LightGBM classifier, gives better accuracy**.

TABLE 6.1
Training accuracy of various models

Model	Accuracy
Naive Bayes	94.06%
Random Forest	100%
XGBoost	100%
LightGBM	100%
Naive bayes + Random Forest	97.75%
Naive Bayes + XGBoost	97.48%
Naive bayes + LightGBM	97.99%

6.6.2 Testing Accuracy

Testing accuracy is a statistic used to assess a ML model's performance, just like training accuracy. Testing accuracy, on the other hand, assesses the model's performance on a different dataset that was not utilized during the training phase, in contrast to the training data. This other dataset, which is sometimes called the test set, functions as an objective assessment of the model's capacity to generalize to fresh, untested data. The formula for testing accuracy is:

$$Testing\ Accuracy = \frac{(Number\ of\ Correct\ Predictions\ in\ Testing\ Set)}{(Total\ Number\ of\ Instances\ in\ Testing\ Set)} \times 100 \quad (6.8)$$

Testing accuracy obtained by various models used in this project is given in Table 6.2.

Inference

- From the testing accuracy score it is clear that all the hybrid models perform well.
- The **hybrid model that combines Naive Bayes and LightGBM classifiers gives better accuracy** compared to the others.

6.6.3 Confusion Matrix

A confusion matrix is a table that evaluates the performance of a classification model by comparing expected and actual values for various classes. It does this by analyzing characteristics such as accuracy, precision, recall, and F1-score.

Confusion matrices of various models used in this project are given in Figures 6.4–6.6.

TABLE 6.2
Testing accuracy of various models

Model	Accuracy
Naive Bayes	94.51%
Random Forest	96.97%
XGBoost	97.06%
LightGBM	97.16%
Naive Bayes + Random Forest	96.31%
Naive bayes + XGBoost	96.41%
Naive bayes + LightGBM	97.16%

Gestational Diabetes Prediction

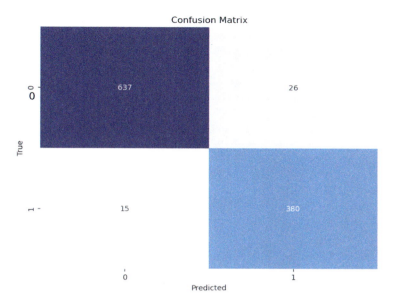

FIGURE 6.4 Confusion matrix for Hybrid Model (Naive Bayes and Random Forest).

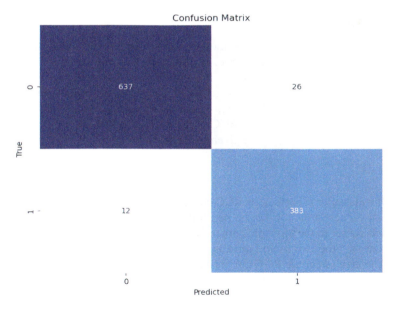

FIGURE 6.5 Confusion matrix for Hybrid Model (Naive Bayes and XGBoost).

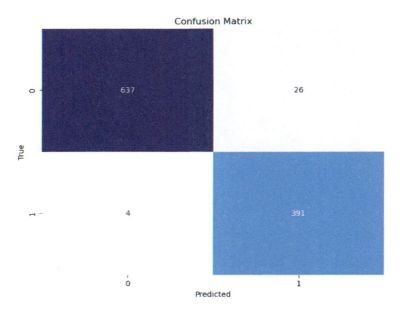

FIGURE 6.6 Confusion matrix for Hybrid Model (Naive Bayes and LightGBM).

6.6.4 ROC Curve

The efficacy of binary classification models is assessed visually using the Receiver Operating Characteristic (ROC) curve. It illustrates how the trade-off between the true positive rate (sensitivity) and the false-positive rate (specificity) is influenced by several factors. A ROC curve that crosses the upper-left corner, signifying great sensitivity and low false-positive rate, characterizes an ideal classifier. Better case distinction is indicated by a higher AUC-ROC score. The ROC curves obtained from the models used in this project are given in Figure 6.7.

Inference

- If a classifier is perfect, its ROC curve will be a vertical line from the bottom-left corner to the top-left corner, followed by a horizontal line from the top-left corner to the top-right corner. Based on the given figure, it can be inferred that the proposed hybrid models are perfect.
- The area under the ROC curve (AUC) is commonly used to evaluate the performance of a classification model. An AUC value above 0.9 indicates excellent model performance. Here, all the three models have AUC > 0.9, which suggests that the models' performance is excellent.
- It is clear from the AUC value that the hybrid model that combines Naive Bayes classifier and LightGBM classifier outperforms the other two models.

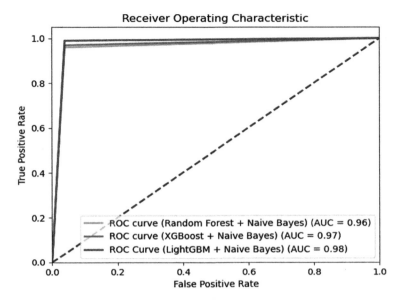

FIGURE 6.7 Comparison of ROC Curve of Hybrid Classifiers.

6.7 CONCLUSION

This project aimed to develop a classification model to classify gestational diabetes by curbing the issue of overfitting in training some of the best performing ML models. The project involved data collection from the online platform Kaggle. Exploratory Data Analysis was done on the collected data, which concluded that some of the features had null values and the class labels were unbalanced. This issue was solved in preprocessing of the data and was given to some of the best performing ML models like Random Forest, XGBoost, and LightGBM. It was observed that there was overfitting in the training set. To solve this problem a hybrid model was introduced that combined the probabilities of these each model with the probability of Naive Bayes classifier. All the hybrid models gave good accuracy, with a combination of Naive Bayes and LightGBM classifiers giving the best accuracy. The training and testing accuracy for this model was 97.99% and 97.16%, respectively. The confusion matrix and ROC curve for each model were plotted for further analysis.

6.8 FUTURE SCOPE

There is a growing need for effective GDM detection at an early stage due to its complications in later stages for both mother and child. The future scope of GDM classification includes combining more models to this hybrid model to get better accuracy and using more features that contribute to better GDM prediction.

REFERENCES

1. Y. Xiong, L. Lin, Y. Chen, S. Salerno, Y. Li, X. Zeng, and H. Li, "Prediction of gestational diabetes mellitus in the first 19 weeks of pregnancy using machine learning techniques," *Journal of Maternal-Fetal & Neonatal Medicine*, vol. 35, no. 13, pp. 2457–2463, 2022.
2. Y.-T. Wu, C.-J. Zhang, B. W. Mol, A. Kawai, C. Li, L. Chen, Y. Wang, J.-Z. Sheng, J.-X. Fan, Y. Shi, et al., "Early prediction of gestational diabetes mellitus in the Chinese population via advanced machine learning," *The Journal of Clinical Endocrinology & Metabolism*, vol. 106, no. 3, pp. e1191–e1205, 2021.
3. Z. Zhang, L. Yang, W. Han, Y. Wu, L. Zhang, C. Gao, K. Jiang, Y. Liu, and H. Wu, "Machine learning prediction models for gestational diabetes mellitus: meta-analysis," *Journal of Medical Internet Research*, vol. 24, no. 3, p. e26634, 2022.
4. Y. Ye, Y. Xiong, Q. Zhou, J. Wu, X. Li, X. Xiao, et al., "Comparison of machine learning methods and conventional logistic regressions for predicting gestational diabetes using routine clinical data: a retrospective cohort study," *Journal of Diabetes Research*, vol. 2020, pp. 4168340, 2020.
5. L. D. Liao, A. Ferrara, M. B. Greenberg, A. L. Ngo, J. Feng, Z. Zhang, P. T. Bradshaw, A. E. Hubbard, and Y. Zhu, "Development and validation of prediction models for gestational diabetes treatment modality using supervised machine learning: a population-based cohort study," *BMC Medicine*, vol. 20, no. 1, p. 307, 2022.
6. I. Gnanadass, "Prediction of gestational diabetes by machine learning algorithms," *IEEE Potentials*, vol. 39, no. 6, pp. 32–37, 2020.
7. H. Liu, J. Li, J. Leng, H. Wang, J. Liu, W. Li, H. Liu, S. Wang, J. Ma, J. C. Chan, et al., "Machine learning risk score for prediction of gestational diabetes in early pregnancy in Tianjin, China," *Diabetes/Metabolism Research and Reviews*, vol. 37, no. 5, p. e3397, 2021.
8. S. S. Reddy, M. Gadiraju, N. M. Preethi, and V. M. Rao, "A novel approach for prediction of gestational diabetes based on clinical signs and risk factors," *EAI Endorsed Transactions on Scalable Information Systems*, vol. 10, no. 3, 2023, p. 4.
9. S. Amarnath, M. Selvamani, and V. Varadarajan, "Prognosis model for gestational diabetes using machine learning techniques.," *Sensors & Materials*, vol. 33, 2021, p. 4.
10. N. Ali, W. Khan, A. Ahmad, M. M. Masud, H. Adam, and L. A. Ahmed, "Predictive modeling for the diagnosis of gestational diabetes mellitus using epidemiological data in the united Arab emirates," *Information*, vol. 13, no. 10, p. 485, 2022.
11. T. A. Buchanan, A. H. Xiang, et al., "Gestational diabetes mellitus," *Journal of Clinical Investigation*, vol. 115, no. 3, pp. 485–491, 2005.
12. L. Breiman, "Random forests," *Machine Learning*, vol. 45, pp. 5–32, 2001.
13. T. Chen and C. Guestrin, "Xgboost: A scalable tree boosting system," in *Proceedings of the 22nd ACM SIGKDD International Conference on Knowledge Discovery and Data Mining*, pp. 785–794, 2016.
14. D. Gupta, A. Choudhury, U. Gupta, P. Singh, M. Prasad, "Computational approach to clinical diagnosis of diabetes disease: a comparative study," *Multimedia Tools and Applications,* pp. 1–26, 2021 Jan 1.
15. G. Ke, Q. Meng, T. Finley, T. Wang, W. Chen, W. Ma, Q. Ye, and T.-Y. Liu, "Lightgbm: A highly efficient gradient boosting decision tree," *Advances in Neural Information Processing Systems*, vol. 30, 2017, p. 8.
16. M. Sharma, S. Deswal, U. Gupta, M. Tabassum, and I. Lawal, eds., *Soft Computing Techniques in Connected Healthcare Systems.* CRC Press, 2023.

17. R. G. Devi, and P. Sumanjani, "Improved classification techniques by combining KNN and Random Forest with naive Bayesian classifier," in *2015 IEEE International Conference on Engineering and Technology (ICETECH)*, pp. 1–4, IEEE, 2015.
18. K. Malik, H. Sadawarti, M. Sharma, U. Gupta, and P. Tiwari, editors. *Computational Techniques in Neuroscience*. CRC Press; 2023 Nov 14.
19. U. Gupta, and D. Gupta, "An improved regularization based Lagrangian asymmetric v-twin support vector regression using pinball loss function," *Applied Intelligence*, vol. 49, pp. 3606–27, 2019 Oct.
20. B. B. Hazarika, D. Gupta, and U. Gupta, "Intuitionistic fuzzy kernel random vector functional link classifier," In *Machine Intelligence Techniques for Data Analysis and Signal Processing: Proceedings of the 4th International Conference MISP 2022*, vol. 1, pp. 881–889. Singapore: Springer Nature Singapore, 2023 May 31.

7 AI in Diagnostics and Disease Prediction
Urolithiasis

Suneel Kumar and Dashrath Singh

7.1 OVERVIEW

In a directive to handle complicated remedial facts and give doctors relevant evidence, AI imitates human social and intellectual behavior using technical skills. [1]. Alan Turing first asked if mainframes might ponder alike people in 1950. Turing is regarded as the founder of AI after passing the so-called "Turing test" [2]. Even though AI was first proposed in the twenty-first century, it continues to be of interest to many scientific disciplines, including medicine. AI is the process of optimizing an algorithm's reward function [3]. AI makes massive dataset analysis possible and efficient, much like the chess analogy method, which gathers and analyzes a vast amount of data from chess games. Mainframe structures can "learn" to identify outlines by utilizing a range of sub-spheres that are created to fulfill a requirement of recognizing and distributing particular forms. Generally, the self-styled artificial neural network (ANN) represents these patterns. An emblematic ANN is made up of several nerve cells, or workstations, that are logically coupled to one another. Environmental sensors stimulate effort nerve cells, and one-sided associates with formerly energetic nerve cells trigger other nerve. Finally, this procedure sets up the behaviour of the corresponding DNN. The goal of the "learning" process is to create corresponding DNN that can do the anticipated action, such operating a device from a distance [4]. A subset of AI, machine learning (ML) is represented by unsupervised learning in Figure 7.1 and is based on inferences made from statistical analysis of massive, multipart neural networks (NNs) [5]. Since ML defines the prospect of a parallel among two or extra variable stars, it chooses to give physicians potential responses rather than "correct" ones. The development of algorithms that can adapt to changing conditions and technologies, high-performance computing, cloud drives for access to personal scientific research, and the use it exposed erasure package are the four main components which guarantee ML functions correctly and flawlessly and serves as a valuable tool for its users [6]. Built on the principles of process stacking, a branch of machine learning known as artificial neural networks (ANNs) enables models to learn autonomously and make independent decisions, embodying a "self-taught" concept (see Figure 7.1) [7]. This approach allows ANNs to layer processes in ways that mimic human learning, progressively refining their understanding and predictive capabilities without explicit programming for each task. Conversely, DL functions

AI in Diagnostics and Disease Prediction   131

autonomously and without the need for outside human involvement, whereas ML usually needs human interaction for upkeep and optimal operation [8]. Convolutional neural networks (CNNs) are a more advanced variety of ANNs that use fully connected, convolution, and pooling layers to integrate spatial motives and features of a picture. They give the user access to data that was previously unavailable, simplify the model, and use less memory and processing power [9]. For instance, double spin in computed tomography (CT) scans could make it more difficult to safely evaluate the results. In order to overcome this disadvantage, CNN concentrates on specific areas of a picture, separates them from nearby buildings, eliminates unwanted noise, and offers previously unobtainable visual data [10].

Urine materials that make decisions, harden like kidney stones, and travel down the urinary tract are known as urolithiasis [11–18]. As a result, drinking water and urolithiasis are intimately associated. Reduced water intake lengthens the period urinary stones remain in the urine, increasing the likelihood that they may form. Additionally, there is a higher occurrence risk in children whose parents have urinary stones because of hereditary influences. In general, it is believed that men are twice as likely as women to develop them, and that the incidence rate is highest in young adults between the ages of 20 and 40. Reduced water consumption leads to extended urinary retention of urinary stones, which is the cause of urolithiasis. Men are more likely than women to experience it, and it happens more frequently in younger age groups. Urinary stone incidence is also significantly influenced by temperature and season. Summertime sweating can lead to concentrated urine and an increased risk of ureteric stones. Kidney stones obstruct urine flow and cause excruciating agony as they travel throughout the renal ureter, ureter, bladder, and urethra. In the process, renal failure, hydropathy, and urinary tract infections may be exacerbated. Failure to treat ureteral stones may result in partial or total blockage of the urethra, which can cause renal failure and death. Stone-related urinary tract infections and localized discomfort are possible side effects. Finally, a patient's symptoms, physical examination, urine test, and radiation test are used to confirm urolithiasis. The patient's clinical symptoms, physical examination, urine examination, and radiation examination are used to diagnose the condition. Simple radiation cannot identify unseen stones, but simple urinary tract imaging can prove the presence of stones. Furthermore, it can be challenging to distinguish ureter stones from other organs or the pelvic bone. Feces can often mask ureter stones. This makes it possible to use CT or urography to check for ureteral stones. The existence of stones is often difficult to prove, although are a crucial component of the clinician's diagnosis. Consequently, the goal of this work is to create a clinical decision support system (CDSS) that can assist in identifying the stone that is most crucial to the urolithiasis diagnosis. In this work, we compare and evaluate AI models in order to determine which one is best at diagnosing urolithiasis, particularly for the creation of AI models that assist in a final decision in CDSS.

7.2 ARTIFICIAL INTELLIGENCE: WHAT IS IT?

Artificial Intelligence (AI) aims to create a self-governing computer that can effectively carry out jobs often done by people. This is achieved by using intricate non-linear mathematical simulation systems with fundamental neural components. The inquiry begins by seeking to determine how the human mind perceives, understands, and executes cognitive processes. The human mind possesses the capacity for intelligence, creativity, pattern recognition, language comprehension, memory, visual perception, logical reasoning, and the ability to establish connections between unrelated pieces of information. AI aims to replicate the aforementioned capabilities in order to perform a range of tasks, ranging from basic and manageable activities such as object recognition to complex tasks such as prediction. In order to enhance the intelligence and efficiency of decision-making, AI strategies involve utilizing existing data without any prejudice, relying solely on statistical models, and predicting future data that is now unknown [19].

The end game of AI is to build a system with enhanced sensory capabilities and the ability to optimize its own performance. Achieving this goal is a complex process involving a variety of subfields of AI, most of which are now applied in medicine and healthcare. Among these subfields you can find ML, ANNs, DL, NLP, computer vision, predictive analytics, genetic and evolutionary computing, expert systems, visual recognition, voice processing, and ANNs. It is necessary to define a few terms in order to move the discussion on the potential clinical effects of AI on the various subspecialties of urology forward.

ML refers to the act of training a computer to accurately predict future events by associating attributes with their corresponding outcome variables in a model that is trained and intended to learn from past data. The fundamental purpose of ML is to allow computers the ability to learn from data on their own. The foundation of deep learning (DL), a subfield of ML, are ANNs. ANNs are structured data processing units that, via the use of synaptic strengths (also called weights), mimic the computing capabilities of the human neurological system and brain. To achieve the best potential input-output mapping accuracy, the learning process involves incrementally adjusting the neuronal weights in a layer-by-layer fashion. A NN that has many layers is called a DL network. Many areas of medicine and healthcare have discovered potential applications for NNs, a highly significant branch of AI. These include sleep apnea, cardiology, electromyography, electroencephalography, and therapeutic medication monitoring.

Decision trees are a popular predictive modeling technique in machine learning, constructed using algorithms that segment the dataset based on specific conditions. This approach involves recursively splitting data into subsets to create a tree-like model of decisions, where each branch represents a possible outcome. By doing so, decision trees can identify patterns and relationships within the data, making them valuable for classification and regression tasks. The healthcare sector frequently uses these algorithms to identify the best option for the patient, such as telemedicine services.

Natural language processing (NLP), a branch of AI that studies how computers and human languages interact, is another area of AI that is vital to the healthcare industry. Managing data that are sparse or poorly detailed—a consequence of data that were once documented in narrative clinical documentation—is the largest problem in clinical research. Analytical, meticulousness remedy, disease investigative imagery, and experimental choice assistance are some of the most promising applications of AI in healthcare.

7.3 AI APPLICATIONS IN UROLOGY

Urology, a field of medicine, has had rapid growth over time. It continues to progress by adopting new technologies to enhance patient results [20]. Since urology is a branch of medicine that primarily treats male and female reproductive systems and urinary tracts, if underlying illnesses in these particular regions are not treated early on, they may worsen. The use of AI in urology is depicted in Figure 7.1. AI has become a common tool in the field of surgery, early diagnosis, and successful treatment planning. AI is becoming more and more prevalent and is assisting doctors in making decisions regarding patients with urological diseases. Studies supporting the safe and efficient use of augmented reality (AR) in urology have emerged throughout the last 5 years. These days, urologists use AR with picture overlay to remotely remove kidneys from patients by means of a machine-like arm with seven notches of autonomy [21].

With AR, information integration into the surgical workflow is greatly improved, reducing the complexity of minimally invasive treatments for surgeons. It is facilitating richer and more participatory experiences by introducing novel methods to surgical operations and medical education. In a similar vein, additional technologies used in conjunction with AI have a significant impact on the area. AI has been utilized in several sub-specialties within urology, such as urologic oncology, reproductive urology, renal transplant, and pediatric urology. Its application has led to breakthroughs in patient care by improving diagnosis, treatment planning, and surgical skill assessment [21]. The utilization of AI in these specific areas is addressed in subsequent sections.

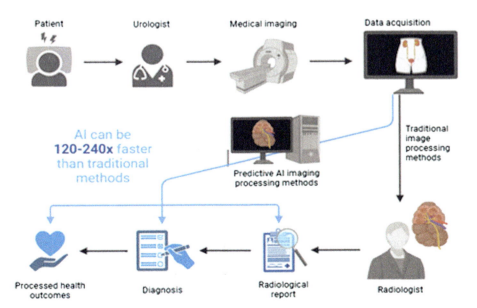

FIGURE 7.1 Augmented reality-guided robotic kidney removal with seven degrees of autonomy.

Creating statistical techniques that can analyze data and predict results for a specific patient is necessary for patient outcome prediction analysis. We have two options: new approaches coming out in AI or statistical modeling techniques. These techniques are well-suited for handling the inherent complexity and variability often present in biological and clinical data. Additionally, AI approaches excel at analyzing massive datasets [25].

7.3.1 Urolithiasis

The management of urolithiasis cases today differs greatly from how they were handled in the past, and AI techniques will have a significant impact on this approach [22]. In this area, AI may eventually offer full urolithiasis management, including diagnosis, treatment, and prevention. In order to aid in the early diagnosis (prevention) of kidney stones, Kazemi et al. [23] presented a unique ensemble learning-based decision support system and described the underlying mechanisms involved in identifying the different types of kidney stones. Using a wide range of AI techniques, such as the Bayesian model, decision trees, ANNs, and rule-based classifiers, this system was able to comprehend the complex biological factors that are involved in the prediction of kidney stones. There was a 97.1% level of accuracy provided by the method. The identification of ureteral stones in high-resolution CT imaging was accomplished by Langkvist et al. [24] by the development of a CNN approach. With an area under the receiver operating characteristic curve (AUC–ROC) of 0.9971 and a false-positive rate of 2.68 per scan, the model was able to accurately classify stones with a specificity of 100%.

7.3.2 Kidney Stone

Shockwave lithotripsy (SWL) and percutaneous nephrolithotomy (PCNL) are two therapeutic techniques that are widely acknowledged for the treatment of urolithiasis. However, the success rates of both approaches vary substantially, and if the therapy is ultimately unsuccessful, it may be necessary to undergo further procedures. Aminsharifi et al. [26] used ANNs to make predictions about the rate of stone-free PCNL with an accuracy of 82.8% and the demand for PCNL repeats with a 97.7% accuracy. The research conducted by Mannil et al. [27] focused on a single patient in order to determine the efficacy of SWL. They took into consideration the skin-to-stone distance, the three-dimensional texture of the stone, and the body mass index (BMI) of the patient. The authors created and assessed five AI systems, each of which utilized distinct 3D textural permutations of patient characteristics in order to achieve AUC values between 0.79 and 0.85. This represents an improvement over the AUC score of 0.58 that was obtained when employing only patient characteristics. Utilizing three-dimensional texture analysis for a different article [28], an estimation was made regarding the quantity of shock waves that are necessary for effective SWL. With an area under the curve (AUC) of 0.838, AI outperformed traditional statistical models when it comes to estimating the required number of shock waves (≥ 72 or ≥ 72). The two experiments that were carried out by Mannil et al. [27, 28] demonstrated that integrating AI with sophisticated textural analysis techniques offers a method that is not only trustworthy but also feasible and reproducible for predicting SWL performance.

7.3.3 ENHANCEMENT OF CT-BASED STONE DISEASE DETECTION

Urinary tract stones have become more commonly identified and diagnosed with abdomen and pelvic CT scans in the past few decades. Since it is the greatest penetrating method for urolithiasis findings, it has now the first choice. The growing trend of unenhanced CT scans has prompted the creation of computer-assisted solutions to help workers expedite this laborious procedure (Table 7.1).

The utilization of AI in the workflow could assist in the prioritization and optimization of patients as the utilization of imaging continues to increase. Li et al. [29] examined the automated segmentation of kidneys and kidney stones in unenhanced CT images using five sophisticated ML algorithms: three-dimensional [3D] U-Net, Res-U-Net, SegNet, DeepLabV3, and UNETR. The segmentation networks were trained using both independent (one-step direct segmentation) and dependent (two-step coarse-to-fine segmentation) approaches. Upon comparing the stated networks, it was found that the Res-U-Net network demonstrated greater performance in comparison to the other networks. Although the two-step segmentation method has significant computational needs and relies on the accuracy of the initial segmentation, it produced better outcomes compared to the one-step strategy. Moreover, dividing larger kidney stones resulted in superior outcomes. Parakh et al. [30] utilized a cascade CNN to analyze the diagnostic accuracy of pretrained models that were enhanced with tagged CT data. This analysis was conducted on non-enhanced CT images obtained from several scanners, with the aim of diagnosing urinary stones. CNN1 and CNN2 were developed for the purpose of urinary stone detection and identification of the urinary system structure, respectively. The CNN architecture used was Inception-v3, which has undergone pretraining on ImageNet. Subsequently, the GrayNet pretrained model, which encompasses a comprehensive assortment of human anatomy images, was employed to enhance the performance of the ImageNet pretrained model. The GrayNet pretrained model was used to initialize the weights of the CNN models for urinary tract recognition and stone detection. The stones in the urinary system were accurately identified with a high level of precision, as evidenced by the AUC of 0.954. Langkvist et al. [31] did a study where even skilled radiologists were able to distinguish between ureteric stones and phleboliths using a DL CNN. The study employed thin-slice CT images from a database of 465 individuals. The CNN model attained a sensitivity of 100% and an average of 3.69 false positives per patient on a test set including 88 scans, even in the absence of segmentation and anatomical information. The study included a probability distribution map that showed the locations of stones. This map resulted in a sensitivity rate of 100% and an average of 2.68 false-positive tests per patient [31]. In their study, Caglayan et al. [32] evaluated the effectiveness of a DL model in accurately detecting kidney stones on unenhanced CT images from different angles, using the stones' size as a determining factor. A retrospective analysis was performed on a cohort of 465 patients who received CT scanning specifically for the purpose of detecting kidney stones. The patients were categorized into three groups according to the size of their renal stones: Group 1 consisted of patients with stones measuring 0 to 1 cm, Group 2 consisted of patients with stones measuring 1 to 2 cm, and Group 3 consisted of patients with stones larger than 2 cm. The accuracy rates of the sagittal plane pictures were the highest among the three groups, with rates of 85%, 89%, and 93%, respectively [16]. Jendeberg et al. [33] employed a 2.5-dimensional CNN to examine 384 pelvic calcifications in an unspecified test dataset. An evaluation was conducted to compare the assessments of seven radiologists, the CNN approach, and a semi-quantitative method. The CNN exhibited superior performance compared to the radiologists. The algorithm had a 92% accuracy rate, surpassing the average accuracy rate of radiologists, which generally stands at 86%. The sensitivity, specificity, and AUC values for differentiating phleboliths from distal ureteric stones were 93%, 89%, and 0.95, respectively [33]. De Perrot et al. [34] assessed the efficacy of a ML model that was trained using radiomics data from a cohort of 369 patients who received low-dose CT scans for acute flank pain. Later, a specific cohort of 43 individuals was utilized to evaluate the ML model's capacity to recognize and differentiate phleboliths from ureteral stones. The ML model attained an AUC of 0.901, an accuracy rate of 85.2%, a positive predictive value (PPV) of 81.5%, and a negative predictive value of 90.0%. They noted that the quantifiable and repeatable attributes of radiomics and ML are some of its benefits. Chak et al. employed a DL network in combination with a support vector machine (SVM) using characteristics from the segmented pictures found to the gray near co-occurrence matrix (GLCM). The region of interest was identified using the K-means segmentation approach after these CT images had previously been processed and visually improved. They stated that they could detect kidney stones in 25 patients' CT scans with a 95% accuracy rate. GP et al. [20] deployed a model called Based on its training in the ImageNet record, exception. The dataset for CT scans involved 1450 pictures of kidneys, and the researchers achieved an accuracy of 96.82% using this particular mix of DL models. In a study conducted by Elton et al. [36], a DL system consisting of a 13-layer CNN classifier demonstrated the ability to accurately detect kidney stones of scan pictures. The system achieved a sensitivity of 86% and a false-positive rate of 0.5 per scan. External validation of the CT image classification system was conducted.

7.3.4 Optimizing Ultrasonography (US) for the Identification of Stone Illness

Krishna et al. [37] suggested a computer-aided detection system using a field programmable gate array to spot anomalies in the kidneys using ultrasound pictures. The US images underwent initial preprocessing to eliminate noise, and the region of interest was manually extracted. Haralick features (GLCM) and intensity histogram properties were extracted from the segmented region of the kidney. The method initially employed a look-up table to differentiate between normal and diseased kidney images. Afterwards, the unusual photos of kidney cysts and stones were identified using a trained SVM with a multi-layer perception (MLP) classifier. The proposed technique precisely detected the specific anomaly exhibited in the renal ultrasound pictures, with a sensitivity of 100%, specificity of 96.8%, and accuracy of 98.1%.

Using an ANN, Balamurugan and Arumugam [38] presented a novel mode to predict renal illness in the United States. Seven hundred and fifty kidney US images were used, 80% for training and 20% for testing. Four categories of photos were used: normal, cyst, stone, and tumor images. The photos from the United States were subjected to preprocessing, the GLCM characteristics were extracted, and the oppositional grasshopper optimization technique was used to choose the key features. Subsequently, an ANN was employed to categorize the photographs as either normal or aberrant. They announced a peak specificity of 97.22% and a peak accuracy of 95.83%.

Selvarani and Rajendran [39] utilized the meta-heuristic support vector networks (SVN) to detect renal stuff of US pictures. The literature extensively reports that speckle noises were reduced using Anisotropic Median Filter (AMM) filter provided by Math Works. A meta-heuristic SVM classifier employed GLCM features that were obtained for classification using standard K-means segmentation. After being trained on 250 US images (150 with stones and 100 without), the system displayed a 98.8% accuracy rate.

Viswanath et al. [40] successfully detected renal stones in US pictures using a deep segmenting reaction-diffusion levels method. For picture processing and quality enhancement, they employed a simple intensity filter, which produced better results than alternative preprocessing techniques. Based on the energy level, advanced wavelet sub-band filters were utilized to extract features from the processed images. With an average image segmentation time of 7.08 seconds, the MLP-backpropagation ANN achieved an accuracy rating of 93.2%. Wavelet decomposition was employed by Akkasaligar and Biradar [41] in 32 US kidney pictures to identify stones. The accuracy of the suggested approach, which makes use of an ANN, was stated to be 96.8%. k nearest neighbor (KNN) and SVM classification were employed in learning.

The purpose of this study was to identify and locate kidney stones in ultrasound images [42]. The Gaussian filter and unsharp masking were initially employed for the analysis and enhancement of the US photos. The region of interest was identified using morphological techniques such as destruction and expansion. Employed to evaluate kidney stone pictures, followed by the application of KNN and SVM classification techniques, the accuracy of KNN stayed initiallyto be 88%, whereas SVM achieved an accuracy of 84%.

7.3.5 Enhancement of X-ray-based Stone Disease Setection

Using kidney-ureter-bladder X-ray imaging as the primary diagnostic method for urolithiasis is not recommended. This is because the method has certain drawbacks, such as the requirement for stones to be visible on the X-ray and the resulting low sensitivity. However, X-rays are still a cost-effective, low-radiation, and easily accessible tool for detecting urinary tract stones. AI can improve the accuracy of X-rays in diagnosing kidney, ureter, and bladder conditions, particularly in identifying stones that are visible on the X-ray (Table 7.1).

TABLE 7.1
An overview of research at the uses in AI for diagnosing stone ailment

Reference	Aim	Observation	AI result	Outcome
Li et al. [29]	Urinary stones detection by CT	Sectional	Finding exactness of 99.95%	Reduced output
Parakh et al. [30]	Urinary Stone Detection Using CT Scans	Sectional	Excellent stone exposure (AUC of 0.953)	Other algorithms with lower performance
La¨ngkvist et al. [31]	Identify urinary stones	Sectional	Optimized accuracy with an AUC of 0.9971	No comparator
Caglayan et al. [32]	Identify urinary stones	Sectional	Precision of 63% to 93%, conditional on stone size class	No comparator
Jendeberg et al. [33]	CT can be used to distinguish between ureteral stones and pelvic phleboliths.	Sectional	Accuracy of 92%	Average radiotherapist precision: 85%; popular division, truth: 92%
De Perrot et al. [34]	CT can be used to distinguish between ureteral stones and pelvic phleboliths.	Sectional	Overall	Other algorithms with lower performance
Chak et al. [35]	Urinary stone detection using CT scans	Sectional	Precision of 96%e98%, Identify urinary stones with CT scans	No comparator

Kobayashi et al. [43] developed a computerized detection system that utilizes DL to be able to detect radio-opaque UTI stones on a standard X-ray. A total of 1016 simple X-radiation depicting radio-opaque superior urination area stones were collected. Out of these, 827 X-rays were utilized for training purposes, whereas the lasting 190 X-rays were assigned for challenging. The X-ray images were uniformly enlarged to dimensions of 1328 by 1328 pixels in order to accommodate variations in picture sizes. Additionally, histogram equalization was employed to enhance the contrast. The study utilized a CNN architecture that consisted of a residual network with 17 layers. The computer's prediction led to the identification and display of a heat map in each input image, where every pixel was assessed for the existence of a stone. The heat map used a color scale, The light red indicates a 100% probability of stone presence, while dark green indicates 0%. Performance was assessed using the F score, which balances sensitivity and PPV via their harmonic mean. The 16-layer residual network achieved a reaction time of 110 milliseconds per X-ray image. Maximum sensitivity and PPV reached 87.2% and 66.2%, respectively, yielding the highest F score of 0.752. Sensitivity and PPV were lowest (92.5% and 87.6%, respectively) for proximal ureter stones, specifically at the mid-ureter during the study.

Aksakalli et al. [44] applied a range of ML and DL techniques to detect kidney stones in X-ray images. They utilized CNN along with several ML algorithms including decision trees (DT), random forests, SVMs, MLP, KNN, Naive Bayes (BernoulliNB), and DNNs. The dataset comprised 221 X-ray images split into 80% for training and 20% for testing. The imaging stayed transformed into grayscale by first being resized to predefined dimensions. Subsequently, the numerical values representing the grey levels in the photographs were extracted in order to generate the data collection. This dataset exhibits imbalanced classes, hence several oversampling and undersampling techniques were employed. This strategy leads to a discernible enhancement in the performance metrics of the methodologies. The F1 score, the harmonic average of precision and recall, played a pivotal role in their analysis. The DT strategy showed superior performance to other methods, attaining the highest F1 score and an 85.3% success rate using synthetic minority over-sampling methodology.

7.4 USING AI TO FORECAST MANAGERIAL RESULTS

7.4.1 Forecasting the Results of Prudent Management

Urinary department visits are primarily made for symptomatic ureteral stones. Spontaneous stone passage (SSP) in this context refers to the natural passage of stones in the body, which poses a therapeutic challenge and disturbs the patient role eminence of lifecycle. Probability of SSP alterations is typically determined by the extent of the stone. In healthcare settings, there is a need for faster and more accurate prediction algorithms to improve patient care by enhancing monitoring and enabling early intervention when necessary. These algorithms can help restore the quality of patient outcomes by identifying potential issues early in the care process, allowing for timely interventions and better management of health conditions (Table 7.2).

Cummings et al. [45] employed a commercially available ANN in 2000 to solve the clinical question of SSP. In the study, 180 patients' worth of data were examined; 125 were used to train the NN and 55 were utilized aimed at trying. For error correction, feedforward backpropagation was utilized. The feature significance analysis assigned the highest weight to the longest duration of symptoms. For those who required intervention, exactness percentage for expecting the SSP remained 76% (42/55) and 100% (25/25), respectively.

A prototype prediction model for the SSP of ureteral stones, using a logistic regression (LR)–ANN–SVM approach was employed by Dal Moro et al. [46]. They asserted that the reproduction's ability to forecast the need for further intervention was enhanced by the addition of SVM. An accuracy rate of 83.9% and a specificity of 86.9% were offered by the SVM-dipend model. Furthermore, the results of the feature importance analysis in this study showed that the three characteristics that were most effective in predicting SSP were stone position, size, and duration of symptoms.

A prototype ANN was employed in a different investigation to evaluate the predicting parameters' effectiveness and estimate the SSP [47]. Three groups—training, validation, and test—were randomly selected from a dataset of 192 patients. The three groups' respective SSP estimation rates for the ANN were 99.2%, 85.5%, and 88.7%. Erythrocyte sedimentation rate (ESR), stone size, and C-reactive protein quantity stayed the prognostic indicators, which showed a stronger correlation with the SSP. For SSP estimation, Park et al. [48] employed an ML and LR model. They examined medical records from 833 patients who had unilateral ureteral stones and visited the emergency room in the past. They assessed the standard statistical approach's (LR) predictive accuracy of SSP using the DL method, which is an MLP built on the Keras context. With a 100% specificity for 5e10 mm stones, MLP outperformed LR in SSP. LR in predicting SSP stayed at 0.860 and 0.848 aimed at ureteral stones smaller than 0.5 mm, and 0.881 and 0.817 for stones larger than 5.5 mm. The LR model's maximum sensitivity for the 5e10 mm stones was 90.5%. They reported that there was no discernible difference between the two models and that image analysis was necessary to enhance the prediction.

7.4.2 Forecasting the Outcome of Extracorporeal Shockwave Lithotripsy

ESWL, a non-invasive treatment for urinary stones ideally under 2 cm, is widely accepted. Several factors influence its success and stone-free rates (SFRs), such as stone size, location, quantity, composition, and density. Numerous mathematical and computational techniques are being researched in an effort to forecast the results and reduce the negative consequences of the operations caused by this unpredictability (Table 7.2)

Poulakis and colleagues [49] conducted an analysis of the medical files of 679 people (constituting 700 kidney elements) who underwent primary shockwave lithotripsy for renal calculi in the lower pole in 2003. Their univariate study examined key variables influencing the clearance of lower pole calculi post-ESWL. To forecast SFR in the test cohort, they utilized an ANN. They documented an area under the AUC–ROC of 0.936 with an accuracy of 92%. The ANN identified urinary transport as the most crucial predictor for predicting SFR in lower pole calculi, as the authors noted. The ANN's performance was also found to be significantly influenced by the following variables: body mass index, caliceal pelvic height, infundibuloureteropelvic angle 2, and others.

An ANN model and an LR model were compared by Gomha et al. [50]. A comparison was made between the performance of two models in predicting the stone-free status (SFS) three months following ESWL, using 10 inputs for ANN and variables for LR. Both models performed adequately, with the ANN outperforming the LR model in accurately predicting individuals who do not respond to ESWL. An ANN was employed by Moorthy and Krishnan [51] to predict the fragmentation of stones based on several features retrieved from CT images that were not improved. ANN was used to compute and assess statistical features such as mean, variance, skewness, and kurtosis using the first-order technique. The findings indicated that the model's prediction displayed a sensitivity of 81.6% when the mean was regarded as the target characteristic. Furthermore, a high accuracy rate of 90% was achieved in correctly identifying both the genuine positive and true negative situations. The specificity of identifying true negative patients is 98.4%.

Choo et al. [52] conducted a study where they introduced a ML approach to detect the SFS (stone-free status) of urinary stones after a single ESWL (extracorporeal shock wave lithotripsy) session. After a period of two weeks following the treatment, they considered a single ESWL session to be successful if there were no fragments larger than 2 mm visible on CT or plain X-ray imaging. The introduction of extra factors resulted in an enhancement of the model's accuracy performance. The decision models were created by considering every conceivable combination of components. When employing DT analysis, the 15-factor model had an accuracy of 92.29% and an average AUC–ROC of 0.951. The stone volume was found to be the most crucial factor in determining the success of a single ESWL session, as per the findings of this study. Seckiner et al. [53] utilized a prototype ANN in a study involving 203 patients who had undergone ESWL. The aim was to predict the SFS and aid in treatment planning. Regression analysis and ANNs were utilized for determination.

The regression analysis utilized identical parameters, with 16 features serving as inputs for the NN. Stone size, number of ESWL sessions, stone placement, infundibulopelvic angle, and skin-to-stone distance (SSD) emerged as significant predictors of stone-free outcomes. Prediction accuracies for the NN were 99.25%, 85.48%, and 88.70% for the training, validation, and test samples, respectively.

Five ML algorithms were used to test 3D texture features within the same context of forecasting the prosperous SFS following ESWL [54]. A purely visual examination could miss distinct, quantifiable differences in stone features that are found by texture analysis (TA). Two important findings were presented in a preliminary investigation by Mannil et al. [54]. Firstly, 3D-TA provides supplementary trustworthy data on the effectiveness of ESWL, along with already established clinical variables such as body mass index (BMI), systemic sclerosis (SSD), and the size of the stone. Furthermore, the correlation between mean CT attenuation values for kidney stones and the success of SWL was not anticipated. When the 3D-TA features were combined with clinical measures, the ability to distinguish improved further. The AUC for 3D-TA features and SSD was 0.84, for 3D-TA features and BMI it was 0.81, and for 3D-TA and stone size it was 0.80.

A ML model was developed using decision tree algorithms on 357 non-contrast CT scans of patients who underwent ESWL for renal and upper ureter stones by Yang et al. [55]. To predict SFS and one-session success, they utilized three DT-based algorithms: random forest, extreme gradient boosting trees (XGBoost), and light gradient boosting method (LightGBM). These algorithms were chosen due to their superior performance in predicting outcomes from small training datasets. According to the authors, the most influential factors affecting procedural outcomes were stone volume and median stone density. The proposed models achieved over 77% accuracy in predicting one-session success and up to 87.8% accuracy in predicting stone-free rates (LightGBM).

An artificial neural network (ANN) was developed by Tsitsiflis et al. [56] to predict and process outcomes based on various features of extracorporeal shock wave lithotripsy (ESWL). They collected medical records from 716 patients, of which 548 were used for training the model, while 166 were set aside for testing. This approach aimed to improve the accuracy of ESWL outcome predictions by leveraging machine learning techniques. Based on the parameter importance analysis, the location of the stone was shown to be the most crucial factor affecting the outcome of the procedure. However, diabetes and hydronephrosis were also found to have a significant predictive value in determining the likelihood of problems occurring. The ANN model achieved a performance rate of 97.73%, a PPV of 82.81%, and an accuracy of 80.42% in accurately predicting complications by the conclusion of the training set.

Handa et al. [57] examined injury lesion volumes in ex vivo kidneys with magnetic resonance imaging and a multi-spectral NN model to predict hemorrhagic injury post-ESWL. They demonstrated that using MRI and the multi-spectral NN classifier allows quick and accurate quantification of renal damage lesion volumes due to ESWL.

7.4.3 Forecasting the Results of Endoscopic Treatments

It is still up for debate on how to predict the SFR and other postoperative metrics for kidney stone patients scheduled for PCNL. To classify renal calculi and estimate SFR, a number of ratings have been established over time. However, in the context of postoperative outcome prediction, AI and its variants have demonstrated results that are comparable to, or in some cases, exceed traditional methods (see Table 7.2). These AI-driven approaches are capable of analyzing complex patterns in patient data, leading to more accurate and reliable predictions for post-surgical recovery.

A model of an ANN was developed by Aminsharifi et al. [58] to forecast four outcomes that occur after PCNL.

The ML model exhibited a sensitivity of over 82% and a success rate of 80% in properly predicting all outcomes of a postoperative assessment. Five ML and DL models were created by Geraghty et al. [59] in order to evaluate the relationship between 11 outcomes and 43 preoperative variables of the patients. Of those outcomes, the majority of patients' post-PCNL infection and transfusion requirements were accurately anticipated. For these two outcomes, the AUCs of all the ML and DL applied models were greater than 0.90 and 0.77, respectively. However, the authors highlighted that these models' incapacity to securely handle incomplete information and extremely uncommon outcomes resulted in subpar forecasts. They came to the conclusion that imaging results might offer AI systems information that is crucial for predicting surgical outcomes. Chen et al. [47] employed a deep neural network (DNN) to calculate and predict the factors contributing to sepsis susceptibility following percutaneous nephrolithotomy (PCNL) or flexible ureteroscopy. The study examined the medical records of 847 patients who met the inclusion criteria. The results indicated that the incidence of sepsis following PCNL or flexible ureteroscopic lithotripsy was assessed to be 5.9%. For every patient, a preoperative CT scan was done. Based on the prediction of preoperative sepsis in patients with ureteral calculi, the DNN model and the performance of the least absolute shrinkage and selection operator model was evaluated. The DNN model demonstrated superior performance compared to the least absolute shrinkage and selection operator model in predicting 25 variables. The AUC for internal validation was 0.873, while for external validation it was 0.919.

TABLE 7.2
Overview of research on the use of AI in forecasting managerial results

Reference	Observation	AI result	Outcome
Cummings et al. [43]	Estimate of SSP	Precision of 77 %	Nonreliable
Dal Moro et al. [44]	Estimate of SSP	Precision 84.5%	Sensitivity specificity
Solakhan et al. [45]	Estimate of SSP	Precision of 92.8%	Further systems with minor concert
Park et al. [46]	Estimate of SSP	Precision AUCs of 0.859	AUC 0.816 (stone of 5.1 mm to 10 mm)
Poulakis et al. [47]	Estimate lower poleclearance after ESWL	Precision of 92%	Nonreliable
Gomha et al. [48]	Estimate of clearance After ESWL for ureteral gravels	Precision of 77.7%, Precision of 93.2%	Nonreliable
Moorthy and Krishnan [49]	Estimate of kidney stone fragmentation after ESWL	Precision of 90%	Nonreliable
Choo et al. [50]	Estimate of consent after ESWL for ureteral	Precision of 92.29%	Nonreliable
Seckiner et al. [51]	Prediction of clearance after ESWL for renal stones	Precision of 88.70%	Nonreliable
Mannil et al. [52]	Estimate of renal stone fragmentation after ESWL	Precision AUC of 0.85%	Further systems with minor concert
Yang et al. [53]	Estimate of clearance after ESWL of Kidney high ureter stones	Precision AUC of 0.85 for stone-free	status 4 weeks; AUC 0.78 stone-free
Tsitsiflis et al. [54]	Estimate of complications after ESWL for renal or ureteral stones	Precision of 81.43%	Nonreliable
Handa et al. [55]	ESWL-induced Experimental Strong correlation between	The precision values are 91.8% and 83%	Nonreliable
Aminsharifi et al. [56]	Estimate of several results later PCNL	Regarding the removal of stones and the requirement for a blood transfusion, the AUC for stone clearance is 0.925.	AUCs of 0.615 and 0.621 for stone clearance by GSS and CROES nomograms

7.5 USING AI TO OPTIMIZE THE SURGICAL PROCESS

The prediction models rely on a wide range of variables that frequently interact with one another, making it difficult to determine the optimal combination of these factors. A crucial step toward efficiently and effectively completing the related procedures is the proper configuration of intraoperative variables. By accurately adjusting these variables, it becomes easier to streamline the process, ensuring better outcomes and reducing potential risks during surgery (Table 7.3).

7.5.1 Enhancement of the ESWL Process

In 2003, Hamid et al. [48] collected data from 81 patients who underwent ESWL treatment for renal stones. This dataset contained a multitude of parameters, such as ESWL settings, radiographic properties of stones, and urine chemical variables. For the purpose of optimizing the prediction of the number of ESWL shockwaves required for attaining ideal fragmentation, the aforementioned data was utilized in the construction and training of an ANN system. All of the patients were identified by the ANN algorithm as requiring a greater quantity of ESWL shockwaves in comparison to the standard treatment plan. Similarly, Goyal et al. [49] employed data from a group of patients who underwent ESWL for the treatment of kidney stones. The training set consisted of 195 patients, while the validation set consisted of 79 patients. For the purpose of training and evaluating the effectiveness of their model, they utilized a technology known as ANN. The research looked at a variety of factors, including the chemical components of urine, the number of ESWL sessions, the size of stones, and the burden they carried. The association between the shockwave strength predicted by the ANN algorithm and the number for optimal stone fragmentation was found to be highly significant (r2=0.8343, r2=0.9328). This connection was found to be highly significant between the observed shockwave intensity and the number of shockwaves that were required during the ESWL procedures.

Mannil et al. [50] conducted a study involving a group of 33 urinary stones in a laboratory setting. The objective was to investigate the relationship between the properties of the stones and the number of shockwaves required to effectively break them apart. This research aimed to provide insights into how different stone characteristics influence the efficiency of shockwave lithotripsy treatment. A machine learning model successfully identified the subset of stones fragmented into smaller bits with fewer than 72 shockwaves, achieving a sensitivity of 94% and a specificity of 59%.

In a recent paper [51] a DL model was developed and trained to generate personalized ESWL protocols. These protocols specify the ESWL settings (like power level, shockwave rate, and total number) for each stage of every ESWL session. The model utilized data based on best practices for ESWL treatments from the International Stone Registry. The DL model demonstrated superiority in comparison to other models and conventional approaches in determining ESWL parameters for each ESWL phase.

In the study by Muller et al. [52] a CNN method was developed to identify kidney stones in images captured during ESWL procedures performed on 11 patients. They next examined the effects employing the aforementioned technique for optical stone

TABLE 7.3
Review of research on the role of AI in enhancing the efficiency of surgical procedures

Reference	Aim	Observation	AI result	Outcome
Hamid et al. [48]	Enhancement of ESWL technique	Partial	75% accuracy predicting shockwave necessity and 99% for patients needing extra shockwaves	Nonreliable
Goyal et al. [49]	Enhancement of ESWL technique	Partial	The correlation coefficient for power level is 0.8343, and the correlation coefficient for shockwave number is 0.9329.	Association factors for rule level and shockwave number 0.0195 and 0.5726, respectively
Mannil et al. [50]	Enhancement of ESWL Technique	Investigational	AUC 0.837 Predicting fragmentation < 71 shockwaves	Alternative different model exhibiting inferior concert
Chen et al. [51]	Enhancement of ESWL Technique	Partial	Prediction accuracy for power level and shockwave rate 98.8%, 98.1%, respectively	Alternative multivariate models exhibiting inferior performance
Muller et al. [52]	Enhancement of ESWL technique	Partial	Effect success frequency of 75.3%	Effect success frequency at 55.1%
Taguchi et al. [53]	Enhancement of PCNL perforation	Investigational	Lesion triumph frequency of 99%; puncture time of 35 s	Nonreliable
Wang et al. [54]	Enhancement of PCNL perforation	Investigational	Normal appreciation care of 78% (SE: 4.5%) for cortex, 84% (SE: 6%) for medulla, and 90% (SE: 5%) for calyx	Nonreliable
Li et al. [55]	Enhancement of PCNL perforation	Partial	ANN perfect realized a restored localization and puncture method comparison to MVRA model and surgeon's expertise	Nonreliable
Jeong et al. [56]	Enhancement of RIRS protection case	Investigational	Gratitude of tissue revelation to laser liveliness with truth of 95% and potential time of 0.5	Nonreliable

detection in directing ESWL shockwaves. Compared to a system where an operator directs the operation, the study discovered that an ESWL approach driven by AI will exhibit a stone strike rate that is 75.3% higher and a mishit rate that is 67.1% lower. To further improve the accuracy of the aforementioned algorithm in steering ESWL shockwaves, it would be beneficial to increase the sample size of patients used for training.

7.5.2 Enhancement of Endoscopic Techniques

Taguchi et al. [53] investigated the probability of using robot-assisted fluoroscopy to controller hole during the PCNL operation. They compared this technique with the traditional method of operator-dependent US-guided wound. A robot-assisted puncture system utilized mainframe idea software and AI to analyze fluoroscopic pictures and provide guidance for hand wound as needed. Robotic-assisted puncture achieved a 100% success rate, whereas operator-dependent puncture had a success rate of 70.6%. The approach significantly decreased the duration required for the entire procedure and the needle puncture.

A different paper [54] introduced another approach for directing PCNL penetration. The system utilized AI with an optical coherence tomography (OCT) scan, a distinctive imaging technique capable of revealing underlying tissue up to several millimeters in depth. Miniature probes that can be customized to match the PCNL puncturing needle can be utilized to administer OCT. In this study, a DL algorithm was developed and trained to distinguish between various OCT patterns in the renal cortex, medulla, and calyx. The mean standard error for these anatomical regions was found to be 79% ± 4%, 85% ± 6%, and 91% ± 5%, respectively. The technique stated above has shown effectiveness in detecting the tissue type in visible of the PCNL needle and guiding the lesion exactly.

In 2016, Li et al. conducted a comprehensive study to identify the optimal tomography mode for targeting lesions during percutaneous nephrolithotomy (PCNL). An artificial neural network (ANN) was developed and trained using data from patients undergoing PCNL surgery. The results from the trained algorithm indicated that using X-ray and ultrasound (US) guidance was particularly advantageous for. However, when it came to treating large or uncomplicated stones, the combination of these approaches did not show significant superiority compared to the other indicated methods. Jeong et al. [56] recently created a monitoring system that analyzes the shockwave form produced when a laser contacts with either soft (tissue) or hard (stone) materials. The technology incorporates AI and aims to reduce the amount of laser energy that soft tissues are exposed to during retrograde intrarenal surgery.

An accelerometer that was modified for the ureteroscope was used to monitor shockwaves, and the results were generated in a surgical simulation setting. In order to enhance the predictive capabilities of a ML model for determining tissue exposure and the resulting laser energy injury, the aforementioned data were added [56]. With a 0.5 s latency time and 95% accuracy, the trained model may alert a surgeon to tissue damage caused by laser radiation. The accuracy of the model would rise with additional optimization, reducing latency time. Furthermore, it appears that the same technology can be applied to other endourological surgical operations.

7.5.3 AI to Clarify the Chemistry and Makeup of Stone Disease

Anticipating the beginning of prime stone illness or the recurrence of nephrolithiasis is considered essential for reducing its occurrence and the resulting impact on kidney role.

AI is considered as a promising method to explicate the intricate relationship amongst many medical and chemical factors, whose deviation is connected through an increased likelihood of developing stone disease (Table 7.4).

7.5.4 Relationship between Blood, Urine Interaction, and Other Medical Variables and the Hazard of Stone Disease

One study employed an artificial neural network (ANN) to analyze data from 118 male patients and 95 control subjects [57]. The study identified calcium oxalate supersaturation and urea concentration in urine as the most influential factors in predicting the development of calcium stones. This finding highlights the potential of ANN in identifying key biomarkers that can help in the prevention and management of kidney stones. A study on the susceptibility to calcium stones utilized ANNs to analyze clinical and biochemical data from 59 female and 119 male patients, along with their respective healthy controls. The study identified 24-hour excreted urea and CaOx (calcium oxide) supersaturation as the two most significant factors contributing to the predisposition to calcium stones [58]. The study emphasized the prevalence of these predisposing factors among male patients, regardless of whether they had a family history of stone disease or not [58].

A recent work utilized ensemble learning to identify relevant features from a dataset consisting of 42 clinical and biochemical indicators. The goal was to establish a prediction model for the development of stone sickness, achieving a correctness rate of 96.2% [59]. A different model utilizing ML algorithms determined that the primary factors contributing to the formation of kidney stones in the specified size category include advanced age, reduced levels of CaOx saturation, greater proportion of protein in the stone structure, and the presence of hypertension, particularly in cases where the kidney stones exceed 20 mm in size [60]. Due to LR, the AI-based prediction in this report was highly satisfactory and precise. While it is established that changes in the levels of urine analytes discharged over a 24-hour period can be used as a dependable interpreter of stone sickness, the specific experimental features responsible for these fluctuations are mostly unknown.

Using data from electronic health records, a sophisticated ensemble model successfully predicted these anomalies with high accuracy [61]. The clinical factors of age, gender, and BMI shown the most robust connection with these anomalies. An ANN technique was employed to assess the likelihood of recurrent stone sickness based on the clinical and biochemical data of 80 patients. A predictive model using various indicators such as serum sodium, serum potassium, urine sodium, urine phosphorus, and urine calcium oxalate was able to accurately predict around 89% of the repetition actions [62]. Genomic predisposition is a standard risk aspect for stone sickness. An examination of the genetic and eco-friendly information of a cluster of patients, as compared to a group of controls, showed that the use of ANNs was

TABLE 7.4
Overview of research on the use of AI in understanding the chemical and compositional aspects of stone diseases

Reference	Observation	Stage	AI result	Outcome
Dussol et al. [57]	Calcium stone Risk factor	Creative level	Accuracy: Stone formers vs. controls: 74.4%	75.8%
Dussol et al. [58]	Calcium stone Risk factor	Creative level	Calcium oxalate supersaturation and 24-hour urinary urea excretion in both men and women with a intimate	Non comparison
Kazemi and Mirroshandel [59]	Renal lithiasis risk	Background	Precision of 97.1%	Alternative classifiers with inferior accuracy
Chen et al. [60]	Chance of creating renal stones of >3 cm	Unit	AUC of 0.68	AUC of 0.75
Kavoussi et al. [61]	Estimate of 24 h urine anomalies related for stone ailment	Unit	Improved precision in forecasting uric acid, and sodium irregularities	Complex precision in guess of pH and citrate anomalies
Caudarella et al. [62]	Danger of stone disease repetition	Unit	Precision of 89.8%	Precision of 67.5%
Chiang et al. [63]	Danger for stone disease	Creative level	Precision of 89%	Precision of 74%
Xiang et al. [64]	Credentials of CaOx summation in urine deposit	Creative level	Precision of 74%	Precision of 74%
Kletzmayr et al. [65]	Credit of illustration embarrassment	Cross-sectional	IP6 analogs effectively inhibit CaOx crystallization	Non comparison
Kriegshauser et al. [66]	Structure of stones using CT scans	Untried	The accuracy rate for UA stones is 97%, whereas for non-UA stones it is 72%.	Alternative multivariate models exhibiting inferior performance

AI in Diagnostics and Disease Prediction 149

Kriegshauser et al. [67]	Structure of stones using CT scans	Partial	100% accuracy for UA stones and 87% for non-UA stones	Alternative multivariate models exhibiting inferior performance
Zhang et al. [68]	Structure of stones using CT scans	Partial	The AUC for UA stones, as opposed to non-UA stones, is 0.965 with a standard deviation (SD) of 0.029	The model utilizing CT TA reached a warmth of 94.4% and a specificity of 93.7%.
Große Hokamp et al. [69]	Structure of stones using CT scans	Partial	Precision of 90.3% on a per-voxel basis; precision of 86.2% to 90.3% on self-reliantly tried procurements	Non comparison
Tang et al. [70]	Structure of stones using CT scans	Partial	Total care of 97%, Inclusive care of 96%	Non comparison
Black et al. [71]	Structure of stones using CT scans	Partial	Truth of 88.3% for COM in its place of non-COM stones (AUCZ0.933)	Non comparison
Lopez et al. [72]	Structure of stones using CT scans	Partial	Estimate precision for each stone structure from 70.48% (struvite) to 94% (COM stones)	Extra multivariate replicas with lesser routine
El Beze et al. [73]	Structure of stones using CT scans	Partial	Precision of 92% of 98%	Depending on stone type
Ochoa-Ruiz et al. [74]	Structure of stones using CT scans	Partial	PPV of 95%e98%, reliant on stone form	Reliant on shingle type

the most effective in distinguishing between the two groups. The ANN accurately classified 89% of the participants [63].

A different study found that the rate of recognizing calcium oxalate gemstones in urine deposit, which are believed to be an aspect in the development of CaOx stones, was improved to 74% in comparison to the conventional identification process that uses microscopy images and a CNN-based identification process [64].

A recent study examined the crystallization dynamics of CaOx in response to potential crystallization inhibitors using electron microscopy images and ML analysis. The results demonstrated the efficacy of myoinositol hexakisphosphate analogues as a preventative measure against CaOx nephropathies or stones [65].

7.5.5 Stone Composition Identification Using CT

In addition to the higher incidence of stone detection, radiographic characteristics in CT exams are a useful. A study utilized a CT approach to examine 32 kidney stones that were removed from the body and had known compositions. This examination provided 52 specific features for each stone. After training with data from these factors, including ANN and SVM algorithms, five multiparametric algorithms achieved a maximum accuracy of 99.5% in individual uric acid (UA) stones from non-UA stones [66]. The precision in distinguishing the non-UA subtitle was 74%.

Researchers conducted a study on 38 renal stones outside of the body using a specific CT method that involved rapidly switching voltages. This method used single-source and dual-energy CT, and provided data on 17 different parameters for each stone. The method for distinguishing between different non-UA subtypes achieved an accuracy of 88%. However, the diversity amongst UA and non-UA stones could be made with 100% accuracy using the same multiparametric techniques [67].

A third study utilized CT TA to differentiate between UA and non-UA stones, employing SVM classifiers to train and construct a corresponding model. The model accurately diagnosed UA stones with an AUC of 0.964, demonstrating a sensitivity of 94.4% and a specificity of 93.7% [68].

Große Hokamp et al. [69] observed 199 renal stones with an identified configuration using a spectrum finder digital scanner, employing several methods. A NN algorithm was developed and verified using the generated data to differentiate between diverse stone structures. The precision rates of the final model were 91.1% per voxel and ranged from 87.1% to 90.4% on independently validated acquisitions. It is intriguing to observe that despite the presence of compound stones, the model accurately predicted the primary component [69].

Tang et al. [70] recently published a study presenting their research findings on the use of an AI model using unenhanced radiography features to differentiate among calcium oxalate monohydrate and non-COM stones. The final model, which consisted of eight selected features from the initial collection of 1218 radiography features, achieved an accuracy of 87.30% (AUC: 0.932; sensitivity: 89.51%; specificity: 84.3%). The aforementioned findings are applicable to the preoperative evaluation of patients as they were obtained from the in vivo radiographic classification of chippings [70].

7.5.6 Identifying the Composition of Stones via Endoscopic Pictures

A competent endourologist can gain valuable visual information about the composition of the stone from intraoperative imaging. This knowledge can then be used to adjust other surgical parameters, such as the energy generator settings for lithotripsy. By applying visual processing with AI techniques, this intraoperative stone characterization could be made more accurate. A CNN algorithm was trained on 63 kidney stones in order to predict stone composition based on macroscopic stone appearance and produced an overall prediction accuracy of 85% for each composition category. The mentioned set included stone samples that were photographed using digital cameras to get detailed images of both the exterior and interior of each stone [71]. Lopez et al. [72] conducted a study to determine the type of stone in surgical procedures. They examined several classification strategies utilizing 87 fragment cross-section images, 90 fragment surface images, and matching algorithms. The most accurate classification technique used a deep CNN algorithm and achieved 98% accuracy for the four stone classifications that were used. Another study used a large number of ureteroscopy-captured pictures of stone surfaces and stone sections to evaluate the classification performance of DL-based approaches with basic algorithms based on texture and color criteria. The six chemical classes of stones depicted in the accompanying images were better predicted by DL-based approaches for four of the lessons mentioned above: UA, struvite, whewellite, and cysteine [73]. In a study released by Ochoa-Ruiz et al. [74], the authors investigated whether it would be possible to anticipate the composition of stones by taking pictures of the stones while doing ureteroscopic procedures. Three deep learning (DL) algorithms and six shallow machine learning (ML) techniques were compared to assess their ability to predict the composition of a virtual breakdown, focusing on the predictive capabilities of these models. The study used a dataset consisting of 93 external photographs and 86 internal images of stones. The results showed that DL algorithms outperformed the other methods, achieving a prediction accuracy of 96%. This indicated that DL models, which rely on texture and color data, provided a slight advantage over traditional ML strategies in predicting stone composition.

According to a recent study, conventional machine learning (ML) models showed inadequate performance when evaluated on a dataset that differed from the one used for model construction and training. This highlights the challenge of model generalization, where algorithms trained on specific datasets may not perform well on new, unseen data, underscoring the need for more robust techniques or cross-domain validation to improve accuracy and reliability in real-world applications. This can result in significantly improved discriminatory performance in comparison to DL [75]. Kim et al. [76] utilized the principal assortment of stone pictures taken using a numeral photographic camera. They presented a total of 1332 stones, each possessing one of 31 unique chemical compositions. A dataset of 965 stones was used to develop and train multiple CNN models. These models were designed to classify the stones into the four most common classes. The highest-performing model achieved an AUC value of at least 0.97, indicating excellent accuracy across all chemical classes it encompassed [76].

7.5.7 STONE COMPOSITION IDENTIFICATION USING FURTHER CUTTING-EDGE TECHNIQUES

Numerous research works presented noteworthy findings regarding the composition of contrasting stones by incorporating data generated by a range of inventive techniques into AI models. The main focus of these cutting-edge techniques was on the chemical and physical characteristics of urethra stone.

Using microtomography, Fitri et al. [77] created a collection of 2430 pictures showing 30 urinary stones from various patients. The aforementioned stone set was divided into three primary components using a CNN algorithm: mixed, UA, or calcium stone. The ultimate model achieved a classification accuracy of 98.52% across the stated classifications. By utilizing the insulator assets of 104 food stones with a given structure, Sa‚clı et al. [78] created a dataset with the appropriate values. An ML-based model for classification was built and trained using the aforementioned dataset, yielding a 98.17% accuracy rate. The method, according to the authors, combines fast processing times and low costs with excellent accuracy in evaluating the precision of the reference techniques used to analyze urinary stones. This approach offers a promising balance between efficiency and effectiveness, making it a valuable tool for improving the analysis and diagnosis of urinary stone composition. Raman spectroscopy is a novel technique that looks at the spectrum properties of the various components to analyze the chemical makeup of the stone. A study reported the use of various Raman spectroscopic measurements to analyze a total of 135 kidney stones, employing machine learning (ML) methods to develop multiple models capable of categorizing the stones into four primary chemical compound groups. The models achieved a high accuracy rate of 96.3%, demonstrating the effectiveness of ML techniques in accurately classifying kidney stones based on their chemical composition [79]. In terms of ease of use and cost-effectiveness, the aforementioned procedure was thought to be advantageous in comparison to current techniques. Onal and Tekgul [80] reported a novel method of identifying the composition of stones by fusing a CNN model with microscope pictures of the stones. A smartphone microscope was used to inspect a group of 37 surgically removed kidney stones, with six separate locations. The classification model was trained using 222 images, and the model successfully predicted the composition of the four primary modules in the learning, namely calcium oxalate, UA, cysteine, and struvite, with an accuracy rate of 88%.

7.6 CONCLUSION

Analysis of the existing publications indicates that the utilization of AI in the treatment of stone disease is an emerging and quickly advancing approach that is currently being evaluated. As expected, the progress of this application corresponds to the development of AI, which is propelled by improvements in both hardware and software. From 2000 to 2010, there were a mere nine reports prepared for publication. But throughout the course of the following ten years, the quantity of publications increased to 35. Within the current decade, a cumulative of 25 papers have already been published between the years 2021 and 2022, demonstrating a noteworthy surge in research endeavors within the related domain. AI has previously focused on the majority of the primary difficulties in diagnosing and treating stone illness in order to improve efficiency, therefore addressing various aspects of stone disease in great detail. Our primary goal is to enhance understanding of the intricate connection between environmental, metabolic, and genetic factors in determining an individual's susceptibility to stone disease. The complex amalgamation of data discussed previously can be employed to generate AI algorithms capable of identifying individuals with a heightened susceptibility to developing kidney stones and devising tailored pharmaceutical therapies to avoid the recurrence of stones. Another area that can be explored is the improvement of lithotripsy procedures by the utilization of AI algorithms to adjust intraoperative factors, such as the parameters of the laser generators or the selection of equipment based on the patient's anatomical and stone characteristics.

The current body of research on the utilization of AI in stone disease is subject to numerous limitations. Several studies just report the results of the AI algorithm in a diagnostic or therapeutic manner, without conducting a comparison to the existing standards in the relevant subfields. Moreover, the results of the research were primarily validated in the same patient cohorts, thereby limiting their relevance to a wider population. The research included in the analysis had an evidence level no higher than level 3. To determine the practical use of AI in managing stone disease in real-life situations, further validation studies, particularly prospective or external validations, are necessary. Out of all the studies included in this review, it is clear that only the diagnostic applications, primarily in the field of radiology, are nearing implementation in urological practice. Nevertheless, numerous publications indicate that the effectiveness of their suggested model could be enhanced by incorporating further data. Ultimately, the variation in study designs and the manner in which the results are presented make it unsuitable to conduct a quantitative analysis and combine the data from the included research.

AI enhances the ability of computer systems to learn, leading to new insights in several scientific fields as well as in everyday life. AI is seen as a crucial aspect in medicine for helping decision-making, particularly in certain areas of healthcare. These areas encompass the clinical domain, which involves the use of AI for diagnosis, prognosis, and prediction. The pharmaceutical field is also included, focusing on the targeted discovery of drugs and the use of computer simulations for clinical trials. Additionally, AI is applied in public health for predicting epidemic outbreaks and shaping public health policies [81]. Furthermore, AI is regarded as a fundamental component for the extensive implementation of precision medicine. This is due to the ongoing accumulation of biomedical data, which is analyzed by ever-advancing AI systems. As a result, new classifications of diseases are anticipated, created on prophets of different types, such as genomic and conservation factors [82]. The aforementioned impact of AI shapes the coming of urogenital medicine, a field; propelled by technology and at the forefront of pioneering advancements. The amount of data connected to urologic problems is growing rapidly, reaching a level known as "Big Data." This term refers to a volume of data that cannot be effectively handled by traditional computer methods but requires the use of AI systems [83]. The integration of "Large Facts" and AI is anticipated to revolutionize urological practice by utilizing computing systems to inform, enhance, and supplement clinicians' decision-making. This will require urologists to acquire familiarity per the submission of AI in their preparation, posing new challenges to their training [84].

A recent assessment found that there are 222 AI-based medical devices that have been authorized in the United States, and 240 in Europe [85]. The authors of the paper state that there has been a significant rise in the availability of AI-based devices in current centuries. Specifically, the number of official policies in the United States increased from 9 in 2015 to 32 in 2017 and further to 77 in 2019. In Europe, the numbers increased from 13 in 2015 to 26 in 2017 and then to 100 in 2019. The field of radiology is the primary medical specialty that utilizes AI-based devices, with the highest number of applications (nZ129). It is followed by cardiology (nZ40) and neurology (nZ21) [85]. This study does not report any urologic device that has been authorized.

The extensive utilization of AI in urology necessitates both the advancement of AI techniques and the doctors' familiarity with AI, which relies on the reliability, comprehensibility, usableness, and photographs of AI procedures [86]. Furthermore, it is imperative for public entities, which have the authority to establish legislative norms in the field of healthcare, to provide a regulatory framework for the implementation of AI. The Food and Drug Administration and European Commission have previously issued the legislative standards and conditions to protect human fundamental rights and offer guidance for the proper utilization of medical AI [87, 88]. According to a survey, the general public had the same level of faith in AI as they had in physicians when it came to diagnostic procedures. Additionally, 94% of the participants were willing to pay for an AI-based assessment of an imaging diagnostic procedure. In contrast, most participants would be uneasy with the idea of automatic robotic surgery. Large multinational firms allocate substantial financial resources for the implementation of AI techniques using "Big Data," whereas governmental financing remains relatively limited [89]. Given patients' interest in AI approaches, particularly for diagnostic purposes, and their highest level of trust in healthcare practitioners, it is incumbent upon the most recent advancements in technology and the public regulatory authority to ensure the provision of dependable and impactful medical AI applications.

REFERENCES

1. Malik P, Pathania M, Rathaur VK. Overview of AI in medicine. *J Fam Med Prim Care* 2019;8:2328e31.
2. Mintz Y, Brodie R. Introduction to artificial intelligence in medicine. *Minim Invasive Ther Allied Technol* 2019;28:73e81.
3. Kueper JK. Primer for artificial intelligence in primary care. *Can Fam Physician* 2021;67:889e93.
4. Schmidhuber J. Deep learning in neural networks: an overview. *Neural Netw* 2015;61:85e117.
5. Frankish K, Ramsey WM. *The Cambridge Handbook of Artificial Intelligence*. Cambridge: Cambridge University Press; 2014. p. 151e66.

6. Rowe M. An introduction to machine learning for clinicians. *Acad Med* 2019;94:1433e6.
7. Choi RY, Coyner AS, Kalpathy-Cramer J, Chiang MF, Campbell JP. Introduction to machine learning, neural networks, and deep learning. *Transl Vis Sci Technol* 2020;9:14. https://doi.org/10.1167/tvst.9.2.14.
8. Rabhi S, Jakubowicz J, Metzger MH. Deep learning versus conventional machine learning for detection of healthcareassociated infections in French clinical narratives. *Methods Inf Med* 2019;58:31e41.
9. Yamashita R, Nishio M, Do RKG, Togashi K. Convolutional neural networks: an overview and application in radiology. *Insights Imaging* 2018;9:611e29.
10. Jin KH, McCann MT, Froustey E, Unser M. Deep convolutional neural network for inverse problems in imaging. *IEEE Trans Image Process* 2017;26:4509e22.
11. Konstantinos-Vaios M, Athanasios O, Ioannis S, Marina K, George M, Evangelia N, et al. Defining voiding dysfunction in women: bladder outflow obstruction versus detrusor underactivity. *Int Neurourol J* 2021;25:244–51.
12. Yu J, Jeong BC, Jeon SS, Lee SW, Lee KS. Comparison of efficacy of different surgical techniques for benign prostatic obstruction. Int Neurourol J 2021;25:252–62.
13. Mytilekas KV, Oeconomou A, Sokolakis I, Kalaitzi M, Mouzakitis G, Nakopoulou E, et al. Defining voiding dysfunction in women: bladder outflow obstruction versus detrusor underactivity. *Int Neurourol J* 2021;25:244–51.
14. Kim HW, Lee JZ, Shin DG. Pathophysiology and management of long-term complications after transvaginal urethral diverticulectomy. *Int Neurourol J* 2021;25:202–9.
15. Jang EB, Hong SH, Kim KS, Park SY, Kim YT, Yoon YE, et al. Catheter-related bladder discomfort: how can we manage it? *Int Neurourol J* 2020;24:324–31.
16. Kwon WA, Lee SY, Jeong TY, Moon HS. Lower urinary tract symptoms in prostate cancer patients treated with radiation therapy: past and present. *Int Neurourol J* 2021;25:119–27.
17. Baser A, Zumrutbas AE, Ozlulerden Y, Alkıs O, Oztekın A, Celen S, et al. Is there a correlation between behçet disease and lower urinary tract symptoms? *Int Neurourol J* 2020;24:150–5.
18. Kim SJ, Choo HJ, Yoon H. Diagnostic value of the maximum urethral closing pressure in women with overactive bladder symptoms and functional bladder outlet obstruction. *Int Neurourol J* 2022;26(Suppl 1):S1–7.
19. Kanagasingam Y, Xiao D, Vignarajan J, Preetham A, Tay-Kearney M.-L, Mehrotra A. Evaluation of artificial intelligence–based grading of diabetic retinopathy in primary care. *JAMA Netw Open* 2018;1: e182665.
20. Venkatramani, V. Urovision 2020: the future of urology. *Indian J. Urol.* 2015;31; 150–155.
21. Porpiglia F, Checcucci E, Amparore D, Piramide F, Volpi G, Granato S, Verri P, Manfredi M, Bellin A, Piazzolla P, et al. Three-dimensional augmented reality robot-assisted partial nephrectomy in case of complex tumours (PADUA ≥ 10): a new intraoperative tool overcoming the ultrasound guidance. *Eur Urol* 2020;78: 229–238.
22. Shah M, Naik N, Somani BK, Hameed BMZ. Artificial intelligence (AI) in urology-Current use and future directions: an iTRUE study. *Türk Urol Derg Turk J Urol* 2020;46: S27–S39.
23. Kazemi Y, Mirroshandel SA. A novel method for predicting kidney stone type using ensemble learning. *Artif Intell Med* 2018;84:117–126.

24. Längkvist M, Jendeberg J, Thunberg P, Loutfi A, Lidén M. Computer aided detection of ureteral stones in thin slice computed tomography volumes using Convolutional Neural Networks. *Comput Biol Med* 2018;97: 153–160.
25. Cosma G, Brown D, Archer M, Khan M, Pockley AG. A survey on computational intelligence approaches for predictive modeling in prostate cancer. *Expert Syst Appl* 2017;70: 1–19.
26. Aminsharifi A, Irani D, Pooyesh S, Parvin H, Dehghani S, Yousofi K, Fazel E, Zibaie F. Artificial neural network system to predict the postoperative outcome of percutaneous nephrolithotomy. *J Endourol* 2017;31: 461–467.
27. Mannil M, Von Spiczak J, Hermanns T, Poyet C, Alkadhi H, Fankhauser CD. Three-dimensional texture analysis with machine learning provides incremental predictive information for successful shock wave lithotripsy in patients with kidney stones. *J Urol* 2018;200: 829–836.
28. Mannil M, Von Spiczak J, Hermanns T, Alkadhi H, Fankhauser CD. Prediction of successful shock wave lithotripsy with CT: a phantom study using texture analysis. *Abdom Radiol* 2017;43: 1432–1438.
29. Li D, Xiao C, Liu Y, Chen Z, Hassan H, Su L, et al. Deep Segmentation Networks for segmenting kidneys and detecting kidney stones in unenhanced abdominal CT images. *Diagnostics* 2022;12:1788. https://doi.org/10.3390/diagnostics12081788.
30. Parakh A, Lee H, Lee JH, Eisner BH, Sahani DV, Do S. Urinary stone detection on CT images using deep convolutional neural networks: evaluation of model performance and generalization. *Radiol Artif Intell* 2019;1:e180066. https://doi.org/10.1148/ryai.2019180066.
31. Längkvist M, Jendeberg J, Thunberg P, Loutfi A, Lidén M. Computer aided detection of ureteral stones in thin slice computed tomography volumes using Convolutional Neural Networks. *Comput Biol Med* 2018;97:153e60.
32. Caglayan A, Horsanali MO, Kocadurdu K, Ismailoglu E, Guneyli S. Deep learning model-assisted detection of kidney stones on computed tomography. *Int Braz J Urol* 2022;48: 830e9.
33. Jendeberg J, Thunberg P, Lidén M. Differentiation of distal ureteral stones and pelvic phleboliths using a convolutional neural network. *Urolithiasis* 2021;49:41e9.
34. De Perrot T, Hofmeister J, Burgermeister S, Martin SP, Feutry G, Klein J, et al. Differentiating kidney stones from phleboliths in unenhanced low-dose computed tomography using radiomics and machine learning. *Eur Radiol* 2019;29:4776e82.
35. Chak P, Navadiya P, Parikh B, Pathak KC. Neural network and SVM based kidney stone based medical image classification. In: *International Conference on Computer Vision and Image Processing*. Singapore: Springer; 2019. p. 158e73.
36. Elton DC, Turkbey EB, Pickhardt PJ, Summers RM. A deep learning system for automated kidney stone detection and volumetric segmentation on noncontrast CT scans. *Med Phys* 2022;49:2545e54.
37. Krishna KD, Akkala V, Bharath R, Rajalakshmi P, Mohammed A, Merchant S, et al. Computer aided abnormality detection for kidney on FPGA based IoT enabled portable ultrasound imaging system. *IRBM* 2016;37:189e97.
38. Balamurugan SP, Arumugam G. A novel method for predicting kidney diseases using optimal artificial neural network in ul- trasound images. *IJIE* 2020;7:37e55.
39. Selvarani S, Rajendran P. Detection of renal calculi in ultra- sound image using meta-heuristic support vector machine. *J Med Syst* 2019;43:1e9.

40. Viswanath K, Anilkumar B, Gunasundari R. Design of deep learning reaction diffusion level set segmentation approach for health care related to automatic kidney stone detection analysis. *Multimed Tool Appl* 2022;81:1e43.
41. Akkasaligar PT, Biradar S. Diagnosis of renal calculus disease in medical ultrasound images. In: *2016 IEEE International Conference on Computational Intelligence and Computing Research (ICCIC)*. IEEE; 2016. p. 1e5. https://doi.org/10.1109/ICCIC.2016.7919642.
42. Verma J, Nath M, Tripathi P, Saini K. Analysis and identification of kidney stone using kth nearest neighbour (KNN) and support vector machine (SVM) classification techniques. *Pattern Recogn Image Anal* 2017;27:574e80.
43. Kobayashi M, Ishioka J, Matsuoka Y, Fukuda Y, Kohno Y, Kawano K, et al. Computer-aided diagnosis with a convolutional neural network algorithm for automated detection of urinary tract stones on plain X-ray. *BMC Urol* 2021;21:1e10.
44. Aksakalli I, Ka¸cdio˘glu S, Hanay YS. Kidney X-ray images classification using machine learning and deep learning methods. *Balkan J Electr Comput Eng* 2021;9: 144e51.
45. Cummings JM, Boullier JA, Izenberg SD, Kitchens DM, Kothandapani RV. Prediction of spontaneous ureteral calculous passage by an artificial neural network. *J Urol* 2000;164:326e8.
46. Dal Moro F, Abate A, Lanckriet G, Arandjelovic G, Gasparella P, Bassi P, et al. A novel approach for accurate prediction of spontaneous passage of ureteral stones: support vector machines. *Kidney Int* 2006;69:157e60.
47. Solakhan M, Seckiner SU, Seckiner I. A neural network-based algorithm for predicting the spontaneous passage of ureteral stones. *Urolithiasis* 2020;48:527e32.
48. Park JS, Kim DW, Lee D, Lee T, Koo KC, Han WK, et al. Development of prediction models of spontaneous ureteral stone passage through machine learning: comparison with conventional statistical analysis. *PLoS One* 2021;16:e0260517. https://doi.org/10.1371/JOURNAL.PONE.0260517.
49. Poulakis V, Dahm P, Witzsch U, De Vries R, Remplik J, Becht E. Prediction of lower pole stone clearance after shock wave lithotripsy using an artificial neural network. *J Urol* 2003;169:1250e6.
50. Gomha MA, Sheir KZ, Showky S, Abdel-Khalek M, Mokhtar AA, Madbouly K. Can we improve the prediction of stone-free status after extracorporeal shock wave lithotripsy for ureteral stones? A neural network or a statistical model? *J Urol* 2004;172:175e9.
51. Moorthy K, Krishnan M. Prediction of fragmentation of kidney stones: a statistical approach from NCCT images. *Can Urol Assoc J* 2016;10:E237e40. https://doi.org/10.5489/cuaj.3674.
52. Choo MS, Uhmn S, Kim JK, Han JH, Kim D-H, Kim J, et al. A prediction model using machine learning algorithm for assessing stone-free status after single session shock wave lithotripsy to treat ureteral stones. *J Urol* 2018;200:1371e7.
53. Seckiner I, Seckiner S, Sen H, Bayrak O, Dogan K, Erturhan S. A neural network-based algorithm for predicting stone-free status after ESWL therapy. *Int Braz J Urol* 2017;43:1110e4.
54. Mannil M, von Spiczak J, Hermanns T, Poyet C, Alkadhi H, Fankhauser CD. Three-dimensional texture analysis with ma- chine learning provides incremental predictive information for successful shock wave lithotripsy in patients with kidney stones. *J Urol* 2018;200:829e36.

55. Yang SW, Hyon YK, Na HS, Jin L, Lee JG, Park JM, et al. Machine learning prediction of stone-free success in patients with urinary stone after treatment of shock wave lithotripsy. *BMC Urol* 2020;20:1e8.
56. Tsitsiflis A, Kiouvrekis Y, Chasiotis G, Perifanos G, Gravas S, Stefanidis I, et al. The use of an artificial neural network in the evaluation of the extracorporeal shockwave lithotripsy as a treatment of choice for urinary lithiasis. *Asian J Urol* 2022;9:132e8.
57. Handa RK, Territo PR, Blomgren PM, Persohn SA, Lin C, Johnson CD, et al. Development of a novel magnetic resonance imaging acquisition and analysis workflow for the quantification of shock wave lithotripsy-induced renal hemorrhagic injury. *Urolithiasis* 2017;45:507e13.
58. Aminsharifi A, Irani D, Pooyesh S, Parvin H, Dehghani S, Yousofi K, et al. Artificial neural network system to predict the postoperative outcome of percutaneous nephrolithotomy. *J Endourol* 2017;31:461e7.
59. Shabaniyan T, Parsaei H, Aminsharifi A, Movahedi MM, Jahromi AT, Pouyesh S, et al. An artificial intelligence-based clinical decision support system for large kidney stone treatment. *Australas Phys Eng Sci Med* 2019;42:771e9.
60. Aminsharifi A, Irani D, Tayebi S, Jafari Kafash T, Shabanian T, Parsaei H. Predicting the postoperative outcome of percutaneous nephrolithotomy with machine learning system: software validation and comparative analysis with Guy's Stone Score and the CROES Nomogram. *J Endourol* 2020;34:692e9.
61. Geraghty R, Finch W, Fowler S, Sriprasad S, Smith D, Dickinson A, et al. Use of internally validated machine and deep learning models to predict outcomes of percutaneous neph-rolithotomy using data from the BAUS PCNL audit. medRxiv 2022. https://doi.org/10.1101/2022.06.16.22276481.
62. Zhao H, Li W, Li J, Li L, Wang H, Guo J. Predicting the stone- free status of percutaneous nephrolithotomy with the machine learning system: comparative analysis with Guy's stone score and the STONE score system. *Front Mol Biosci* 2022;9:880291. https://doi.org/10.3389/fmolb.2022.880291.
63. Xiang H, Chen Q, Wu Y, Xu D, Qi S, Mei J, et al. Urine calcium oxalate crystallization recognition method based on deep learning. In: *2019 international conference on automation, computational and technology management, ICACTM 2019*; 2019. p. 30e3. https://doi.org/10.1109/ICACTM.2019.8776769.
64. Kletzmayr A, Mulay SR, Motrapu M, Luo Z, Anders HJ, Ivarsson ME, et al. Inhibitors of calcium oxalate crystallization for the treatment of oxalate nephropathies. *Adv Sci* 2020;7:1903337. https://doi.org/10.1002/advs.201903337.
65. Kriegshauser JS, Silva AC, Paden RG, He M, Humphreys MR, Zell SI, et al. Ex vivo renal stone characterization with single- source dual-energy computed tomography: a multiparametric approach. *Acad Radiol* 2016;23:969e76.
66. Kriegshauser JS, Paden RG, He M, Humphreys MR, Zell SI, Fu Y, et al. Rapid kV-switching single-source dual-energy CT ex vivo renal calculi characterization using a multiparametric approach: refining parameters on an expanded dataset. *Abdom Radiol (NY)* 2018;43:1439e45.
67. Zhang GMY, Sun H, Shi B, Xu M, Xue HD, Jin ZY. Uric acid versus non-uric acid urinary stones: differentiation with single energy CT texture analysis. *Clin Radiol* 2018;73:792e9.
68. Große Hokamp N, Lennartz S, Salem J, Pinto dos Santos D, Heidenreich A, Maintz D, et al. Dose independent characterization of renal stones by means of dual energy computed tomography and machine learning: an ex-vivo study. *Eur Radiol* 2020;30:1397e404.

69. Tang L, Li W, Zeng X, Wang R, Yang X, Luo G, et al. Value of artificial intelligence model based on unenhanced computed tomography of urinary tract for preoperative prediction of calcium oxalate monohydrate stones in vivo. *Ann Transl Med* 2021;9:1129. https://doi.org/10.21037/atm-21-965.
70. Black KM, Law H, Aldoukhi A, Deng J, Ghani KR. Deep learning computer vision algorithm for detecting kidney stone composition. *BJU Int* 2020;125:920e4.
71. Lopez F, Varelo A, Hinojosa O, Mendez M, Trinh DH, ElBeze Y, et al. Assessing deep learning methods for the identification of kidney stones in endoscopic images. *Annu Int Conf IEEE Eng Med Biol Soc* 2021;2021:2778e81.
72. El Beze J, Mazeaud C, Daul C, Ochoa-Ruiz G, Daudon M, Eschwe`ge P, et al. Evaluation and understanding of auto- mated urinary stone recognition methods. *BJU Int* 2022;130: 786e98.
73. Ochoa-Ruiz G, Estrade V, Lopez F, Flores-Araiza D, Beze JE, Trinh DH, et al. On the in vivo recognition of kidney stones using machine learning. 2022. arXiv preprint arXiv:2201.08865.
74. Mendez-Ruiz M, Lopez-Tiro F, El-Beze J, Estrade V, Ochoa- Ruiz G, Hubert J, et al. On the generalization capabilities of FSL methods through domain adaptation: a case study in endoscopic kidney stone image classification. 2022. arXiv preprint arXiv:2205.00895.
75. Kim US, Kwon HS, Yang W, Lee W, Choi C, Kim JK, et al. Prediction of the composition of urinary stones using deep learning. *Investig Clin Urol* 2022;63:441e7.
76. Fitri LA, Haryanto F, Arimura H, YunHao C, Ninomiya K, Nakano R, et al. Automated classification of urinary stones based on microcomputed tomography images using convolutional neural network. *Phys Med* 2020;78:201e8.
77. Sa¸clı B, Aydınalp C, Cansız G, Joof S, Yilmaz T, C¸ayo¨ren M, et al. Microwave dielectric property based classification of renal calculi: application of a kNN algorithm. *Comput Biol Med* 2019;112:103366. https://doi.org/10.1016/j.compbiomed.2019.103366.
78. Cui X, Zhao Z, Zhang G, Chen S, Zhao Y, Lu J. Analysis and classification of kidney stones based on Raman spectroscopy. *Biomed Opt Express* 2018;9:4175e83.
79. Onal EG, Tekgul H. Assessing kidney stone composition using smartphone microscopy and deep neural networks. *BJUI Compass* 2022;3:310e5.
80. Noorbakhsh-Sabet N, Zand R, Zhang Y, Abedi V. Artificial intelligence transforms the future of health care. *Am J Med* 2019;132:795e801.
81. Denny JC, Collins FS. Precision medicine in 2030dseven ways to transform healthcare. *Cell* 2021;184:1415e9.
82. Hameed BZ, Dhavileswarapu S, Naik N, Karimi H, Hegde P, Rai BP, et al. Big data analytics in urology: the story so far and the road ahead. *Ther Adv Urol* 2021;13:1756287221998134. https://doi.org/10.1177/1756287221998134.
83. John-Charles R, Tsanas A, Singh S. Rise of the machines: will artificial intelligence replace the urologist? www.urologynews.uk.com/features/features/post/rise-of-the-machines-will-artificial-intelligence-replace-the-urologist. [Accessed 29 March 2021].
84. Muehlematter UJ, Daniore P, Vokinger KN. Approval of artificial intelligence and machine learning-based medical devices in the USA and Europe (2015e20): a comparative analysis. *Lancet Digit Health* 2021;3:e195e203. https://doi.org/10.1016/S2589-7500(20)30292-2.
85. Cutillo CM, Sharma KR, Foschini L, Kundu S, Mackintosh M, Mandl KD. Machine intelligence in health cared perspectives on trustworthiness, explain ability, usability, and transparency. *NPJ Digit Med* 2020;3:1e5.

86. Food and Drug Administration. Artificial Intelligence/Machine Learning (AI/ML)-Based Software as a Medical Device (SaMD) Action Plan. www.fda.gov/media/145022/download. [Accessed July 5 2021].
87. Sto¨ger K, Schneeberger D, Holzinger A. Medical artificial intelligence: the European legal perspective. *Commun ACM* 2021;64:34e6.
88. Stai B, Heller N, McSweeney S, Rickman J, Blake P, Vasdev R, et al. Public perceptions of artificial intelligence and robotics in medicine. *J Endourol* 2020;34:1041e8.
89. Schoenthaler M, Boeker M, Horki P. How to compete with Google and Co.: big data and artificial intelligence in stones. *Curr Opin Urol* 2019;29:135e42.

8 Unveiling Cheminformatics for Accelerated Drug Discovery and Development
A Computational-Guided Approach

Amanpreet Kaur and Debasish Mandal

8.1 INTRODUCTION

Cheminformatics is a growing field that integrates computer science, statistics, mathematics, and chemistry. It is at the cutting edge of the push to transform drug discovery and delivery (Gasteiger and Engel 2003; Varnek and Tropsha 2008; Brown 2009). This area is very important for creating, improving, and testing possible drug options because it gives us a solid way to understand how chemical compounds work (Ekins et al. 2019; Lounkine et al. 2012; Walters et al. 2011; Leeson and Springthorpe 2007).

Traditional methods of making drugs face a lot of problems, including long development times, huge costs, low output, and the creation of dangerous by-products (DiMasi et al. 2016; Scannell et al. 2012; Paul et al. 2010). These problems often lead to low success rates because the targets are not clear enough and the effects are not where they are supposed to be (Kola and Landis 2004; Arrowsmith 2011; Munos 2009; Proudfoot 2002). The growth of computational techniques in cheminformatics has caused a big change in drug discovery. These techniques have solved problems and made drug development more efficient overall (Schneider et al. 2020; Chen et al. 2018; Xu et al. 2019; Narayanan et al. 2014).

Cheminformatics is a broad field of techniques used to deal with the complicated nature of chemical data (Bender et al. 2007; Cortés-Ciriano and Bender 2019). Various techniques, like molecular modeling, virtual screening, quantitative structure-activity relationship (QSAR) studies, mutations and machine learning (ML) techniques help medicinal chemists to make smart decisions about how to make novel drugs wisely. (Gilson and Zhou 2007; Cheng et al. 2009; Lavecchia and Cerchia 2015).

Chemistry and biology data integration and analysis using cutting-edge informatics methods form the basis of cheminformatics (Gaulton et al. 2012; Wang et al. 2020). Researchers utilize databases that contain chemical structures and their respective biological activities as indispensable assets (Gaulton et al. 2012; Wang et al. 2020). Databases like PubChem, CHEMBL, ZINC, etc., contain millions of molecules and their experimental data. This allows scientists to explore and analyze through the whole chemical space (Wang et al. 2020; Sun et al. 2016).

Furthermore, cheminformatics also includes data mining and ML techniques (Lavecchia and Cerchia 2015; Gómez-Bombarelli et al. 2018) that help scientists to find hidden patterns and correlations in huge datasets by using various algorithms. These help them to plan focused experiments and compound libraries (Lavecchia and Cerchia 2015; Cortés-Ciriano and Bender 2019; Vieth et al. 2004).

In recent years, adding artificial intelligence (AI) has also massively changed cheminformatics by making computational drug development faster and more accurate (Gómez-Bombarelli et al. 2018). Deep learning networks and other AI-driven algorithms are very good at predicting molecular properties and finding good drug options from huge chemical libraries (Gómez-Bombarelli et al. 2018; Cortés-Ciriano and Bender 2019). Some well-known drugs that have been found using cheminformatics techniques are discussed in the following.

Firstly, Oseltamivir, also known as Tamiflu, was found by using structure-based drug design and virtual screening to target influenza neuraminidase (Kim et al. 1997). Secondly, Raloxifene (Evista) was found using computer methods to have selective estrogen receptor modulation qualities, which led to its use in treating osteoporosis (Black et al. 1994). Thirdly, Raltegravir (Isentress) was created using molecular docking and QSAR methods to block HIV integrase. This is a milestone in antiretroviral medication (Summa et al. 2008). Other examples include the development of imatinib (Gleevec) for chronic myeloid leukemia, a good example of how cheminformatics techniques helped to find and improve kinase inhibitors, which subsequently led to a successful drug (Capdeville et al. 2002; Druker et al. 2001; Goldman and Melo 2003). In the same way, computer methods that helped scientists to understand the structure of HIV protease inhibitors like saquinavir sped up the process of finding new drugs that could block them (Krohn et al. 1991; Wlodawer and Vondrasek 1998; Perola et al. 2004). Thus, cheminformatics is at the cutting edge of finding novel drugs as it connects chemical data to new treatments (Bajorath 2002; Gilson and Zhou 2007). It helps to quickly find and improve bioactive compounds using computational methods and various software that help them to explore complicated chemical spaces (Oprea and Matter 2004; Gaulton et al. 2012 for examples). As technology keeps getting better, cheminformatics will play a bigger part in finding new drugs. This will open up new ways to meet unmet medical needs and make healthcare better around the world (Bender et al. 2007; Lavecchia and Cerchia 2015).

A number of popular techniques such as pharmacophore modeling, docking, QSAR, Virtual Screening, Molecular Dynamics (MD), Mutations, and ML algorithms have had made a great impact on cheminformatics. The detailed techniques along with their case studies are discussed later in this chapter.

8.2 PHARMACOPHORE MODELING

Paul Ehrlich came up with pharmacophore modeling in 1909, and it is now one of the most popular ways to figure out what properties a molecule needs to have in order to have drug-like behavior (Ehrlich 1909; Balakin et al. 2006; Yang 2010; Koshland 1958). To put it simply, a pharmacophore is a picture of the molecular details a biological macromolecule needs to recognize a drug (Wolber and Langer 2005; Klebe 2000; Schneider 2010). It is based on finding the arrangements of functional groups that interact with certain target sites (Schaller et al. 2011; Lin et al. 2020; Langer and Hoffmann 2001; McGaughey et al. 2000). This allowed researchers to make new drugs that have the required biological effects.

Pharmacophore modeling is an important part of drug development because it helps find possible drug candidates by focusing on the molecular features that are needed for activity (Lipinski and Hopkins 2004; Scior et al. 2007; Teague et al. 1999). Scientists can find novel molecules that can work well with their biological targets by mapping these pharmacophoric traits (Leach et al. 2010; Lionta et al. 2014; Schneider 2010). Some of the available pharmacophore modeling software includes LigandScout, ZINCPharmer MOE (Molecular Operating Environment), Pharmit, Phase (Schrödinger), Catalyst (BIOVIA), SYBYL, PharmaGist,, Discovery Studio, and ICM (MolSoft).

The steps involved in the pharmacophore search are given in Figure 8.1.

FIGURE 8.1 Various Steps involved in the workflow of Pharmacophore Modeling in drug discovery.

8.2.1 CASE STUDY OF RALTEGRAVIR USING PHARMACOPHORE MODELING

An important example of good pharmacophore modeling is raltegravir, which blocks HIV integrase. Scientists explored specific pharmacophore features that needed to stop integrase. This helped them to find an improved raltegravir, a drug that stops viral DNA from integrating into the host genome (Hazuda et al. 2004; Shimura et al. 2008; Espeseth et al. 2000). This crucial step forward in treating HIV shows how significant pharmacophore modeling is for finding molecular interactions for drug development and its effectiveness.

8.3 MOLECULAR DOCKING

Molecular docking is a crucial technique used to determine the optimal manner in which two molecules can bind together (Figure 8.2). The compounds formed by the binding of receptors and ligands are referred to as receptor-ligand complexes (Trott and Olson 2010; Morris et al. 2009; Pagadala et al. 2017; Lengauer and Rarey 1996). Molecular docking depends upon chemical recognition to determine the best optimal orientation of a ligand when binding to a protein receptor (Halperin et al., 2002; Meng et al., 2011; Pinzi and Rastelli, 2019; Taylor et al. 2002). This methodology facilitates our comprehension of molecular interconnections and holds significant relevance for lead optimization (Kroemer 2007; Friesner et al. 2004; Sousa et al. 2006). Different docking software commonly used are AutoDock, Schrödinger Glide, MOE (Molecular Operating Environment), Vina, DOCK, GOLD, PLANTS, LigandScout, PyRx, and OpenEye FRED.

Researchers can enhance the ability of potential drug candidates to attach to the receptor and increase their selectivity by conducting docking studies, which provide insights into the molecular interactions between medications and their targets (Yuriev and Ramsland 2013; Vilar et al. 2008; Brooijmans and Kuntz 2003). This computerized approach reduces the necessity for several laboratory tests, hence accelerating the drug development process and enhancing its efficiency (Sousa et al.

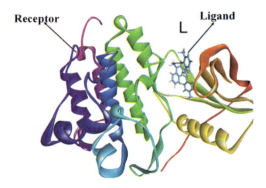

FIGURE 8.2 Molecular docking technique showing docked complex containing receptor (ribbon shaped) and ligand (ball and stick representation).

Unveiling Cheminformatics 165

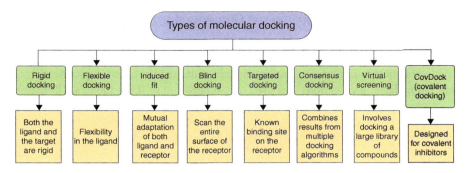

FIGURE 8.3 Different types of docking approaches offering insights into binding affinities, binding mechanisms, modes, and potential efficacy of drug molecules.

2006; Moitessier et al. 2008; Sliwoski et al. 2014; Gupta et al., 2024). Some of the popular types of docking used are shown in Figure 8.3.

8.3.1 CASE STUDY OF ZANAMIVIR

Zanamivir, an antiviral medication, utilized for the treatment of influenza was developed using extensive docking trials. Researchers used docking techniques to find the best binding interactions between potential inhibitors and the active site of the influenza neuraminidase enzyme. The binding investigations found that zanamivir exhibits potent inhibitory effects by effectively preventing viral replication and blocking the enzyme's activity (von Itzstein et al., 1993; Kim et al., 1997; Varghese et al., 1992). This discovery showed that using docking studies helps to accelerate the development of efficacious antiviral medications.

8.4 QUANTITATIVE STRUCTURE ACTIVITY RELATIONSHIP

QSAR models uses the structural features of chemical compounds to forecast their activity thereby helping in assessing the effectiveness of potential drugs (Cherkasov et al., 2014; Hansch and Fujita, 1964; Todeschini and Consonni, 2009). QSAR provides insights into the molecular characteristics that influence the pharmacological effects of a substance by linking its chemical structure to its biological activity. This aids researchers in developing more effective and less harmful pharmaceuticals (Patel et al., 2014; Pires et al., 2015; Bender, 2010; Llinàs and Goodman 2008). QSAR investigations involve the creation of mathematical models to demonstrate the relationship between chemical structure and biological activity. This facilitates the prediction of the mechanism of action of novel medications (Roy et al. 2015; Katritzky et al. 2010; Bajorath 2001). The models play a crucial role in the discovery of new medications as they aid in the identification and enhancement of lead molecules that possess the desired pharmacological properties (Gonzalez and Helguera 2011; Puzyn et al. 2010; Duchowicz et al. 2013). Some of the software used to build QSAR include MOE (Molecular Operating Environment), ADMET Predictor, KNIME, Schrödinger's QikProp, QSAR Toolbox, Discovery Studio, Dragon, and ChemOffice.

8.4.1 CASE STUDY: SUBSTITUTES FOR CHLOROQUINE

The discovery and development of chloroquine compounds for the treatment of malaria was explored by QSAR investigations. To predict the efficacy of recently developed chloroquine analogues, an analysis of the structural characteristics of present antimalarial medications was used as a foundation of these models. Ultimately, the discovery of novel antimalarial pharmaceuticals was facilitated by the development and enhancement of molecules that exhibited superior efficacy and safety in these models (Egan et al. 2002; Ghose et al. 1998; Melagraki et al. 2010).

8.5 VIRTUAL SCREENING

Virtual screening is a computational technique used by the pharmaceutical sector to identify prospective drugs. It is performed by assessing large numbers of small molecules in order to evaluate their affinity for an identified target, such as an enzyme or protein receptor (Shoichet 2004; Kitchen et al. 2004; Lavecchia and Di Giovanni 2013). This technique saves a lot of time, money, and labor on extensive experimental screening, allow for the speedy identification of successful pharmaceutical drugs (McInnes 2007; Green and Kuntz 1992; Alvarez 2004).

Computer programs can be utilized to search through lists of chemical compounds and identify those that are believed to have a high affinity for a specific target protein (Irwin and Shoichet 2005; Walters et al. 1998; Huang et al. 2006). This approach is particularly advantageous in the first stages of medication development since it enables researchers to promptly identify compounds that require further examination (Lavecchia and Di Giovanni 2013; Schneider 2010). A number of software tools are used for virtual screening are AutoDock, MOE, LigandScout, PLANTS, GOLD, Schrödinger Glide, Vina, PyRx, and OpenEye FRED.

8.5.1 CASE STUDY: BACE1 INHIBITORS FOR ALZHEIMER'S DISEASE TREATMENT

The identification of BACE1 inhibitors used in treatment of Alzheimer's disease to identify small molecules capable of inhibiting BACE1, an enzyme involved in the production of amyloid-beta peptides, was discovered using virtual screening. Through the utilization of this approach, a number of potent BACE1 inhibitors were discovered and are currently under investigation as potential therapies for Alzheimer's disease (Ghosh et al., 2012; Vassar, 2014; Wang et al., 2012; Malik et al., 2023).

8.6 MOLECULAR DYNAMICS

MD models play a crucial role in cheminformatics since they provide a very detailed representation of the temporal evolution of molecular systems (Karplus and McCammon 2002; Dror et al. 2012). Scientists employ MD approaches to construct models that depict the motion of atoms and molecules. These models provide insights into the interactions and dynamic changes in the form of molecules over time (Karplus

Unveiling Cheminformatics

and McCammon 2002). Scientists can investigate the arrangement and energy of molecular assemblies by solving Newton's equations of motion for a collection of particles using this computational technique (Karplus and McCammon 2002; Dror et al. 2012; Merz 2010). General steps to build a stable system for MD simulation include parametrization, preparation of system, energy minimization, equilibration, production, and trajectory analysis (Figure 8.4).

Understanding the thermodynamics and kinetics of ligand binding is crucial for medication development. MD can enhance understanding of these processes by simulating atomic-scale binding (Hollingsworth and Dror 2018; Swanson et al. 2011).

The investigation of the interaction between proteins and ligands is a critical application of MD simulations. This facilitates the identification of the mechanism by which medication alternatives interact and the extent of their binding affinities (Hollingsworth and Dror 2018; Swanson et al. 2011).

MD simulations can be used to identify possible binding sites on target proteins and demonstrate the changes in their shapes when a ligand binds to them (Dror et al., 2012; Shaw et al., 2010). This information is extremely advantageous for improving lead compounds, allowing scientists to create treatments with enhanced effectiveness and selectivity (Shaw et al., 2010). Popular MD software used for simulation are GROMACS, CHARMM, OpenMM, AMBER, NAMD, Desmond, LAMMPS, and TINKER.

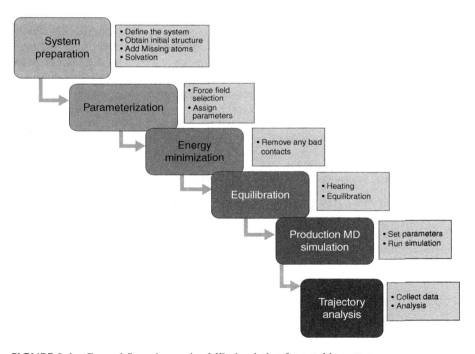

FIGURE 8.4 General Steps in running MD simulation for a stable system.

8.6.1 CASE STUDY OF GLEEVEC (IMATINIB)

The development of Imatinib, a novel pharmaceutical agent, has significantly improved the treatment outcomes for chronic myeloid leukemia, thanks to the utilization of MD models. Researchers employed molecular dynamics (MD) to investigate the binding mechanism of imatinib to the BCR-ABL kinase. The researchers discovered significant interactions that maintained the medication in its active state (Lin et al., 2013). The understanding of this information enhanced the capacity of imatinib to attach to and selectively target particular cells, resulting in its efficacy as a targeted therapy for cancer (Lin et al. 2013; Druker et al. 2001).

8.7 MUTATIONS

Cheminformatics is particularly effective in understanding how mutations affect the effectiveness and tolerance of drugs. Drug resistance can occur as a result of changes in target proteins, making it extremely difficult to treat disorders including cancer and infectious diseases (Gonzalez et al., 2015; Swainston and Jansen, 2019). Molecular modeling and MD simulations are computer methods used to forecast the effects of mutations on the structure and function of specific proteins. As a result, these changes have the ability to modify the interaction between drugs and these proteins (Gonzalez et al., 2015; Pickett et al. 2011). Various mutations are found that can change the function of DNA and protein synthesis (Sharma et al., 2023). The types are shown in Figure 8.5.

Researchers can determine probable pathways via which pharmaceuticals may be ineffective in binding to mutant receptors and can produce therapies that maintain their efficacy by examining the structural implications of mutations (Swainston and Jansen, 2019; Ali et al., 2018). The ability to predict is essential in the development of sophisticated medications that may overcome resistance and offer long-lasting health advantages (Ali et al., 2018). Some of the commonly used software for mutations are PROVEAN, CADD, ANNOVAR, VEP, SIFT, PolyPhen, MutationTaster, GEMINI, FATHMM, and MutationAssessor.

8.7.1 A CASE STUDY ON HIV PROTEASE INHIBITORS FOR MUTATIONS

Saquinavir and other HIV protease inhibitors were initially developed using computer technology to address drug resistance issues. MD simulations and structural analysis were employed by researchers to investigate modifications in the HIV protease enzyme. This made it possible to develop inhibitors that remain effective against isolates that have developed resistance (Krohn et al. 1991; Ali et al. 2018; Peet et al. 1998).

8.8 MACHINE LEARNING

By using extensive datasets available in libraries, the drug discovery process has been transformed by the integration of ML into cheminformatics (Ekins et al., 2019; Vamathevan et al., 2019). In order to improve the efficacy of lead molecules,

Unveiling Cheminformatics 169

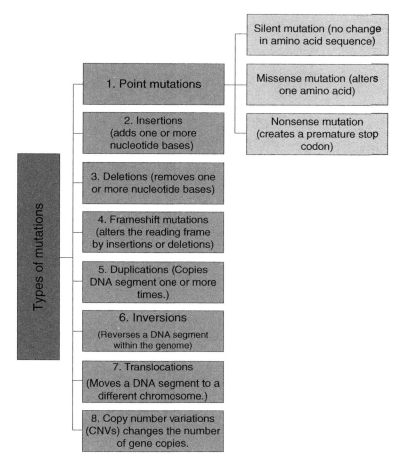

FIGURE 8.5 Classification of mutations along with their brief description.

identify potential therapies, and forecast chemical characteristics, ML techniques are implemented (Ekins et al., 2019). These algorithms analyze data from literature in order to acquire knowledge and provide predictions. The authors Vamathevan et al. (2019) identify trends that may be difficult to detect using conventional methods.

ML in cheminformatics has changed the way drugs are made by making it possible to analyze and understand large datasets (Ekins et al., 2019; Vamathevan et al., 2019). ML methods are used to predict chemical properties, find possible treatment approaches, and improve the functionality of basic molecules (Ekins et al., 2019). In order to refine more and make predictions, these algorithms look at facts from the previous studies. They find trends that might be hard to find with normal methods (Vamathevan et al., 2019).

ML has demonstrated significant potential in the field of cheminformatics. These have been categorized into supervised and unsupervised ML models The ML models falling under these category include J48, Naïve Bayes, SMO, IBk, **Multilayer**

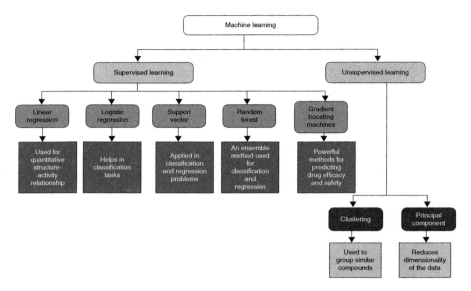

FIGURE 8.6 Some key ML algorithms to build models in drug discovery.

Perceptron, JRip, Random Forest, Linear Regression, and SimpleKMeans (Gómez-Bombarelli et al., 2018; Vamathevan et al., 2019) (Figure 8.6). Due to their ability to process intricate and multi-dimensional data, these tools are highly valuable for tasks such as virtual screening and QSAR modeling (Gómez-Bombarelli et al. 2018). Software used to build ML model include TensorFlow, Chemprop, RDKit, DataRobot, Scikit-learn, KNIME, DeepChem, H2O.ai, PyTorch, and Weka.

8.8.1 Case Study: Drug-Induced Liver Injury

ML techniques have been used to predict the likelihood of liver damage caused by drug molecules. ML models have been created to detect compounds that may cause liver damage by analyzing large collections of chemical and biological data. This adds to the improved safety of recently produced drugs (Xu et al., 2019; Chen et al., 2018 Yang et al. 2009).

8.9 CONCLUSION

In summary, cheminformatics has transformed the field of drug development by integrating the disciplines of chemistry, computer science, mathematics, and statistics. The implementation of this multidisciplinary strategy efficiently tackles the drawbacks associated with conventional drug development approaches, including exorbitant expenses, prolonged timeframes, and low rates of success. Advanced computational methods such as pharmacophore modeling, molecular docking, QSAR, virtual screening, MD, and ML have greatly improved the effectiveness and accuracy of drug development. These techniques offer deep understanding of molecular interactions, forecast biological activities, and enhance lead compounds, thus accelerating the drug development process. The discovery of Tamiflu and Gleevec drugs exemplify the significant influence of cheminformatics in developing efficacious treatments and drugs. With the ongoing advancement of technology, the field of cheminformatics will become more and more important in solving medical demands that have not been discovered and enhancing healthcare results worldwide. Cheminformatics plays a leading role in current medicinal chemistry by providing creative ideas and improving the process of finding safer and more efficient drugs.

REFERENCES

Egan T.J., Hunter R., Basilico N., Parapini S., Taramelli D., Pasini E., and Monti D. (2002). "Structure-activity relationships in 4-aminoquinoline antiplasmodials. The role of the group at the 7-position." *Journal of Medicinal Chemistry* 45(16), 3531–3539. doi: 10.1021/jm020858u. PMID: 12139464.

Ali, A., et al. (2018). "Investigating the Impact of Resistance Mutations on HIV-1 Protease Inhibitors." *Biochemical and Biophysical Research Communications*, 495(4), 2020–2024.

Alvarez, J.C. (2004). "High-throughput docking as a source of novel drug leads." *Current Opinion in Chemical Biology*, 8(4), 365–370.

Arrowsmith, J. (2011). "Trial watch: phase II failures: 2008-2010." *Nature Reviews Drug Discovery*, 10(5), 328–329.

Bajorath, J. (2001). "Integration of virtual and high-throughput screening." *Nature Reviews Drug Discovery*, 1(11), 882–894.

Bajorath, J. (2002). "Integration of virtual and high-throughput screening." *Nature Reviews Drug Discovery*, 1(11), 882–894.

Balakin, K.V., et al. (2006). "Pharmacophore modeling: advances, limitations, and current utility." *Molecular Pharmacology*, 70(1), 1–8.

Bender, A. (2010). "Evaluation of molecular descriptors for the prediction of biological activity using unstructured data." *Journal of Chemical Information and Modeling*, 50(1), 18–24.

Bender, A., et al. (2007). "Chemoinformatics in drug discovery." *Molecular Informatics*, 26(2), 102–116.

Black, L.J., et al. (1994). "Raloxifene (LY139481 HCl) prevents bone loss and reduces serum cholesterol without causing uterine hypertrophy in ovariectomized rats." *Journal of Clinical Endocrinology & Metabolism*, 79(5), 1428–1436.

Brooijmans, N., and Kuntz, I.D. (2003). "Molecular recognition and docking algorithms." *Annual Review of Biophysics and Biomolecular Structure*, 32, 335–373.

Brown, N. (2009). "Cheminformatics – an introduction." *Wiley Interdisciplinary Reviews: Computational Molecular Science*, 1(4), 318–322.

Capdeville, R., et al. (2002). "Glivec (STI571, imatinib), a rationally developed, targeted anticancer drug." *Nature Reviews Drug Discovery*, 1(7), 493–502.

Chen, M., et al. (2018). "Machine learning in cheminformatics and drug discovery." *Drug Discovery Today*, 23(8), 1538–1546.

Cheng, T., et al. (2009). "Structure-based virtual screening for drug discovery: a problem-centric review." *AAPS Journal*, 14(1), 133–141.

Cherkasov, A., et al. (2014). "QSAR modeling: where have you been? Where are you going to?" *Journal of Medicinal Chemistry*, 57(12), 4977–5010.

Cortés-Ciriano, I., and Bender, A. (2019). "Artificial intelligence in drug discovery: recent advances and future perspectives." *Expert Opinion on Drug Discovery*, 14(8), 793–806.

DiMasi, J.A., et al. (2016). "Innovation in the pharmaceutical industry: New estimates of R&D costs." *Journal of Health Economics*, 47, 20–33.

Dror, R.O., et al. (2012). "Biomolecular simulation: a computational microscope for molecular biology." *Annual Review of Biophysics*, 41, 429–452.

Druker, B.J., et al. (2001). "Activity of a specific inhibitor of the BCR-ABL tyrosine kinase in the blast crisis of chronic myeloid leukemia and acute lymphoblastic leukemia with the Philadelphia chromosome." *New England Journal of Medicine*, 344(14), 1038–1042.

Duchowicz, P.R., and Castro, E.A. (2013). "The importance of the QSAR-QSPR methodology to the theoretical study of pesticides." *International Journal of Chemical Modeling*, 5(1), 35–50.

Ehrlich, P. (1909). "Experimental research on immunity." *Journal of Immunology*, 4(1), 1–16.

Ekins, S., et al. (2019). "Exploiting machine learning for end-to-end drug discovery and development." *Nature Materials*, 18(5), 435–441.

Espeseth, A.S., et al. (2000). "HIV integrase inhibitors that compete with the target DNA substrate define a unique strand transfer conformation for integrase." *Proceedings of the National Academy of Sciences*, 97(21), 11244–11249.

Friesner, R.A., et al. (2004). "Glide: a new approach for rapid, accurate docking and scoring. 1. Method and assessment of docking accuracy." *Journal of Medicinal Chemistry*, 47(7), 1739–1749.

Gasteiger, J., and Engel, T. (2003). *"Cheminformatics: A Textbook."* Wiley-VCH.

Gaulton, A., et al. (2012). "ChEMBL: a large-scale bioactivity database for drug discovery." *Nucleic Acids Research*, 40(D1), D1100–D1107.

Ghose, A.K., Viswanadhan, V.N., and Wendoloski, J.J. (1998). "Prediction of hydrophobic (lipophilic) properties of small organic molecules using fragmental methods: an analysis of ALOGP and CLOGP methods." *The Journal of Physical Chemistry A*, 102(21), 3762–3772.

Ghosh, A.K., et al. (2012). "Structure-based design, synthesis, and biological evaluation of potent inhibitors of human beta-secretase." *Bioorganic & Medicinal Chemistry Letters*, 22(8), 2976–2980.

Gilson, M.K., and Zhou, H.X. (2007). "Calculation of protein-ligand binding affinities." *Annual Review of Biophysics and Biomolecular Structure*, 36, 21–42.

Goldman, J.M., and Melo, J.V. (2003). "Chronic myeloid leukemia – advances in biology and new approaches to treatment." *New England Journal of Medicine*, 349(15), 1451–1464.

Gonzalez, M.W., and Helguera, A.M. (2011). "Novel molecular descriptors applied to the QSAR analysis of hepatitis C virus NS5B inhibitors." *Molecular Informatics*, 30(7), 638–646.

Gonzalez, M.W., et al. (2015). "The impact of mutations on protein structure and function: Self-learning module." *Biochemical and Biophysical Research Communications*, 456(4), 725–729.

Gómez-Bombarelli, R., et al. (2018). "Design of efficient molecular organic light-emitting diodes by a high-throughput virtual screening and experimental approach." *Nature Materials*, 17(10), 974–979.

Green, J.R., and Kuntz, I.D. (1992). "A novel method for identifying potential drug candidates." *Journal of Medicinal Chemistry*, 35(5), 859–870.

Gupta, U., et al. (2024). "The Contribution of Artificial Intelligence to Drug Discovery: Current Progress and Prospects for the Future." In: Aditya Khamparia, Babita Pandey, Devendra Kumar Pandey, and Deepak Gupta (eds.), *Microbial Data Intelligence and Computational Techniques for Sustainable Computing*, 1–23. Springer.

Halperin, I., et al. (2002). "Principles of docking: An overview of search algorithms and a guide to scoring functions." *Proteins: Structure, Function, and Genetics*, 47(4), 409–443.

Hansch, C., and Fujita, T. (1964). "ρ-σ-π Analysis. A method for the correlation of biological activity and chemical structure." *Journal of the American Chemical Society*, 86(8), 1616–1626.

Hazuda, D.J., et al. (2004). "Raltegravir (MK-0518): inhibition of HIV-1 integrase, a novel target for antiretroviral therapy." *Antiviral Research*, 65(1), A49.

Hollingsworth, S.A., and Dror, R.O. (2018). "Molecular dynamics simulation for all." *Neuron*, 99(6), 1129–1143.

Huang, N., et al. (2006). "Novel strategies for compound selection in structure-based virtual screening." *Journal of Molecular Graphics and Modelling*, 25(1), 81–96.

Irwin, J.J., and Shoichet, B.K. (2005). "ZINC—a free database of commercially available compounds for virtual screening." *Journal of Chemical Information and Modeling*, 45(1), 177–182.

Karplus, M., and McCammon, J.A. (2002). "Molecular dynamics simulations of biomolecules." *Nature Structural Biology*, 9(9), 646–652.

Katritzky, A.R., et al. (2010). *Comprehensive Descriptors for Structural and Statistical Analysis*." Volume 1. Elsevier.

Kim, C.U., et al. (1997). "Influenza neuraminidase inhibitors: design, synthesis, and biological activity of a novel hydroxyethylene benzoic acid inhibitor (GS 4104)." *Journal of Medicinal Chemistry*, 40(21), 3975–3977.

Kitchen, D.B., et al. (2004). "Docking and scoring in virtual screening for drug discovery: methods and applications." *Nature Reviews Drug Discovery*, 3(11), 935–949.

Klebe, G. (2000). "Recent developments in structure-based drug design." *Journal of Molecular Medicine*, 78(5), 269–281.

Kola, I., and Landis, J. (2004). "Can the pharmaceutical industry reduce attrition rates?" *Nature Reviews Drug discovery*, 3(8), 711–716.

Koshland, D.E. (1958). "Application of a theory of enzyme specificity to protein synthesis." *Proceedings of the National Academy of Sciences*, 44(2), 98–104.

Kroemer, R.T. (2007). "Structure-based drug design: docking and scoring." *Current Protein and Peptide Science*, 8(4), 312–328.

Krohn, A., Redshaw, S., Ritchie, J.C., Graves, B.J., and Hatada, M.H. (1991). "Novel binding mode of highly potent HIV-proteinase inhibitors incorporating the (R)-hydroxyethylamine isostere." *Journal of Medicinal Chemistry*, 34(11), 3340–3342.

Langer, T., and Hoffmann, R.D. (2001). "Virtual screening an effective tool for lead structure discovery." *Current Pharmaceutical Design*, 7(7), 509–527.

Lavecchia, A., and Cerchia, L. (2015). "In silico methods to improve drug discovery." *Current Medicinal Chemistry*, 22(5), 699–714.

Lavecchia, A., and Di Giovanni, C. (2013). "Virtual screening strategies in drug discovery: a critical review." *Current Medicinal Chemistry*, 20(23), 2839–2860.

Leach, A.R., et al. (2010). "Structure-Based Drug Design: A Historical Perspective." In: Jiangang Chen and K. N. Houk (eds.), *"Molecular Modeling: Principles and Applications,"* 2nd ed. Pearson Education Limited, pp. 481–512.

Leeson, P.D., and Springthorpe, B. (2007). "The influence of drug-like concepts on decision-making in medicinal chemistry." *Nature Reviews Drug Discovery*, 6(11), 881–890.

Lengauer, T., and Rarey, M. (1996). "Computational methods for biomolecular docking." *Current Opinion in Structural Biology*, 6(3), 402–406.

Lin, A., et al. (2020). "Recent advances in AI-driven drug discovery." *Drug Discovery Today*, 25(10), 1838–1846.

Lionta, E., Spyrou, G., K Vassilatis, D., and Cournia, Z. (2014). "Structure-based virtual screening for drug discovery: principles, applications and recent advances." *Current Topics in Medicinal Chemistry*, 14(16), 1923–1938.

Lipinski, C., and Hopkins, A. (2004). "Navigating chemical space for biology and medicine." *Nature*, 432(7019), 855–861.
Lipinski, C.A., et al. (2001). "Experimental and computational approaches to estimate solubility and permeability in drug discovery and development settings." *Advanced Drug Delivery Reviews*, 46(1–3), 3–26.
Llinàs, A., and Goodman, J.M. (2008). "Drug–drug interaction screening: A role for chemistry and computer-aided drug design." *Drug Discovery Today*, 13(17–18), 749–755.
Lounkine, E., Keiser, M. J., Whitebread, S., Mikhailov, D., Hamon, J., Jenkins, J. L., ... and Urban, L. (2012). "Large-scale prediction and testing of drug activity on side-effect targets." *Nature*, 486(7403), 361–367.
Malik, M.I., Iqbal, M.A., and Kaur, T. (2023). "Recent advances in the development of BACE1 inhibitors for Alzheimer's disease therapy: a review." *Current Medicinal Chemistry*, 30(12), 1556–1580.
McGaughey, G.B., et al. (2000). "Nature, occurrence, and hydrogen bonding of NH/π interactions: An in vitro and database study." *Journal of Biological Chemistry*, 275(37), 1343–1348.
McInnes, C. (2007). "Virtual screening strategies in drug discovery." *Current Opinion in Chemical Biology*, 11(5), 494–502.
Melagraki, G., Sopasakis, P., Afantitis, A., and Sarimveis, H. (2010, September). "Consensus QSAR modeling and domain of applicability: An integrated approach." In *18th European Symposium on Quantitative Structure-Activity Relationships, Rhodes, Greece*.
Meng, X.Y., Zhang, H.X., Mezei, M., and Cui, M. (2011). "Molecular docking: a powerful approach for structure-based drug discovery." *Current Computer-Aided Drug Design*, 7(2), 146–157.
Merz, K.M. (2010). "Insight into the role of molecular dynamics in drug design." *Advances in Protein Chemistry and Structural Biology*, 79, 103–131.
Moitessier, N., Englebienne, P., Lee, D., Lawandi, J., and Corbeil, A.C. (2008). "Towards the development of universal, fast and highly accurate docking/scoring methods: a long way to go." *British Journal of Pharmacology*, 153(S1), S7–S26.
Morris, G.M., Huey, R., Lindstrom, W., Sanner, M.F., Belew, R.K., Goodsell, D.S., and Olson, A.J. (2009). "AutoDock4 and AutoDockTools4: automated docking with selective receptor flexibility." *Journal of Computational Chemistry*, 30(16), 2785–2791.
Munos, B. (2009). "Lessons from 60 years of pharmaceutical innovation." *Nature Reviews Drug Discovery*, 8(12), 959–968.
Narayanan, D., et al. (2014). "Computational Drug Discovery: The Road Ahead." In: D. K. Srivastava (ed.), *"Computational Drug Discovery."* Springer, pp. 1–17.
Oprea, T.I. (2000). *"Cheminformatics in Drug Discovery."* Wiley-VCH.
Pagadala, N.S., Syed, K., and Tuszynski, J. (2017). "Software for molecular docking: a review." *Biophysical Reviews*, 9, 91–102.
Patel, H.M., Noolvi, M.N., Sharma, P., Jaiswal, V., Bansal, S., Lohan, S., ... and Bhardwaj, V. (2014). "Quantitative structure–activity relationship (QSAR) studies as strategic approach in drug discovery." *Medicinal Chemistry Research*, 23, 4991–5007.
Paul, S.M., Mytelka, D.S., Dunwiddie, C.T., Persinger, C.C., Munos, B.H., Lindborg, S.R., & Schacht, A.L. (2010). "How to improve R&D productivity: the pharmaceutical industry's grand challenge." *Nature Reviews Drug Discovery*, 9(3), 203–214.
Peet, N.P., et al. (1998). "Discovery of the first anti-HIV agent targeting the viral integrase." *Journal of Medicinal Chemistry*, 41(11), 2029–2043.
Perola, E., et al. (2004). "A Guide to Pharmaceutical Lead Optimization." In: Hans Gerhard Vogel (ed.), *"Drug Discovery and Evaluation: Pharmacological Assays."* Springer, pp. 49–68.

Pickett, S.D., et al. (2011). "Computational Approaches to the Prediction of Toxicity." In: John B. Taylor and David J. Triggle (eds.), *"Comprehensive Medicinal Chemistry II."* Volume 4. Elsevier, pp. 629–652.

Pinzi, L., and Rastelli, G. (2019). "Molecular docking: shifting paradigms in drug discovery." *International Journal of Molecular Sciences*, 20(18), 4331.

Pires, D.E., Blundell, T.L., and Ascher, D.B. (2015). "pkCSM: predicting small-molecule pharmacokinetic and toxicity properties using graph-based signatures." *Journal of Medicinal Chemistry*, 58(9), 4066–4072.

Proudfoot, J.R. (2002). "The evolution of synthetic oral drug properties." *Bioorganic & Medicinal Chemistry Letters*, 12(12), 1647–1650.

Puzyn, T., Gajewicz, A., Leszczynska, D., and Leszczynski, J. (2010). "Nanomaterials–the next great challenge for QSAR modelers." In Tomasz Puzyn, Jerzy Leszczynski and Mark T.D. Cronin (eds.), *"Recent Advances in QSAR Studies: Methods and Applications"*. CRC, pp. 383–409.

Roy, K., Kar, S., and Ambure, P. (2015). "On a simple approach for determining applicability domain of QSAR models." *Chemometrics and Intelligent Laboratory Systems*, 145, 22–29.

Scannell, J.W., Blanckley, A., Boldon, H., and Warrington, B. (2012). "Diagnosing the decline in pharmaceutical R&D efficiency." *Nature Reviews Drug Discovery*, 11(3), 191–200.

Schneider, G. (2010). "Virtual screening: an endless staircase?" *Nature Reviews Drug Discovery*, 9(4), 273–276.

Schneider, G., and Fechner, U. (2005). "Computer-based de novo design of drug-like molecules." *Nature Reviews Drug Discovery*, 4(8), 649–663.

Schneider, P., Walters, W.P., Plowright, A.T., Sieroka, N., Listgarten, J., Goodnow Jr, R. A., ... and Schneider, G. (2020). "Rethinking drug design in the artificial intelligence era." *Nature Reviews Drug Discovery*, 19(5), 353–364.

Scior, T., Bernard, P., and Medina-Franco, J.L. (2007). "Large compound databases for structure-activity relationships studies in drug discovery." *Mini Reviews in Medicinal Chemistry*, 7(8), 851–860.

Sharma, M., et al. (2023), eds. *Soft Computing Techniques in Connected Healthcare Systems.* CRC Press.

Shaw, D.E., Maragakis, P., Lindorff-Larsen, K., Piana, S., Dror, R.O., Eastwood, M. P., ... and Wriggers, W. (2010). "Atomic-level characterization of the structural dynamics of proteins." *Science*, 330(6002), 341–346.

Shimura, K., Kodama, E., Sakagami, Y., Matsuzaki, Y., Watanabe, W., Yamataka, K., ... and Matsuoka, M. (2008). "Broad antiretroviral activity and resistance profile of the novel human immunodeficiency virus integrase inhibitor elvitegravir (JTK-303/GS-9137)." *Journal of Virology*, 82(2), 764–774.

Shoichet, B.K. (2004). "Virtual screening of chemical libraries." *Nature*, 432(7019), 862–865.

Sliwoski, G., Kothiwale, S., Meiler, J., and Lowe, E.W. (2014). "Computational methods in drug discovery." *Pharmacological Reviews*, 66(1), 334–395.

Sousa, S.F., Fernandes, P.A., and Ramos, M.J. (2006). "Protein–ligand docking: current status and future challenges." *Proteins: Structure, Function, and Bioinformatics*, 65(1), 15–26.

Summa, V., Petrocchi, A., Bonelli, F., Crescenzi, B., Donghi, M., Ferrara, M., ... and Rowley, M. (2008). "Discovery of raltegravir, a potent, selective orally bioavailable HIV-integrase inhibitor for the treatment of HIV-AIDS infection." *Journal of Medicinal Chemistry*, 51(18), 5843–5855.

Sun, H., et al. (2016). "A review of the recent advances in the chemical biology of small molecules." *Chemistry & Biology*, 23(5), 584–593.

Swainston, N.H., and Jansen, J. (2019). "Cheminformatics approaches to address drug resistance in cancer and infectious diseases." *Frontiers in Pharmacology*, 10, 1221.

Swanson, J.M., Kuo, R.C., and Wang, S. (2011). "The role of molecular dynamics simulations in drug discovery: Insights into binding and unbinding processes." *Journal of Chemical Information and Modeling*, 51(5), 1038–1046.

Taylor, R.D., et al. (2002). "A fast flexible docking method using an incremental construction algorithm." *Journal of Computational Chemistry*, 23(5), 584–590.

Teague, S.J., Davis, A.M., Leeson, P.D., and Oprea, T. (1999). "The design of leadlike combinatorial libraries." *Angewandte Chemie International Edition*, 38(24), 3743–3748.

Todeschini, R., and Consonni, V. (2009). *Molecular Descriptors for Chemoinformatics*, Volume I. John Wiley & Sons.

Trott, O., and Olson, A.J. (2010). "AutoDock Vina: improving the speed and accuracy of docking with a new scoring function, efficient optimization, and multithreading." *Journal of Computational Chemistry*, 31(2), 455–461.

Vamathevan, J., Clark, D., Czodrowski, P., Dunham, I., Ferran, E., Lee, G., ... & Zhao, S. (2019). "Applications of machine learning in drug discovery and development." *Nature Reviews Drug Discovery*, 18(6), 463–477.

Varghese, J.N., McKimm-Breschkin, J.L., Caldwell, J.B., Kortt, A.A., and Colman, P.M. (1992). "The structure of the complex between influenza virus neuraminidase and sialic acid, the viral receptor." *Proteins: Structure, Function, and Bioinformatics*, 14(3), 327–332.

Varnek, A., and Tropsha, A. (Eds.). (2008). *Chemoinformatics Approaches to Virtual Screening*. Royal Society of Chemistry.

Vassar, R. (2014). "BACE1 inhibitor drugs in clinical trials for Alzheimer's disease." *Alzheimer's Research & Therapy*, 6(9), 89.

Vieth, M., et al. (2004). "Current and future trends in structure-based drug design." *Journal of Medicinal Chemistry*, 47(11), 2819–2832.

Vilar, S., Cozza, G., & Moro, S. (2008). "Medicinal chemistry and the molecular operating environment (MOE): application of QSAR and molecular docking to drug discovery." *Current Topics in Medicinal Chemistry*, 8(18), 1555–1572.

Von Itzstein, M., Wu, W. Y., Kok, G. B., Pegg, M. S., Dyason, J. C., Jin, B., ... and Penn, C.R. (1993). "Rational design of potent sialidase-based inhibitors of influenza virus replication." *Nature*, 363(6428), 418–423.

Walters, W.P., et al. (1998). "Virtual screening—an overview." *Drug Discovery Today*, 3(4), 160–178.

Walters, W.P., Green, J., Weiss, J.R., and Murcko, M.A. (2011). "What do medicinal chemists actually make? A 50-year retrospective." *Journal of Medicinal Chemistry*, 54(19), 6405–6416.

Wang, Y., Huang, Y., and Li, H. (2012). "Discovery of novel BACE1 inhibitors using virtual screening and molecular dynamics simulations." *Bioorganic & Medicinal Chemistry Letters*, 22(24), 7563–7568.

Wang, Y., Yang, H., and Zhang, W. (2020). "Integrating chemistry and biology for drug discovery: advances in cheminformatics methods." *Molecular Informatics*, 39(5), 1900136.

Wlodawer, A., & Vondrasek, J. (1998). "Inhibitors of HIV-1 protease: a major success of structure-assisted drug design." *Annual Review of Biophysics and Biomolecular Structure*, 27(1), 249–284.

Wolber, G., and Langer, T. (2005). "LigandScout: 3-D pharmacophores derived from protein-bound ligands and their use as virtual screening filters." *Journal of Chemical Information and Modeling*, 45(1), 160–169.

Xu, Y., Lin, K., Wang, S., Wang, L., Cai, C., Song, C., ... and Pei, J. (2019). "Deep learning for molecular generation." *Future Medicinal Chemistry*, 11(6), 567–597.

Yang, S.Y. (2010). "Pharmacophore modeling and applications in drug discovery: challenges and recent advances." *Drug Discovery Today*, 15(11-12), 444–450.

Yang, Y., et al. (2009). "Protein–ligand docking: A review of recent developments." *Journal of Computational Chemistry*, 30(5), 898–907.

Yuriev, E., and Ramsland, P.A. (2013). "Latest developments in molecular docking: 2010–2011 in review." *Journal of Molecular Recognition*, 26(5), 215–239.

9 Transformative Applications of AI and Machine Learning in Bioinformatics for Healthcare Systems

Arooj Fatima Tul Zahra, Mujahid Tabassum, Nabiea Shehma, Moeza Anam, and Tripti Sharma

9.1 INTRODUCTION

Artificial intelligence (AI) and bioinformatics combine to provide sophisticated tools for studying and interpreting complicated biological data, revolutionizing the healthcare industry (C. Liu et al., 2023). Bioinformatics, a field that utilizes computational methods to comprehend and handle biological data, is progressively utilizing AI and machine learning (ML) to improve data processing skills and extract significant discoveries (Sharieff & Sameer, 2023). These breakthroughs are expediting research and fueling innovations in diagnostics, customized medicine, and public health (Athanasopoulou et al., 2022).

9.1.1 Bioinformatics and AI: Revolutionizing Healthcare

The modern healthcare landscape is fundamentally transforming due to the integration of AI and bioinformatics. Conventional approaches to data processing are typically inadequate when dealing with the large amount and intricate nature of biological data produced by advanced technologies like next-generation sequencing, mass spectrometry, and single-cell RNA sequencing (Athanasopoulou et al., 2022). AI and ML with their capacity to analyze extensive datasets and detect complex patterns provide effective answers to these difficulties (Karim et al., n.d.) AI plays a crucial role in the analysis of genetic data, making it one of the most important contributions of AI in the field of bioinformatics (Akhtar et al., 2021). AI algorithms, namely those utilizing deep learning (DL) techniques, can scan vast amounts of genomic data, detect genetic abnormalities, and forecast their impact on an individual's health (Min et al., 2017). CNNs have been utilized to identify malignant mutations by analyzing genomic sequences, resulting in notable enhancements in the precision and efficiency of diagnosis (Rana, 2019). Early discovery of mutations is essential in the field of oncology as it can significantly improve treatment regimens and enhance patient outcomes (Noorbakhsh-Sabet et al., 2019).

AI and ML models play a crucial role in customizing treatments for specific patients in the field of customized medicine. These models utilize genetic, environmental, and lifestyle variables to forecast the likelihood of developing a disease and the potential effectiveness of different treatments. This individualized strategy improves the efficacy of therapies and reduces the occurrence of adverse side effects (Mehta, 2023). ML algorithms have the capability to recognize biomarkers for early disease identification, which allows for prompt therapies that enhance patient outcomes (Qiu et al., 2023). An exemplary instance involves the utilization of AI to scrutinize genetic information in order to forecast individual reactions to chemotherapy. This enables physicians to tailor treatment strategies and mitigate the likelihood of unfavorable consequences ((Karim et al., n.d.)). AI has a significant influence on the process of discovering and developing drugs by targeting the shortcomings of conventional approaches. The conventional method of drug discovery is widely known for its lengthy duration and high costs, typically requiring more than ten years and billions of dollars to introduce a new medicine to the market successfully (Jiang et al., 2017). AI algorithms can enhance this process by forecasting the biological activity of substances, pinpointing prospective medication targets, and refining drug design (Lysaght et al., 2019). DL models have been employed to forecast the interactions between various medications and target proteins. This has resulted in the discovery of potential drug candidates and the utilization of current drugs for novel therapeutic purposes. This strategy not only expedites the process of discovery but also substantially decreases costs (Biswas et al., 2023).

Furthermore, the utilization of AI-driven predictive analytics is transforming the field of public health by facilitating more precise illness forecasting and monitoring. AI models have the capability to examine epidemiological data in order to forecast disease outbreaks and monitor the transmission of contagious diseases. This enables prompt implementation of public health treatments (S. Z. Rahman et al., 2023). These models have proven to be highly useful in effectively handling global health emergencies, such as the COVID-19 pandemic. AI technologies have been employed to forecast the transmission of the virus and evaluate the efficacy of containment strategies. ML algorithms were utilized to assess real-time data from various sources, such as social media and news reports, in order to forecast COVID-19 hotspots and provide guidance for public health interventions (Syrowatka et al., 2021).

Furthermore, AI and ML are also improving healthcare operations. AI systems can aid in patient triage, streamline repetitive activities, and offer clinical decision assistance, enabling healthcare workers to concentrate on more intricate elements of patient care. AI algorithms can utilize electronic health records (EHRs) to assess and identify individuals who are at a high risk of developing issues. This allows for preventive care and helps to decrease the number of hospital readmissions (Kaur et al., 2021). AI-powered chatbots and virtual assistants also offer patients customized health guidance and assistance, enhancing their involvement and contentment (Jeyaraj & Narayanam, 2023).

However, the incorporation of AI in bioinformatics poses difficulties. Extensive and higher-quality datasets are necessary for the training of precise models. Ensuring the confidentiality and protection of data is of the utmost importance, mainly when dealing with sensitive patient information. In order to safeguard patient data and retain trust, the integration of AI in healthcare must adhere to standards such as the General Data Protection Regulation (GDPR) in Europe and the Health Insurance Portability and Accountability Act (HIPAA) in the United States (Bhattamisra et al., 2023). AI model interpretability is a significant challenge. Although DL models can attain great levels of accuracy, their inherent "black box" nature poses challenges in comprehending individual predictions. The absence of transparency can impede the acceptance of AI in therapeutic environments, where comprehending the reasoning behind a choice is essential. Current endeavors are focused on the development of explainable AI strategies that offer elucidation on the decision-making processes of these models (Pavan Kumar et al., 2022).

Deploying AI in healthcare is heavily influenced by ethical considerations. Imbalanced training data can result in biased AI models, which in turn can provide unfair outcomes. It is crucial to confront these prejudices in order to guarantee that AI systems offer fair and impartial care to all groups of patients. Furthermore, it is crucial to thoroughly contemplate the ethical ramifications of AI decision-making in healthcare, including issues of accountability and the necessity for human supervision (Puaschunder & Feierabend, 2019).

To summarize, the revolutionary impact of AI and ML in bioinformatics is improving healthcare systems by strengthening the processing and interpretation of intricate biological data. AI technologies are making tremendous improvements in healthcare, ranging from genetic data analysis and customized medicine to medication discovery and public health management. However, fully harnessing the capabilities of AI in bioinformatics necessitates tackling obstacles pertaining to the quality of data, the interpretability of models, and ethical considerations. As these obstacles are successfully addressed, AI is positioned to assume a more pivotal role in advancing biomedical research and enhancing healthcare outcomes (Parums, 2023).

9.1.2 Objectives and Scope of the Chapter

This chapter seeks to investigate the profound influence of AI and ML in the field of bioinformatics, specifically in healthcare systems. This will emphasize the ways in which these technologies are transforming the examination of intricate biological data, enhancing diagnostics, customizing therapies, optimizing drug discovery, and improving the management of public health.

Firstly, a historical overview of AI in bioinformatics will be presented, highlighting notable technological progress and methodologies that have enabled its integration into the field. Comprehending this context is crucial in order to grasp existing capabilities and future potentials fully. Following that, we will analyze different ML methodologies, encompassing supervised, unsupervised, and reinforcement learning. The concepts and implementations of each technique will be examined through case studies to illustrate their efficacy in the field of bioinformatics. A specialized segment will be devoted to DL, focusing on its profound influence, notably through neural networks such as CNNs and recurrent neural networks (RNNs). The discussion will focus on the utilization of these technologies in the analysis of genomic data, medical imaging, and healthcare data.

This chapter will also explore the field of natural language processing (NLP) as it pertains to the interpretation of biological texts and the integration of genomic data. Practical applications and case studies will demonstrate how NLP is utilized in the field of bioinformatics. This text will investigate the function of AI in genomic analysis, specifically focusing on its applications in DNA sequencing and predictive modeling. Genomic research success stories will showcase the progress made in genetics through the use of AI. In addition, the chapter will explore the impact of AI on precision medicine and personalized healthcare. AI greatly enhances patient outcomes by facilitating the creation of individualized treatment plans and predictive analytics. The chapter will also discuss the influence of AI on drug research and development, demonstrating how AI expedites the drug discovery process and its involvement in drug repurposing. Case studies will illustrate practical applications in the field of pharmaceutical research. Furthermore, we will discuss the utilization of AI in the forecast of diseases and the administration of public health. The discussion will focus on AI models used to anticipate disease outbreaks and improve public health surveillance, highlighting their significance in the field of global health.

Finally, this discussion will address the challenges and ethical concerns associated with incorporating AI into the fields of bioinformatics and healthcare. The examination will focus on data privacy, security, ethical implications, and bias in AI systems, as well as solutions to address these concerns. This chapter seeks to offer a thorough comprehension of the revolutionary capacity of AI and ML in the field of bioinformatics. It provides essential insights for researchers, healthcare professionals, and politicians who are navigating the changing field of AI and bioinformatics, and achieves this by emphasizing current research, practical applications, and future directions.

9.2 HISTORICAL CONTEXT AND TECHNOLOGICAL INTEGRATION

The application of AI and ML in bioinformatics is a relatively new but swiftly developing domain. To fully grasp the current accomplishments and future potentials, it is essential to comprehend the historical backdrop and the integration of technology. This section explores the significant historical events that have influenced the incorporation of AI in bioinformatics and analyzes the crucial technological progress that has enabled this merging.

9.2.1 Evolution of AI in Bioinformatics

The advent of AI in bioinformatics commenced in the latter part of the 20th century, propelled by the rapid expansion of biological data and the necessity for sophisticated computing tools to effectively handle and analyze this data (Gauthier et al., 2019a). At first, bioinformatics heavily depended on conventional computing methods to study biological sequences and structures. Nevertheless, the shortcomings of these systems in managing extensive and intricate datasets quickly became evident (Akhtar et al.,

2021). The incorporation of AI and ML into the field of bioinformatics represented a substantial transformation (Saifi et al., 2023). During the 1990s, the emergence of algorithms like hidden Markov models and neural networks introduced novel methods for forecasting protein structures and gene sequences (Narayanan et al., 2002a). These initial applications showcased the capacity of AI to reveal patterns and insights that were previously unachievable using conventional methods. The completion of the Human Genome Project (HGP) in 2003 was a significant achievement that highlighted the essential role of AI in the field of bioinformatics. The study produced an unparalleled quantity of genomic data, requiring advanced methods for data processing. AI and ML methods were utilized to annotate the human genome, ascertain gene functions, and comprehend genetic variances (Lai et al., 2019). The incorporation of AI in bioinformatics during the HGP era established the foundation for subsequent progress.

In the following years, the introduction of high-throughput sequencing technology significantly increased the amount of biological data (Cai et al., 2020). There was an increasing urgency for data analysis tools that could be easily expanded and performed effectively. AI techniques, including ML, were progressively utilized to handle and analyze this overwhelming amount of data (Cai et al., 2020). Support vector machines (SVMs) and clustering algorithms were employed to categorize gene expression data and detect biomarkers (Ezziane, 2006). DL became increasingly prevalent in the field of bioinformatics during the 2010s. CNNs and RNNs have been increasingly used in bioinformatics for applications such as medical diagnoses through image analysis and genomics through sequence analysis. DL models, due to their capacity to acquire hierarchical representations, have offered unparalleled precision and understanding, hence reinforcing the significance of AI in the field of bioinformatics (Tabassum et al., 2021).

The incorporation of AI into bioinformatics has been driven by recent breakthroughs in processing capacity and the accessibility of extensive, annotated datasets (Sharieff & Sameer, 2023b). Currently, researchers are investigating the use of transfer learning and reinforcement learning methods to improve predictive modeling and optimize bioinformatics workflows (Terranova et al., 2021). In addition, the development of explainable AI seeks to tackle the opaque aspect of DL models, enhancing their transparency and reliability in therapeutic contexts. The development of AI in bioinformatics demonstrates an ongoing effort to enhance the precision, effectiveness, and scalability of biological data analysis. AI has revolutionized bioinformatics, turning it into a dynamic and rapidly progressing science, from its early algorithmic applications to its advanced DL models (Sharieff & Sameer, 2023b). To comprehend the current capabilities and predict future developments that will further transform healthcare systems through bioinformatics, it is crucial to understand this evolution.

9.2.2 Key Technological Milestones and Methodologies

The field of bioinformatics has seen significant technological advancements and the use of novel approaches that have contributed significantly to its expansion and integration with AI and ML (Athanasopoulou et al., 2022).

A significant development occurred with the introduction of high-throughput sequencing technology in the early 2000s. These technologies have transformed genomic research by greatly enhancing the speed and reducing the expense of sequencing, resulting in the production of extensive genetic data that requires sophisticated computer tools for processing (Gauthier et al., 2019). High-throughput sequencing has facilitated extensive investigations of genomes, transcriptomes, and epigenomes, allowing for the execution of large-scale initiatives such as the 1000 Genomes Project and the Cancer Genome Atlas (Gauthier et al., 2019). The emergence and execution of microarray technology signified notable progress. Microarrays enabled the concurrent examination of gene expression in numerous genes, offering a potent instrument for comprehending gene functionality, detecting disease indicators, and investigating the impacts of genetic variations. These technologies played a crucial role in enhancing our comprehension of intricate diseases and aiding in the discovery of therapeutic targets (Dopazo, 2006).

Simultaneously, the emergence of cloud computing has had a profound impact on the field of bioinformatics. Cloud platforms such as Amazon Web Services (AWS), Google Cloud, and Microsoft Azure offer flexible infrastructure for storing, processing, and analyzing extensive biological datasets. These platforms provide robust computational resources and bioinformatics tools that allow researchers to conduct intricate investigations without requiring significant local computational equipment (Masulli & Tagliaferri, 2011). Another noteworthy approach is the utilization of ensemble learning techniques. Ensemble approaches, which amalgamate the forecasts of several models to enhance accuracy and resilience, have been effectively utilized in the field of bioinformatics. Random forests and gradient-boosting machines have been employed to forecast disease outcomes, categorize biological samples, and discern significant genetic characteristics from intricate datasets (Bacciu et al., 2018).

The establishment of specialized databases with resources has also proven indispensable. Genetic and protein information is stored in databases like GenBank, EMBL, and the Protein Data Bank (PDB), which serve as extensive repositories. These databases enable the sharing of data and promote collaborative research endeavors. These resources have played a crucial role in numerous bioinformatics applications, ranging from sequence alignment to structural biology (Akhtar et al., 2021). Recently, the incorporation of AI and ML techniques, such as transfer learning and federated learning, has become a significant development. Transfer learning enables the refinement of models that have been trained on big datasets, specifically for bioinformatics tasks that have minimal data. This process enhances performance and minimizes the requirement for extensive labeled datasets. Federated learning allows for the training of models using decentralized datasets while still ensuring data privacy. This addresses essential challenges in the field of biomedical research (Sharieff & Sameer, 2023b).

9.2.3 BIG DATA INTEGRATION AND ITS IMPACT

The incorporation of big data into the field of bioinformatics has had significant and extensive effects, fundamentally altering the scope of biological study and healthcare. Big data in bioinformatics comprises a wide range of data types, such as genomic, proteomic, transcriptomic, and clinical data. The incorporation of these datasets has facilitated a comprehensive comprehension of biological systems and disorders (Athanasopoulou et al., 2022b). A notable consequence of integrating big data is the progress made in the field of personalized medicine. Researchers can use large-scale genomic data and patient health records to detect genetic differences that impact disease risk and drug response. This enables the development of more personalized and efficient treatments (Gauthier et al., 2019).

The integration of big data has also improved the study of systems biology. Researchers can create comprehensive models of biological pathways and networks by integrating data from several biological levels, including genes, proteins, and metabolites. This methodology offers a valuable understanding of the intricate interplay involved in biological processes and disease mechanisms, hence aiding in the discovery of new therapeutic targets and biomarkers (Ouzounis, 2012). Big data utilization has also propelled progress in predictive analytics. ML algorithms have the ability to examine large datasets in order to detect patterns and make predictions, such as the course of diseases, how patients will respond to therapy, and the likelihood of infectious disease outbreaks. One instance of this is the development of prediction models that utilize both genomic and clinical data to predict cancer prognosis and provide guidance for treatment options (Masulli & Tagliaferri, 2011).

Cloud computing has been significant in enabling the incorporation of large volumes of data. Cloud platforms provide the necessary computational power and storage capacity to manage extensive datasets, as well as capabilities for data analysis and collaboration (Perumal et al., 2022). These platforms have made bioinformatics resources accessible to everyone, enabling researchers from around the world to do advanced analyses without needing extensive local infrastructure (Bacciu et al., 2018). Nevertheless, the incorporation of big data in bioinformatics also brings about difficulties, specifically with the protection and safeguarding of data privacy and security. Safeguarding confidential biological and health information while maintaining its availability for research purposes is of utmost importance. Regulatory frameworks like the GDPR and the HIPAA have been created to deal with these problems. Continued endeavors are required to achieve a harmonious equilibrium between the accessibility of data and the preservation of privacy and security (Mushegian, 2011).

Big data integration has transformed bioinformatics by facilitating comprehensive and accurate studies, promoting advancements in personalized medicine, systems biology, and predictive analytics. Efficient handling and examination of large datasets will be essential for future advancements in healthcare and biological research as the area continues to develop (Jindal et al., 2021).

9.3 MACHINE LEARNING TECHNIQUES IN BIOINFORMATICS

ML (ML) is crucial in the field of bioinformatics as it enables the analysis of extensive biological datasets to identify patterns and make accurate predictions (Shastry & Sanjay, 2020). The applications of this technology encompass a wide range of subfields, including genomics, proteomics, systems biology, and drug development

(Larrañaga et al., 2006). This section explores the fundamental machine-learning techniques employed in the field of bioinformatics.

9.3.1 SUPERVISED LEARNING: CONCEPTS AND APPLICATIONS

Supervised learning is a foundational technique in ML that entails training a model using a dataset that has been labeled (Mitchell, 2014). The model acquires the ability to establish a correlation between input characteristics and a predetermined output, enabling it to make forecasts on novel, unobserved data (Dittman et al., 2013). This approach has been extensively utilized in the field of bioinformatics for tasks such as disease classification, gene expression analysis, and protein function prediction. A prominent use of supervised learning is in forecasting disease outcomes. Support Vector Machines (SVMs) have been employed to categorize malignant tissues using gene expression profiles as a basis(Cheng et al., 2008). Tan and Gilbert conducted a study that showed Support Vector Machines (SVMs) to have a high level of accuracy in accurately differentiating between various forms of cancer (Choon & Gilbert, 2003). Decision trees have also been used to detect critical genetic markers linked to particular diseases, assisting in the early detection and customized treatment strategies. Protein function prediction is another important use. Supervised learning algorithms, such as neural networks, have undergone training using protein sequences to forecast their functional roles (Shastry & Sanjay, 2020). These models employ labeled datasets that contain known protein activities to acquire patterns and generate precise predictions for new sequences (Mohammed & Mohammed, 2014). This method has proven to be especially valuable in labeling proteins in recently sequenced genomes, where it is often difficult to do experimental verification.

Additionally, supervised learning has been applied in the field of genomics for variant calling and annotation. Random Forests and Gradient Boosting Machines algorithms are trained using labeled genomic data to detect genetic variations and forecast their possible influence on gene function and susceptibility to diseases (Auslander et al., 2021). These models have greatly enhanced the precision and effectiveness of genomic investigations, making it easier to identify variations related to diseases (Mitchell, 2014).

9.3.2 UNSUPERVISED LEARNING: DISCOVERING HIDDEN PATTERNS

Unsupervised learning involves working with data that does not have any labels, as opposed to supervised learning. The objective is to uncover concealed patterns or inherent structures within the data. Clustering and dimensionality reduction are prevalent unsupervised learning methods employed in the field of bioinformatics (Shastry & Sanjay, 2020).

Clustering methods, such as k-means & hierarchical clustering, are utilized to categorize comparable biological samples according to their characteristics (Larrañaga et al., 2006). Clustering has been employed to detect different categories of cancer by examining gene expression data (Dittman et al., 2013). This method facilitates the comprehension of the diverse composition of tumors at a molecular level, which is essential for the advancement of specific treatment approaches (Narayanan et al., 2002b). Another use case involves the examination of protein-protein interaction networks. Unsupervised learning techniques can group proteins into functional modules, providing valuable information on their involvement in biological processes. For instance, Kasturi and Acharya introduced an unsupervised approach to detect gene clusters by leveraging integrated data from promoter sequences, gene ontologies, and location data. This method successfully identified genes that are linked and their possible interactions (Y. Liu, 2004).

Dimensionality reduction methods, such as Principal Component Analysis (PCA) as well as t-distributed Stochastic Neighbor Embedding (t-SNE), are employed to simplify high-dimensional biological data (Yang et al., 2021). These techniques aid in the visualization of data patterns and the identification of crucial characteristics that lead to variations in the dataset (Dong & Pan, 2020). PCA has been used to decrease the dimensionality of gene expression data, making it easier to identify principle components that capture most of the variation in the data. Unsupervised learning is essential for conducting exploratory data analysis(Alexander et al., 2023). Elucidating inherent clusters and connections within the data assists researchers in formulating hypotheses and directing subsequent experimental inquiries.

9.3.3 Reinforcement Learning: Optimizing Bioinformatics Processes

Reinforcement learning (RL) is a form of ML in which an agent acquires the ability to make decisions by engaging with its surroundings and obtaining incentives for activities that lead to desired results. Reinforcement learning (RL) is well-suited for improving intricate bioinformatics procedures that require making decisions in a specific order.

An application of Reinforcement Learning (RL) in the field of bioinformatics involves optimizing the process of DNA fragment assembling. Bocicor et al. introduced a model based on reinforcement learning to address the fragment assembly problem. In this approach, the agent learns to arrange DNA fragments in the correct sequence by maximizing overlap scores. This strategy demonstrated decreased computing complexity and enhanced efficiency in comparison to conventional methods. Reinforcement learning (RL) has also been used to maximize the coordinated movement of groups of cells, a crucial process in activities such as wound healing and the spread of cancer. Hou et al. employed deep reinforcement learning to simulate and expedite the coordinated motion of cells, enhancing comprehension of cellular actions and facilitating the advancement of therapeutic approaches.

In addition, reinforcement learning (RL) is employed in the development of adaptive clinical trials, where the trial procedures are dynamically modified in response to accumulating data. This methodology facilitates the expedited identification of the most efficacious medicines with a reduced patient cohort. Reinforcement learning (RL) improves the efficiency and ethical conduct of clinical trials by teaching optimal strategies for patient allocation and dosage changes. Reinforcement learning provides a robust framework for addressing intricate optimization problems in bioinformatics, leading to progress in DNA assembly, cellular modeling, and clinical trial design.

9.4 DEEP LEARNING APPROACHES

9.4.1 NEURAL NETWORKS IN GENOMIC DATA ANALYSIS

Neural networks are now indispensable tools in the analysis of genetic data, as they can effectively capture intricate and non-linear correlations in extensive datasets (Muzio et al., 2021). Deep neural networks (DNNs) excel in their ability to accurately forecast gene expression levels, detect regulatory elements, and provide annotations for genomic sequences. For instance, deep neural networks (DNNs) have been employed to forecast the levels of gene expression based on DNA sequences (Silva et al., 2020). This is achieved by capturing the complex relationships between genetic sequences and their corresponding expression patterns (Demetci et al., 2020). This methodology has resulted in substantial advancements in comprehending the mechanics of gene regulation. Moreover, neural networks have been utilized to detect regulatory regions in the genome, such as promoters and enhancers, which play a vital role in gene regulation (Demetci et al., 2020). These technological improvements have made it easier to make precise annotations of genomic sequences, which helps in identifying functional elements and understanding their functions in maintaining health and causing diseases (X. M. Zhang et al., 2021).

9.4.2 CONVOLUTIONAL NEURAL NETWORKS FOR MEDICAL IMAGING

CNNs are very suitable for assessing medical images because they possess the capability to automatically extract hierarchical characteristics from raw image data(Sarvamangala & Kulkarni, 2022). CNNs have been widely used in many medical imaging applications, such as picture categorization, partitioning, and identification of abnormalities (Bir & Balas, 2020). CNNs have proven effective in accurately categorizing lung nodules in CT scans, hence improving the ability to detect lung cancer at an early stage (Xu et al., 2021). Utilizing CNNs in this field has led to increased precision rates in comparison to conventional techniques, thereby minimizing incorrect identifications and enhancing diagnostic certainty (Bir & Balas, 2020; Sarvamangala & Kulkarni, 2022). In addition, CNNs have been utilized to segment brain tumors in MRI scans, enabling accurate identification and classification of tumors. This capability is essential for the development of treatment strategies. In addition, CNNs have been utilized to detect diabetic retinopathy from retinal pictures. They are capable of accurately classifying the severity of retinopathy, which enables early intervention and effective disease management(Ghosal et al., 2019).

9.4.3 RECURRENT NEURAL NETWORKS IN HEALTHCARE DATA

RNNs are specifically intended to process sequential data, which makes them well-suited for evaluating time-series data in the healthcare field (Lalapura et al., 2021). RNNs have different versions, one of which is long short-term memory (LSTM) networks. These networks are very good at collecting long-term dependencies, which are crucial for various healthcare applications (Y. Chen & Li, 2021; Neves et al., 2021). An important use of RNNs is to forecast patient outcomes using EHRs. RNNs can forecast forthcoming health occurrences, such as hospital readmissions or the initiation of chronic illnesses, by examining sequences of medical events and patient history (Ho et al., 2021). This assists in proactive healthcare management. In addition, RNNs have been employed to diagnose diseases, like as heart failure, at an early stage (Ledbetter et al., 2021). This is achieved by continuously monitoring changes in patient data over time, which allows for prompt interventions and has the potential to enhance patient outcomes. RNNs are utilized in healthcare for various NLP activities, including the anonymization of patient notes and the extraction of significant information from clinical narratives (Ackerson et al., 2021). This program facilitates enhanced data management and enables more informed clinical decision-making (Ackerson et al., 2021; A. Rahman et al., 2021).

9.5 ADVANCED AI APPLICATIONS IN BIOINFORMATICS

9.5.1 Natural Language Processing and Genomics

9.5.1.1 NLP for Biological Text Mining

NLP is becoming an essential technology in bioinformatics for extracting relevant information from the extensive collection of biological literature. Due to the rapid increase in scientific publications, researchers are able to automate the extraction and analysis of biological data using NLP approaches. For instance, NLP models like PubTator Central and BioBERT have been created to annotate biomedical texts and extract significant biological elements and relationships. NLP approaches are widely employed for extracting associations between genes and diseases. NLP systems can analyze scientific literature to detect and emphasize references to genes and diseases, as well as their connections. The implementation of this automated procedure has greatly expedited the detection of genetic indicators for different illnesses, enhancing the productivity and efficacy of biomedical investigation. The integration of NLP with genomic data improves the comprehension of genetic functions and disease mechanisms by establishing connections between textual information extracted from scientific literature and genomic datasets. The integration of clinical notes with genomic data in EHRs is particularly beneficial for identifying genetic variants that are related to specific clinical characteristics. For example, NLP approaches have been employed to remove identifying information from patient notes and extract clinical data, which is subsequently connected to genomic data for comprehensive study.

This systematic approach is also advantageous for annotating genes using information derived from books. Researchers can enhance the functional annotation of genes by merging data from sources such as Ensembl and NCBI. This can assist in identifying gene regulation mechanisms and possible targets for therapeutic interventions.

9.5.2 AI in Genomic Research

9.5.2.1 AI in DNA Sequencing

AI has greatly improved the process of DNA sequencing by enhancing the precision and efficiency of sequence analysis (Mumtaz et al., 2023). DeepVariant, a DL model created by Google, demonstrates this by utilizing CNNs to enhance the precision of variant calling from sequencing data (Afanasiev et al., 2021; Giudice et al., 2021). This model can differentiate between authentic genetic variations and errors that occur during the sequencing process, resulting in more dependable sequencing outcomes.

In addition, AI has been utilized in nanopore sequencing, a technology renowned for its ability to generate data in real time but with significant mistake rates (Jena & Pathak, 2023). AI models, which have been taught using large sequencing datasets, can promptly rectify mistakes, hence enhancing the suitability of nanopore sequencing for clinical purposes. These findings have substantial ramifications for areas such as customized treatment and genetic research (Kacew et al., 2021).

9.5.2.2 Predictive Modeling and Success Stories

AI-driven predictive modeling has emerged as a fundamental aspect of genomic research, facilitating the anticipation of disease vulnerability and treatment outcomes through the analysis of genetic information (Kosvyra et al., 2020). ML methods, such as Random Forests and Gradient Boosting Machines, combine diverse biological data, including genomic, epigenomic, and transcriptome information, to generate precise predictions on illness outcomes (Srinivasu et al., 2022). An exemplary achievement in AI-powered genomic research is the application of predictive modeling in the field of cancer genomics. AI models have been employed to detect mutations in the BRCA1 and BRCA2 genes, which are linked to a heightened susceptibility to breast and ovarian malignancies (Wang et al., 2020). These models enable prompt identification and individualized treatment strategies, leading to substantial enhancements in patient results (Du et al., 2023).

Furthermore, AI has been influential in finding genetic variations linked to uncommon medical conditions. Through the analysis of extensive genomic datasets, AI algorithms can identify uncommon genetic variations that conventional methods may overlook (Robson & Ioannidis, 2023). This ability assists in the identification and management of rare genetic illnesses. DL methods have enhanced the identification of harmful variations in genes linked to neurodevelopmental disorders, thereby offering a fresh understanding of the genetic foundation of these problems (Haga et al., 2020; Stein et al., 2022).

9.5.2.3 Hereditary Fructose Intolerance

Hereditary Fructose Intolerance (HFI) is an uncommon genetic condition resulting from mutations in the ALDOB gene, responsible for producing the enzyme aldolase B (Coffee et al., 2010). This enzyme is essential for the process of fructose metabolism, and if there is a lack of it, harmful metabolites will build up in the body when fructose is consumed. The utilization of AI and ML has been employed to augment the diagnosis and comprehension of HFI through the analysis of genetic data and the prediction of the consequences of specific mutations (Ferri et al., 2012). The latest progress in AI has made it possible to identify harmful variations in the ALDOB gene by examining sequencing data (Esposito et al., 2004). The utilization of AI-driven methods has significantly enhanced the precision of genetic testing for HFI, facilitating prompt detection and more effective treatment of the illness. For example, predictive models can anticipate the probable consequences of particular ALDOB mutations, aiding clinicians in comprehending the extent of the condition in individual patients (Adamowicz et al., 2007).

Moreover, the integration of AI with EHRs enables the detection of patients displaying symptoms that suggest HFI, even in cases when genetic testing is not readily accessible (Beyzaei et al., 2023). Natural language processing (NLP) methods can retrieve pertinent clinical data from patient records, thereby aiding in the prompt identification and treatment of HFI (Ferri et al., 2012; Gunduz et al., 2021).

9.6 AI IN HEALTHCARE AND PUBLIC HEALTH

AI is transforming healthcare and public health through its sophisticated powers in data processing, pattern recognition, and predictive analysis. This transformation has a significant impact on precision medicine and personalized healthcare. AI plays a crucial role in tailoring medicines according to individual genetic profiles and improving public health management through predictive analytics.

9.6.1 Precision Medicine and Personalized Healthcare

Precision medicine, along with personalized healthcare, aims to customize medical treatments based on individual patient characteristics, in contrast to conventional techniques that are designed to fit all patients uniformly (Sisk et al., 2020). AI is essential in facilitating this transition by allowing for the incorporation and examination of intricate data to enhance patient results and treatment approaches.

9.6.1.1 AI's Role in Precision Medicine

AI is becoming essential in precision medicine, specifically in genomics and the management of genetic illnesses. ML algorithms have the capability to analyze large quantities of genomic data in order to find genetic changes that are associated with diseases (Takei et al., 2018). This ability helps in the advancement of targeted medicines. AI systems excel in identifying mutations linked to particular types of cancer, facilitating timely detection and tailored treatment strategies. This is particularly apparent in the field of oncology, where AI assists in customizing treatments according to the genetic characteristics of tumors, hence improving effectiveness and reducing adverse effects (Girotti et al., 2016; Kurnaz & Loaiza-Bonilla, 2019). AI plays a crucial role in discovering biomarkers for different genetic illnesses. Biomarkers play a vital role in the diagnosis and monitoring of diseases such as Alzheimer's, Parkinson's, and metabolic disorders (Bahado-Singh et al., 2022).

AI has been employed to detect genetic changes in individuals diagnosed with thalassemia, a blood disorder resulting from abnormalities in the HBB gene for beta-thalassemia and the HBA1/HBA2 genes for alpha-thalassemia (Takei et al., 2018). This level of accuracy enables more efficient administration and therapy of the illness. AI can utilize genetic data to forecast patient reactions to dietary fructose in cases of HFI. HFI is a result of genetic abnormalities in the ALDOB gene, which causes a shortage of aldolase B (Petersson et al., 2023). AI can assist in the early detection of people with this ailment, enabling dietary adjustments that can prevent the occurrence of severe symptoms such as liver and kidney damage (Sisk et al., 2020).

9.6.1.2 Developing Personalized Treatment Plans

Personalized treatment plans require the integration of data from multiple sources, such as lifestyle data and genetic test results (Güvenç Paltun et al., 2021). AI systems evaluate this data to uncover patterns and connections that provide insights for personalized treatment plans. Pharmacogenomics is a notable field where AI is used to forecast the effectiveness of drugs and possible adverse effects by analyzing a patient's genetic makeup (Gkouvas, 2022). This approach minimizes the need for trial and error when administering pharmaceuticals. AI can assist in customizing blood transfusion schedules and chelation therapy for individual patients with hereditary illnesses such as thalassemia (Spaulding & Deogun, 2011). This individualized strategy enhances patient outcomes and enhances quality of life (Gifari et al., 2021). AI can provide dietary recommendations for genetic fructose intolerance that eliminates fructose, sucrose, and sorbitol. This helps prevent severe metabolic crises and improves the management of the condition (Rezayi et al., 2022).

AI-driven decision support systems are also transforming clinical workflows. These technologies offer healthcare practitioners evidence-based therapy recommendations by assessing patient data in real-time. For instance, AI can utilize patient data to effectively manage metabolic illnesses such as glycogen storage disease type III (GSDIII) (Winter & Hahn, 2020). It can improve treatment plans, track the evolution of the condition, and make necessary adjustments to medications. Adopting this proactive strategy aids in the management of symptoms and enhances patient outcomes (Schork, 2019; Wei et al., 2023).

9.6.1.3 Predictive Analytics and Case Studies

Predictive analytics use AI to predict health outcomes and identify populations that are at risk, which is crucial for early diagnosis of diseases, assessing risks, and promoting preventative healthcare (Houfani et al., 2022). AI-powered predictive models examine data from several sources to deliver precise health forecasts, facilitating prompt interventions and enhancing patient outcomes.

AI models in chronic illness management utilize pattern recognition in health data to accurately forecast the occurrence of ailments such as diabetes and cardiovascular diseases in patients who are at high risk (Hernandez & Zhang, 2017). For instance, AI can examine glucose monitor data to anticipate episodes of hypoglycemia in individuals with diabetes, enabling proactive interventions that improve patient safety and overall well-being (Mukherjee, 2019). AI has demonstrated efficacy in forecasting the advancement of diseases in the field of oncology. AI models can utilize clinical trial data and patient records to predict the course of cancer and the outcomes of treatment, enabling physicians to customize treatment programs with greater precision (Frownfelter et al., 2019; Iqbal et al., 2020). This method enhances the rates of survival and the overall quality of life for individuals diagnosed with cancer.

Moreover, AI is employed in medical imaging to augment diagnostic precision. CNNs and generative adversarial networks (GANs) are utilized in the analysis of medical pictures for the purposes of anomaly detection, disease classification, and anatomical structure segmentation (Houfani et al., 2022). These AI-driven techniques automate operations that were previously prone to human mistakes and inconsistency, thereby enhancing diagnostic accuracy and efficiency.

Another significant application of predictive analytics is in the field of public health management, where AI models forecast the propagation of infectious diseases and aid in the optimal allocation of resources (Hernandez & Zhang, 2017). During influenza outbreaks, AI may utilize data from health records and social media to forecast the transmission and intensity of the virus (Dhar, 2014). This information can then be used to inform public health initiatives, including vaccine campaigns and resource distribution. To summarize, the incorporation of AI into precision medicine and personalized healthcare signifies revolutionary progress in the field of medical science. Healthcare professionals may leverage AI's analytical capabilities to create personalized treatment regimens and employ predictive analytics to enhance patient outcomes and public health measures. The ongoing development and acceptance of AI technologies hold the potential for additional advancements in healthcare provision, ultimately improving patient care and the management of public health (Axelrod & Vogel, 2003).

9.6.2 Disease Prediction and Public Health Management

AI and ML are revolutionizing the way health systems anticipate, respond to, and manage disease outbreaks in the context of public health management and disease prediction. Public health officials can utilize AI and ML models to obtain valuable information about prospective epidemics, boost the functionality of epidemiological apps, and strengthen public health surveillance in order to prevent and manage diseases more efficiently.

9.6.2.1 AI Models for Predicting Disease Outbreaks

AI and ML models are more and more utilized to forecast disease outbreaks by examining extensive datasets from many sources, including healthcare records, social media, climatic data, and travel patterns (M. Chen et al., 2017; Shinde et al., 2022). These models utilize advanced algorithms to discover patterns and trends that human analysts may miss, enabling prompt identification and response to possible outbreaks (Abdullahi & Nitschke, 2021; Thapen et al., 2016).

During the COVID-19 epidemic, AI models such as BlueDot and HealthMap were important in forecasting the transmission of the virus. These methods utilized data from airline ticketing systems, press stories, and social media to predict the occurrence and dissemination of COVID-19, facilitating governments and health organizations in optimizing resource allocation and implementing focused interventions (Allam et al., 2020; Santosh, 2020). In addition, ML algorithms are utilized to forecast seasonal flu epidemics (Dai & Bikdash, 2016; Volkova et al., 2017). Through the examination of past flu season data, weather patterns, and vaccination rates, ML models can predict the timing and intensity of future flu seasons. This enables health systems to make early preparations by ensuring the availability of required vaccines and antiviral drugs (Rastogi & Keshtkar, 2020; Volkova et al., 2017; Wankhede et al., 2022).

9.6.2.2 Applications in Epidemiology

AI and ML boost the field of epidemiology by enhancing the capability to monitor and evaluate disease patterns across different populations. These technologies can analyze extensive volumes of epidemiological data in order to detect risk factors and patterns of disease transmission (Boman & Gillblad, 2014; Wiens & Shenoy, 2018). This is crucial for comprehending the spread of diseases and devising methods to manage them.

For example, AI and ML models are employed to simulate the spread patterns of vector-borne diseases such as malaria and dengue fever. By integrating statistics regarding mosquito populations, environmental conditions, and human mobility patterns, these models have the ability to forecast outbreaks and direct public health actions to regions with the greatest susceptibility (Hamilton et al., 2021). This aids in the optimization of the allocation of preventive measures such as pesticides and bed nets (Supriya & Chattu, 2021). In the field of cancer epidemiology, AI and ML are used to aid in the identification of risk factors and patterns linked to different types of cancer (Goldenberg et al., 2019). Through the examination of patient data, genetic information, and lifestyle factors, these models can identify populations with an elevated risk and suggest specific screening programs. By adopting this proactive strategy, there is an earlier identification and management of medical conditions, resulting in better patient results and a decrease in the overall healthcare load (Bhattamisra et al., 2023).

AI and ML significantly improve public health surveillance by offering immediate monitoring and analysis of health data, which is essential for identifying and addressing new health risks. Conventional surveillance systems frequently depend on manual reporting and data collection, which can be time-consuming and inadequate (Wiemken & Kelley, 2019). On the other hand, AI and ML systems can constantly examine data from many sources, offering prompt and precise information to public health authorities (Miller & Brown, 2018).

AI-driven platforms such as IBM's Watson Health and Google's DeepMind Health analyze health data to detect anomalies that could potentially suggest the occurrence of an outbreak (Yu et al., 2018). These systems utilize hospital records, laboratory results, and social media data to identify abnormal patterns and notify health authorities. This facilitates faster reactions to potential outbreaks and aids in controlling the spread of diseases (Ramkumar et al., 2021). Furthermore, ML techniques are employed in syndromic monitoring, a method that tracks the prevalence of symptoms among communities in order to identify developing health risks (Shameer et al., 2018). ML models can detect abnormal surges in particular symptoms, such as fever or respiratory distress, which may suggest the occurrence of an outbreak (Scott et al., 2021). This is achieved by evaluating data obtained from emergency department visits, medication sales, and web searches (Ellahham, 2020). This enables public health officials to examine and address possible hazards before they intensify (Supriya & Chattu, 2021). Ultimately, the incorporation of AI and ML in disease forecasting and public health administration is fundamentally transforming the way healthcare systems predict and address disease epidemics. AI and ML utilize sophisticated algorithms and extensive datasets to offer significant insights that boost epidemiological applications and public health surveillance, ultimately leading to improved public health outcomes and preparedness (Bhattamisra et al., 2023).

9.7 FUTURE DIRECTIONS AND ETHICAL CONSIDERATIONS

Healthcare systems could undergo a massive transformation as bioinformatics and AI advances. It is imperative to consider ethical and regulatory factors, as well as emerging trends and innovations in these disciplines, in order to ensure responsible development and execution.

9.7.1 Emerging Trends and Innovations

The fields of AI and bioinformatics are progressing quickly, propelled by enhancements in processing capacity, data accessibility, and algorithmic complexity (Goldenberg et al., 2019). A significant development is the incorporation of multi-omic data. This convergence allows for a thorough comprehension of biological processes and disease mechanisms, hence permitting more precise forecasts and tailored treatments (Tabassum et al., 2021). The emergence of digital twins is yet another revolutionary trend. These virtual patient replicas utilize real-time data to mimic and predict health outcomes, thereby optimizing treatment strategies and improving preventative care (Akhtar et al., 2021). AI-driven digital twins are utilized in Alzheimer's research to monitor cognitive deterioration and detect individuals who are susceptible to getting the disease (Goldenberg et al., 2019).

The progress in NLP is also revolutionizing the field of bioinformatics. NLP enables the extraction of significant insights from extensive unstructured biological data, including research papers and clinical notes (Sharieff & Sameer, 2023b; Welivita et al., 2018). This process enhances the speed and efficiency of knowledge discovery and decision-making in the healthcare industry. In the future, the use of AI in drug discovery will speed up the process of finding possible drugs, forecast how well they will work and how safe they are, and improve clinical trials (Sharieff & Sameer, 2023b). This can significantly diminish the duration and expenses linked to introducing novel pharmaceuticals to the market. AI will augment the accuracy of forecasts on individual reactions to treatments in customized medicine, customizing cancer therapy according to the genetic composition of tumors. In addition, the utilization of AI-driven remote monitoring tools as well as telehealth platforms would provide ongoing health monitoring and immediate interventions, hence enhancing the availability of healthcare services and enhancing patient health outcomes (Goldenberg et al., 2019).

9.7.2 Ethical and Regulatory Considerations

As AI systems manage progressively more sensitive health data, ensuring the confidentiality and safety of information becomes of utmost importance (Vuori, 1977). To safeguard patient information from breaches and illegal access, it is crucial to employ strong encryption techniques, reliable data storage systems, and strict access controls

(Sharma et al., 2023). Adhering to standards such as the GDPR and the HIPAA is of utmost importance (Murdoch, 2021). The ethical ramifications of AI in the healthcare sector are complex and have many different aspects. A significant issue revolves around the concepts of openness and explainability. Transparency is essential for AI systems since it allows healthcare professionals and patients to understand and trust the judgments made by these systems. It is crucial to develop algorithms that offer explicit and comprehensible justifications for their forecasts and suggestions (Jeyaraman et al., 2023). Furthermore, it is imperative to establish comprehensive informed consent procedures to guarantee that patients receive complete information regarding the utilization of their data and the consequences of AI-driven decisions on their healthcare (Murdoch, 2021).

Another crucial ethical aspect is the need to tackle bias and ensure fairness in AI systems. In order to avoid biases, it is crucial to train AI models using different datasets that accurately represent the population (Challen et al., 2019). In order to achieve fair healthcare outcomes, it is crucial to guarantee that underrepresented groups are sufficiently included. Regular and ongoing surveillance and assessment of AI systems are essential to identify and rectify biases, ensuring fairness and precision in predictions and recommendations (Ellahham et al., 2020). To summarize, the potential of AI and bioinformatics in healthcare is quite promising. However, harnessing this potential necessitates a meticulous examination of developing patterns, inventive uses, and ethical and legal dilemmas. Through careful consideration of these factors, AI has the potential to revolutionize the healthcare industry by providing tailored, effective, and fair treatment to every individual.

9.7.3 Conclusion

The incorporation of AI alongside ML into bioinformatics is entirely revolutionizing healthcare systems by augmenting the capacity to analyze intricate biological data, enhancing diagnostics, tailoring treatments, optimizing drug discovery, and fortifying public health management. These technologies have shown exceptional promise in different aspects of healthcare, leading to the development of creative solutions and progress. The integration of AI and ML in the field of bioinformatics has resulted in substantial discoveries and progress. The utilization of AI and ML techniques enables more advanced data analysis, enabling the processing of large quantities of biological data with higher precision and comprehensiveness. These technologies can detect patterns and connections that are frequently overlooked by conventional methods, resulting in a more profound comprehension of genetic variants and disease causes. In addition, AI-powered models have transformed the field of diagnostics by allowing for the detection of illness signs and the forecasting of disease advancement. This development has created opportunities for personalized medicine, which involves customizing treatment programs based on the genetic characteristics of each patient. As a result, this approach enhances the effectiveness of treatments and minimizes adverse side effects. AI is crucial in speeding up the process of identifying novel drug candidates and repurposing existing medications for new therapeutic purposes in the field of drug research. AI models can forecast the biological activity of compounds and enhance the process of medication design, resulting in a substantial reduction in the time and expenses associated with introducing new drugs to the market. In addition, AI and ML have become extremely important in the field of epidemiology. They offer powerful tools for predicting diseases, detecting outbreaks, and monitoring public health. These technologies provide prompt reactions to emerging health concerns and effective allocation of assets during healthcare emergencies.

The incorporation of AI and ML in the field of bioinformatics has significant implications for the future of healthcare. An important consequence is the transition towards data-driven decision-making in healthcare, resulting in more accurate and efficient treatments. AI will persistently improve its capacity to evaluate and understand intricate biological data, hence stimulating advancements in diagnostics and tailored therapy. Moreover, the automation functionalities of AI systems will optimize repetitive operations in healthcare, such as patient prioritization and administrative procedures. This would enable healthcare workers to concentrate on more intricate patient care tasks, hence enhancing efficiency and elevating the overall standard of care. With the increasing use of AI in healthcare, ethical and legal frameworks must adapt to tackle concerns around data privacy, algorithmic bias, and transparency. It is imperative to prioritize the fairness, accountability, and explainability of AI systems in order to uphold trust and attain equal healthcare results. Ongoing focus is necessary to address the ethical considerations related to the use of AI in healthcare in order to ensure responsible and beneficial application for all patients.

The potential of bioinformatics and AI in healthcare is highly promising, with continuous breakthroughs anticipated to transform the area further. Novel technologies like digital twins, multi-omics integration, and explainable AI will offer a more profound understanding of illness causes and treatment responses. The integration of bioinformatics, AI, and clinical practice will result in enhanced and tailored healthcare solutions, ultimately enhancing patient outcomes and public health. Nevertheless, it is crucial to confront the obstacles related to data accuracy, the ability to interpret models, and ethical concerns in order to fully exploit the capabilities of AI in the field of bioinformatics. Ongoing research and development, along with strong regulatory supervision, will guarantee the responsible and efficient use of these technologies. This will facilitate the emergence of a new era of precision medicine and sophisticated healthcare systems in which AI and bioinformatics collaborate synergistically to improve the quality and efficiency of healthcare services.

REFERENCES

Abdullahi, T., & Nitschke, G. (2021). Predicting Disease Outbreaks with Climate Data. *2021 IEEE Congress on Evolutionary Computation (CEC)*, 989–996. https://doi.org/10.1109/CEC45853.2021.9504740

Ackerson, J. M., Dave, R., & Seliya, J. (2021). Applications of Recurrent Neural Network for Biometric Authentication & Anomaly Detection. *ArXiv, abs/2109.05701*(7). https://doi.org/10.3390/INFO12070272

Adamowicz, M., Płoski, R., Rokicki, D., Morava, E., Gizewska, M., Mierzewska, H., Pollak, A., Lefeber, D. J., Wevers, R. A., & Pronicka, E. (2007). Transferrin Hypoglycosylation in Hereditary Fructose Intolerance: Using the Clues and Avoiding the Pitfalls. *Journal of Inherited Metabolic Disease*, *30*(3), 407. https://doi.org/10.1007/S10545-007-0569-Z

Afanasiev, O., Berghout, J., Brenner, S., Bulyk, M. L., Crawford, D. C., Chen, J. H., Daneshjou, R., & Kidzinski, L. (2021). Computational Challenges and Artificial Intelligence in Precision Medicine. *Pacific Symposium on Biocomputing. Pacific Symposium on Biocomputing*, *26*, 166–171. https://doi.org/10.1142/9789811232701_0016

Akhtar, M. N., Abbas, G., & Khan, M. (2021). AI in Bioinformatics. *International Journal of Sciences: Basic and Applied Research*, *56*(1), 301–311.

Alexander, T. A., Irizarry, R. A., & Bravo, H. C. (2023). Capturing Discrete Latent Structures: Choose LDs Over PCs. *Biostatistics*, *24*(1), 1–16. https://doi.org/10.1093/BIOSTATISTICS/KXAB030

Allam, Z., Dey, G., & Jones, D. S. (2020). Artificial Intelligence (AI) Provided Early Detection of the Coronavirus (COVID-19) in China and Will Influence Future Urban Health Policy Internationally. *AI*, *1*(2), 156–165. https://doi.org/10.3390/AI1020009

Athanasopoulou, K., Daneva, G. N., Adamopoulos, P. G., & Scorilas, A. (2022a). Artificial Intelligence: The Milestone in Modern Biomedical Research. *BioMedInformatics*, *2*(4), 727–744. https://doi.org/10.3390/BIOMEDINFORMATICS2040049

Athanasopoulou, K., Daneva, G. N., Adamopoulos, P. G., & Scorilas, A. (2022b). Artificial Intelligence: The Milestone in Modern Biomedical Research. *BioMedInformatics*, *2*(4), 727–744. https://doi.org/10.3390/BIOMEDINFORMATICS2040049

Auslander, N., Gussow, A. B., & Koonin, E. V. (2021). Incorporating Machine Learning into Established Bioinformatics Frameworks. *International Journal of Molecular Sciences*, *22*(6), 1–19. https://doi.org/10.3390/IJMS22062903

Axelrod, R. C., & Vogel, D. (2003). Predictive Modeling in Health Plans. *Disease Management & Health Outcomes*, *11*(12), 779–787. https://doi.org/10.2165/00115677-200311120-00003

Bacciu, D., Lisboa, P., Martín, J. D., Stoean, R., & Vellido, A. (2018). Bioinformatics and Medicine in the Era of Deep Learning. *The European Symposium on Artificial Neural Networks*.

Bahado-Singh, R. O., Vishweswaraiah, S., Aydas, B., Yilmaz, A., Saiyed, N. M., Mishra, N. K., Guda, C., & Radhakrishna, U. (2022). Precision Cardiovascular Medicine: Artificial Intelligence and Epigenetics for the Pathogenesis and Prediction of Coarctation in Neonates. *Journal of Maternal-Fetal & Neonatal Medicine*, *35*(3), 457–464. https://doi.org/10.1080/14767058.2020.1722995

Beyzaei, Z., Ezgu, F., Imanieh, M. H., Haghighat, M., Dehghani, S. M., Honar, N., & Geramizadeh, B. (2023). Identification of a Novel Mutation in the ALDOB Gene in Hereditary Fructose Intolerance. *Journal of Pediatric Endocrinology and Metabolism*, *36*(3), 331–334. https://doi.org/10.1515/JPEM-2022-0566

Bhattamisra, S. K., Banerjee, P., Gupta, P., Mayuren, J., Patra, S., & Candasamy, M. (2023). Artificial Intelligence in Pharmaceutical and Healthcare Research. *Big Data and Cognitive Computing*, *7*(1), 10. https://doi.org/10.3390/BDCC7010010

Bir, P., & Balas, V. E. (2020). A Review on Medical Image Analysis with Convolutional Neural Networks. *2020 IEEE International Conference on Computing, Power and Communication Technologies (GUCON)*, 870–876. https://doi.org/10.1109/GUCON48875.2020.9231203

Biswas, A., Kumari, A., Gaikwad, D. S., & Pandey, D. K. (2023). Revolutionizing Biological Science: The Synergy of Genomics in Health, Bioinformatics, Agriculture, and Artificial Intelligence. *Omics: A Journal of Integrative Biology*, *27*(12), 550–569. https://doi.org/10.1089/OMI.2023.0197

Boman, M., & Gillblad, D. (2014). Learning Machines for Computational Epidemiology. *2014 IEEE International Conference on Big Data (Big Data)*, 1–5. https://doi.org/10.1109/BIGDATA.2014.7004419

Cai, Y., Dong, Q., & Li, A. (2020). Application and Research Progress of Machine Learning in Bioinformatics. *2020 International Conference on Computer Vision, Image and Deep Learning (CVIDL)*, 369–374. https://doi.org/10.1109/CVIDL51233.2020.00-69

Challen, R., Denny, J., Pitt, M., Gompels, L., Edwards, T., & Tsaneva-Atanasova, K. (2019). Artificial Intelligence, Bias and Clinical Safety. *BMJ Quality & Safety*, 28(3), 231–237. https://doi.org/10.1136/BMJQS-2018-008370

Chen, M., Hao, Y., Hwang, K., Wang, L., & Wang, L. (2017). Disease Prediction by Machine Learning Over Big Data From Healthcare Communities. *IEEE Access*, 5, 8869–8879. https://doi.org/10.1109/ACCESS.2017.2694446

Chen, Y., & Li, J. (2021). Recurrent Neural Networks Algorithms and Applications. *2021 2nd International Conference on Big Data & Artificial Intelligence & Software Engineering (ICBASE)*, 38–43. https://doi.org/10.1109/ICBASE53849.2021.00015

Cheng, J., Tegge, A. N., & Baldi, P. (2008). Machine Learning Methods for Protein Structure Prediction. *IEEE Reviews in Biomedical Engineering*, 1, 41–49. https://doi.org/10.1109/RBME.2008.2008239

Choon, A., & Gilbert, D. (2003). *An Empirical Comparison of Supervised Machine Learning Techniques in Bioinformatics*. http://www.cs.waikato.ac.nz/~ml/weka/

Coffee, E. M., Yerkes, L., Ewen, E. P., Zee, T., & Tolan, D. R. (2010). Increased Prevalence of Mutant Null Alleles that Cause Hereditary Fructose Intolerance in the American Population. *Journal of Inherited Metabolic Disease*, 33(1), 33–42. https://doi.org/10.1007/S10545-009-9008-7

Dai, X., & Bikdash, M. (2016). Distance-Based Outliers Method for Detecting Disease Outbreaks Using Social Media. *SoutheastCon 2016*, 2016-July, 1–8. https://doi.org/10.1109/SECON.2016.7506752

Demetci, P., Cheng, W., Darnell, G., Zhou, X., Ramachandran, S., & Crawford, L. (2020). Multi-Scale Inference of Genetic Trait Architecture Using Biologically Annotated Neural Networks. *BioRxiv*. https://doi.org/10.1101/2020.07.02.184465

Dhar, V. (2014). Big Data and Predictive Analytics in Health Care. *Big Data*, 2(3), 113–116. https://doi.org/10.1089/BIG.2014.1525

Dittman, D. J., Khoshgoftaar, T. M., Wald, R., & Napolitano, A. (2013). Simplifying the Utilization of Machine Learning Techniques for Bioinformatics. *2013 12th International Conference on Machine Learning and Applications*, 2, 396–403. https://doi.org/10.1109/ICMLA.2013.155

Dong, H., & Pan, J. (2020). Cascaded Dimensionality Reduction Method and Its Application in Spectral Classification. *Journal of Physics: Conference Series*, 1624(3), 32017. https://doi.org/10.1088/1742-6596/1624/3/032017

Dopazo, J. (2006). Bioinformatics and Cancer: An Essential Alliance. *Clinical and Translational Oncology*, 8(6), 409–415. https://doi.org/10.1007/S12094-006-0194-6

Du, D., Zhong, F., & Liu, L. (2023). Enhancing Recognition and Interpretation of Functional Phenotypic Sequences through Fine-Tuning Pre-Trained Genomic Models. *BioRxiv*. https://doi.org/10.1101/2023.12.05.570173

Ellahham, S. (2020). Artificial Intelligence in Diabetes Care. *The American Journal of Medicine*, 133(8), 895–900. https://doi.org/10.1016/J.AMJMED.2020.03.033

Ellahham, S., Ellahham, N., & Simsekler, M. C. E. (2020). Application of Artificial Intelligence in the Health Care Safety Context: Opportunities and Challenges. *American Journal of Medical Quality*, 35(4), 341–348. https://doi.org/10.1177/1062860619878515

Esposito, G., Santamaria, R., Vitagliano, L., Ieno, L., Viola, A., Fiori, L., Parenti, G., Zancan, L., Zagari, A., & Salvatore, F. (2004). Six Novel Alleles Identified in Italian Hereditary Fructose Intolerance Patients Enlarge the Mutation Spectrum of the Aldolase B Gene. *Human Mutation*, *24*(6), 534. https://doi.org/10.1002/HUMU.9290

Ezziane, Z. (2006). Applications of Artificial Intelligence in Bioinformatics: A Review. *Expert Systems Applications*, *30*(1), 2–10. https://doi.org/10.1016/J.ESWA.2005.09.042

Ferri, L., Caciotti, A., Cavicchi, C., Rigoldi, M., Parini, R., Caserta, M., Chibbaro, G., Gasperini, S., Procopio, E., Donati, M. A., Guerrini, R., & Morrone, A. (2012). Integration of PCR-Sequencing Analysis with Multiplex Ligation-Dependent Probe Amplification for Diagnosis of Hereditary Fructose Intolerance. *JIMD Reports*, *6*, 31–37. https://doi.org/10.1007/8904_2012_125

Frownfelter, J., Blau, S., Page, R. D., Showalter, J., Miller, K., Kish, J., Valley, A. W., & Nabhan, C. (2019). Artificial Intelligence (AI) to Improve Patient Outcomes in Community Oncology Practices. *Journal of Clinical Oncology*, *37*(15_suppl), e18098–e18098. https://doi.org/10.1200/JCO.2019.37.15_SUPPL.E18098

Gauthier, J., Vincent, A. T., Charette, S. J., & Derome, N. (2019a). A Brief History of Bioinformatics. *Briefings in Bioinformatics*, *20*(6), 1981–1996. https://doi.org/10.1093/BIB/BBY063

Gauthier, J., Vincent, A. T., Charette, S. J., & Derome, N. (2019b). A Brief History of Bioinformatics. *Briefings in Bioinformatics*, *20*(6), 1981–1996. https://doi.org/10.1093/BIB/BBY063

Ghosal, P., Reddy, S., Sai, C., Pandey, V., Chakraborty, J., & Nandi, D. (2019). A Deep Adaptive Convolutional Network for Brain Tumor Segmentation from Multimodal MR Images. *TENCON 2019 – 2019 IEEE Region 10 Conference (TENCON)*, *2019-October*, 1065–1070. https://doi.org/10.1109/TENCON.2019.8929402

Gifari, M. W., Samodro, P., & Kurniawan, D. (2021). Artificial Intelligence Toward Personalized Medicine. *Pharmaceutical Sciences and Research*, *8*(2), 1. https://doi.org/10.7454/PSR.V8I2.1199

Girotti, M. R., Gremel, G., Lee, R., Galvani, E., Rothwell, D., Viros, A., Mandal, A. K., Lim, K. H. J., Saturno, G., Furney, S. J., Baenke, F., Pedersen, M., Rogan, J., Swan, J., Smith, M., Fusi, A., Oudit, D., Dhomen, N., Brady, G., … Marais, R. (2016). Application of Sequencing, Liquid Biopsies, and Patient-Derived Xenografts for Personalized Medicine in Melanoma. *Cancer Discovery*, *6 3*(3), 286–299. https://doi.org/10.1158/2159-8290.CD-15-1336

Giudice, M. Del, Peirone, S., Perrone, S., Priante, F., Varese, F., Tirtei, E., Fagioli, F., & Cereda, M. (2021). Artificial Intelligence in Bulk and Single-Cell RNA-Sequencing Data to Foster Precision Oncology. *International Journal of Molecular Sciences*, *22*(9), 4563. https://doi.org/10.3390/IJMS22094563

Gkouvas, N. (2022). Precision Medicine & Pharmacogenomics: Personalized Medication in Neuropsychiatric Disorders using AI and Telepsychiatry. *European Psychiatry*, *65*(S1), 678–678. https://doi.org/10.1192/J.EURPSY.2022.1745

Goldenberg, S. L., Nir, G., & Salcudean, S. E. (2019). A New Era: Artificial Intelligence and Machine Learning in Prostate Cancer. *Nature Reviews Urology*, *16*(7), 391–403. https://doi.org/10.1038/S41585-019-0193-3

Gunduz, M., Ünal-Uzun, Ö., Koç, N., Ceylaner, S., Özaydln, E., & Kasapkara, Ç. S. (2021). Molecular and Clinical Findings of Turkish Patients with Hereditary Fructose Intolerance. *Journal of Pediatric Endocrinology and Metabolism*, *34*(8), 1017–1022. https://doi.org/10.1515/JPEM-2021-0303

Güvenç Paltun, B., Mamitsuka, H., & Kaski, S. (2021). Improving Drug Response Prediction by Integrating Multiple Data Sources: Matrix Factorization, Kernel and Network-Based

Approaches. *Briefings in Bioinformatics*, 22(1), 346–359. https://doi.org/10.1093/BIB/BBZ153

Haga, H., Sato, H., Koseki, A., Saito, T., Okumoto, K., Hoshikawa, K., Katsumi, T., Mizuno, K., Nishina, T., & Ueno, Y. (2020). A Machine Learning-Based Treatment Prediction Model Using Whole Genome Variants of Hepatitis C Virus. *PLoS ONE*, 15(11 November), e0242028. https://doi.org/10.1371/JOURNAL.PONE.0242028

Hamilton, A. J., Strauss, A. T., Martinez, D. A., Hinson, J. S., Levin, S., Lin, G., & Klein, E. Y. (2021). Machine Learning and Artificial Intelligence: Applications in Healthcare Epidemiology. *Antimicrobial Stewardship & Healthcare Epidemiology: ASHE*, 1(1), e28. https://doi.org/10.1017/ASH.2021.192

Hernandez, I., & Zhang, Y. (2017). Using Predictive Analytics and Big Data to Optimize Pharmaceutical Outcomes. *American Journal of Health-System Pharmacy*, 74 18(18), 1494–1500. https://doi.org/10.2146/AJHP161011

Ho, L. V., Aczon, M., Ledbetter, D., & Wetzel, R. (2021). Interpreting a Recurrent Neural Network's Predictions of ICU Mortality Risk. *Journal of Biomedical Informatics*, 114, 103672. https://doi.org/10.1016/J.JBI.2021.103672

Houfani, D., Slatnia, S., Kazar, O., Saouli, H., & Merizig, A. (2022). Artificial Intelligence in Healthcare: A Review on Predicting Clinical Needs. *International Journal of Healthcare Management*, 15(3), 267–275. https://doi.org/10.1080/20479700.2021.1886478

Iqbal, U., Celi, L. A., & Li, Y. C. J. (2020). How Can Artificial Intelligence Make Medicine More Preemptive? *Journal of Medical Internet Research*, 22(8). https://doi.org/10.2196/17211

Jena, M. K., & Pathak, B. (2023). Development of an Artificially Intelligent Nanopore for High-Throughput DNA Sequencing with a Machine-Learning-Aided Quantum-Tunneling Approach. *Nano Letters*, 23(7), 2511–2521. https://doi.org/10.1021/ACS.NANOLETT.2C04062

Jeyaraj, P., & Narayanan, T. S. (2023). Role of Artificial Intelligence in Enhancing Healthcare Delivery. *International Journal of Innovative Science and Modern Engineering*, 11(12), 1–13. https://doi.org/10.35940/IJISME.A1310.12111223

Jeyaraman, M., Balaji, S., Jeyaraman, N., & Yadav, S. (2023). Unraveling the Ethical Enigma: Artificial Intelligence in Healthcare. *Cureus*, 15(8), e43262. https://doi.org/10.7759/CUREUS.43262

Jiang, F., Jiang, Y., Zhi, H., Dong, Y., Li, H., Ma, S., Wang, Y., Dong, Q., Shen, H., & Wang, Y. (2017). Artificial Intelligence in Healthcare: Past, Present and Future. *Stroke and Vascular Neurology*, 2(4), 230–243. https://doi.org/10.1136/SVN-2017-000101

Jindal, S., Marriwala, D. N. K., Sharma, A., & Bhatia, R. (2021). Methodological Analysis with Informative Science in Bioinformatics. In Nikhil Marriwala, C. C. Tripathi, & Shruti Jain (Eds.), *Soft Computing for Intelligent Systems* (pp. 49–57). Springer.. https://doi.org/10.1007/978-981-16-1048-6_5

Kacew, A. J., Strohbehn, G. W., Saulsberry, L., Laiteerapong, N., Cipriani, N. A., Kather, J. N., & Pearson, A. T. (2021). Artificial Intelligence Can Cut Costs While Maintaining Accuracy in Colorectal Cancer Genotyping. *Frontiers in Oncology*, 11, 630953. https://doi.org/10.3389/FONC.2021.630953/PDF

Karim, Md. R., Islam, T., Beyan, O., Lange, C., Cochez, M., Rebholz-Schuhmann, D., & Decker, S. (n.d.). Explainable AI for Bioinformatics: Methods, Tools, and Applications. *Briefings in Bioinformatics, 24*, bbad236. https://doi.org/10.48550/ARXIV.2212.13261

Kaur, I., Kumar, Y., & Sandhu, A. K. (2021). A Comprehensive Survey of AI, Blockchain Technology and Big Data Applications in Medical Field and Global Health. *2021 International Conference on Technological Advancements and Innovations (ICTAI)*, 593–598. https://doi.org/10.1109/ICTAI53825.2021.9673285

Kosvyra, A., Maramis, C., & Chouvarda, I. (2020). A Data-Driven Approach to Build a Predictive Model of Cancer Patients' Disease Outcome by Utilizing Co-Expression Networks. *Computers in Biology and Medicine, 125*, 103971. https://doi.org/10.1016/J.COMPBIOMED.2020.103971

Kurnaz, S., & Loaiza-Bonilla, A. (2019). SYNERGY-AI: Artificial Intelligence-Based Precision Oncology Clinical Trial Matching and Registry. *Journal of Global Oncology, 5*(suppl), 22–22. https://doi.org/10.1200/JGO.2019.5.SUPPL.22

Lai, K., Twine, N., O'Brien, A., Guo, Y., & Bauer, D. (2019). Artificial Intelligence and Machine Learning in Bioinformatics. *Encyclopedia of Bioinformatics and Computational Biology, 1–3*, 272–286. https://doi.org/10.1016/B978-0-12-809633-8.20325-7

Lalapura, V. S., Amudha, J., & Satheesh, H. S. (2021). Recurrent Neural Networks for Edge Intelligence. *ACM Computing Surveys (CSUR), 54*(4), 1–38. https://doi.org/10.1145/3448974

Larrañaga, P., Calvo, B., Santana, R., Bielza, C., Galdiano, J., Inza, I., Lozano, J. A., Armañanzas, R., Santafé, G., Pérez, A., & Robles, V. (March 2006). Machine Learning in Bioinformatics. *Briefings in Bioinformatics, 7*, 86–112. https://doi.org/10.1093/bib/bbk007

Ledbetter, D. R., Laksana, E., Aczon, M., & Wetzel, R. (2021). Improving Recurrent Neural Network Responsiveness to Acute Clinical Events. *IEEE Access, 9*, 106140–106151. https://doi.org/10.1109/ACCESS.2021.3099996

Liu, C., Liu, X., Shangguan, H., Wen, S., & Zheng, F. (2023). Review on the Application of Artificial Intelligence in Bioinformatics. *Highlights in Science, Engineering and Technology, 30*, 209–214. https://doi.org/10.54097/HSET.V30I.4978

Liu, Y. (2004). Active Learning with Support Vector Machine Applied to Gene Expression Data for Cancer Classification. *Journal of Chemical Information and Computer Sciences, 44* 6(6), 1936–1941. https://doi.org/10.1021/CI049810A

Lysaght, T., Lim, H. Y., Xafis, V., & Ngiam, K. Y. (2019). AI-Assisted Decision-making in Healthcare. *Asian Bioethics Review, 11*(3), 299–314. https://doi.org/10.1007/S41649-019-00096-0

Masulli, F., & Tagliaferri, R. (2011). Advances in Computational Intelligence and Bioinformatics. *Soft Computing, 15*(8), 1457–1458. https://doi.org/10.1007/S00500-010-0595-X

Mehta, V. (2023). Artificial Intelligence in Medicine: Revolutionizing Healthcare for Improved Patient Outcomes. *Journal of Medical Research and Innovation, 7*(2), e000292. https://doi.org/10.32892/JMRI.292

Miller, D. D., & Brown, E. W. (2018). Artificial Intelligence in Medical Practice: The Question to the Answer? *American Journal of Medicine, 131* 2(2), 129–133. https://doi.org/10.1016/J.AMJMED.2017.10.035

Min, S., Lee, B., & Yoon, S. (2017). Deep Learning in Bioinformatics. *Briefings in Bioinformatics, 18*(5), 851–869. https://doi.org/10.1093/BIB/BBW068

Mitchell, J. B. O. (2014). Machine Learning Methods in Chemoinformatics. *Wiley Interdisciplinary Reviews. Computational Molecular Science, 4*(5), 468–481. https://doi.org/10.1002/WCMS.1183

Mohammed, J., & Mohammed, J. (2014). *Machine Learning in Bioinformatics.* Springer. https://doi.org/10.1007/SPRINGERREFERENCE_65255

Mukherjee, S. (2019). Predictive Analytics and Predictive Modeling in Healthcare. *Demand & Supply in Health Economics.* https://doi.org/10.2139/SSRN.3403900

Mumtaz, H., Saqib, M., Jabeen, S., Muneeb, M., Mughal, W., Sohail, H., Safdar, M., Mehmood, Q., Khan, M. A., & Ismail, S. M. (2023). Exploring Alternative Approaches to Precision Medicine Through Genomics and Artificial Intelligence – A Systematic

Review. *Frontiers in Medicine, 10*, 1227168. https://doi.org/10.3389/FMED.2023.1227168/PDF

Murdoch, B. (2021). Privacy and Artificial Intelligence: Challenges for Protecting Health Information in a New Era. *BMC Medical Ethics, 22*(1), 1–5. https://doi.org/10.1186/S12910-021-00687-3

Mushegian, A. (2011). Grand Challenges in Bioinformatics and Computational Biology. *Frontiers in Genetics, 2*(SEP), 60. https://doi.org/10.3389/FGENE.2011.00060/PDF

Muzio, G., O'Bray, L., & Borgwardt, K. (2021). Biological Network Analysis with Deep Learning. *Briefings in Bioinformatics, 22*(2), 1515–1530. https://doi.org/10.1093/BIB/BBAA257

Narayanan, A., Keedwell, E., & Olsson, B. (2002a). Artificial Intelligence Techniques for Bioinformatics. *Applied Bioinformatics, 1*, 191–222.

Neves, G. F., Chaudron, J. B., & Dion, A. (2021). Recurrent Neural Networks Analysis for Embedded Systems. *International Joint Conference on Computational Intelligence, 1*, 374–383. https://doi.org/10.5220/0010715700003063

Noorbakhsh-Sabet, N., Zand, R., Zhang, Y., & Abedi, V. (2019). Artificial Intelligence Transforms the Future of Health Care. *American Journal of Medicine, 132*(7), 795–801. https://doi.org/10.1016/J.AMJMED.2019.01.017

Ouzounis, C. A. (2012). Rise and Demise of Bioinformatics? Promise and Progress. *PLoS Computational Biology, 8*(4), e1002487. https://doi.org/10.1371/JOURNAL.PCBI.1002487

Parums, D. V. (2023). Editorial: Infectious Disease Surveillance Using Artificial Intelligence (AI) and its Role in Epidemic and Pandemic Preparedness. *Medical Science Monitor, 29*, 941209–1. https://doi.org/10.12659/MSM.941209

Pavan Kumar, I., Mahaveerakannan, R., Praveen Kumar, K., Basu, I., Anil Kumar, T. C., & Choche, M. (2022). A Design of Disease Diagnosis based Smart Healthcare Model using Deep Learning Technique. *2022 International Conference on Electronics and Renewable Systems (ICEARS)*, 1444–1449. https://doi.org/10.1109/ICEARS53579.2022.9752063

Perumal, S., Tabassum, M., Sharma, M., & Mohanan, S. (2022). Next Generation Communication Networks for Industrial Internet of Things Systems. In S. Perumal, M. Tabassum, M. Sharma, & S. Mohanan (Eds.), *Next Generation Communication Networks for Industrial Internet of Things Systems*. CRC Press. https://doi.org/10.1201/9781003355946

Petersson, L., Svedberg, P., Nygren, J. M., & Larsson, I. (2023). Healthcare Leaders' Perceptions of the Usefulness of AI Applications in Clinical Work: A Qualitative Study. *Studies in Health Technology and Informatics, 302*, 678–679. https://doi.org/10.3233/SHTI230235

Puaschunder, J. M., & Feierabend, D. (2019). Artificial Intelligence in the Healthcare Sector. *Scientia Moralitas-International Journal of Multidisciplinary Research, 4*(2), 1–4. https://doi.org/10.2139/SSRN.3469423

Qiu, J., Li, L., Sun, J., Peng, J., Shi, P., Zhang, R., Dong, Y., Lam, K., Lo, F. P. W., Xiao, B., Yuan, W., Wang, N., Xu, D., & Lo, B. (2023). Large AI Models in Health Informatics: Applications, Challenges, and the Future. *IEEE Journal of Biomedical and Health Informatics, 27*(12), 6074–6087. https://doi.org/10.1109/JBHI.2023.3316750

Rahman, A., Chang, Y., & Rubin, J. (2021). Interpretable Additive Recurrent Neural Networks For Multivariate Clinical Time Series. *ArXiv.Org*.

Rahman, S. Z., Senthil, R., Ramalingam, V., & Gopal, R. (2023). Predicting Infectious Disease Outbreaks with Machine Learning and Epidemiological Data. *Journal of Advanced Zoology, 44*(S4), 110–121. https://doi.org/10.17762/JAZ.V44IS4.2177

Ramkumar, P. N., Kunze, K. N., Haeberle, H. S., Karnuta, J. M., Luu, B. C., Nwachukwu, B. U., & Williams, R. J. (2021). Clinical and Research Medical Applications of Artificial Intelligence. *Arthroscopy*, *37*(5), 1694–1697. https://doi.org/10.1016/J.ARTHRO.2020.08.009

Rana, V. (2019). Role of Artificial Intelligence in Bioinformatics. *Indian Journal of Pure & Applied Biosciences*, *7*(6), 317–321. https://doi.org/10.18782/2582-2845.7918

Rastogi, N., & Keshtkar, F. (2020). Using BERT and Semantic Patterns to Analyze Disease Outbreak Context over Social Network Data. *HEALTHINF 2020 – 13th International Conference on Health Informatics, Proceedings; Part of 13th International Joint Conference on Biomedical Engineering Systems and Technologies, BIOSTEC 2020*, 854–863. https://doi.org/10.5220/0009375908540863

Rezayi, S., R Niakan Kalhori, S., & Saeedi, S. (2022). Effectiveness of Artificial Intelligence for Personalized Medicine in Neoplasms: A Systematic Review. *BioMed Research International*, *2022*, 7842556. https://doi.org/10.1155/2022/7842566

Robson, E. S., & Ioannidis, N. M. (2023). GUANinE v1.0: Benchmark Datasets for Genomic AI Sequence-to-Function Models. *BioRxiv*. https://doi.org/10.1101/2023.10.12.562113

Saifi, I., Bhat, B. A., Hamdani, S. S., Bhat, U. Y., Lobato-Tapia, C. A., Mir, M. A., Dar, T. U. H., & Ganie, S. A. (2023). Artificial Intelligence and Cheminformatics Tools: a Contribution to the Drug Development and Chemical Science. *Journal of Biomolecular Structure and Dynamics*, *42*, 6523–41. https://doi.org/10.1080/07391102.2023.2234039

Santosh, K. C. (2020). AI-Driven Tools for Coronavirus Outbreak: Need of Active Learning and Cross-Population Train/Test Models on Multitudinal/Multimodal Data. *Journal of Medical Systems*, *44*(5), 93. https://doi.org/10.1007/S10916-020-01562-1

Sarvamangala, D. R., & Kulkarni, R. V. (2022a). Convolutional Neural Networks in Medical Image Understanding: A Survey. *Evolutionary Intelligence*, *15*(1), 1–22. https://doi.org/10.1007/S12065-020-00540-3

Sarvamangala, D. R., & Kulkarni, R. V. (2022b). Convolutional Neural Networks in Medical Image Understanding: A Survey. *Evolutionary Intelligence*, *15*(1), 1–22. https://doi.org/10.1007/S12065-020-00540-3

Schork, N. J. (2019). Artificial Intelligence and Personalized Medicine. *Cancer Treatment and Research*, *178*, 265–283. https://doi.org/10.1007/978-3-030-16391-4_11

Scott, I., Cook, D., & Coiera, E. (2021). Evidence-Based Medicine and Machine Learning: A Partnership with a Common Purpose. *BMJ Evidence-Based Medicine*, *26*(6), 290–294. https://doi.org/10.1136/BMJEBM-2020-111379

Shameer, K., Johnson, K. W., Glicksberg, B. S., Dudley, J. T., & Sengupta, P. P. (2018). Machine Learning in Cardiovascular Medicine: Are We There Yet? *Heart*, *104*(14), 1156–1164. https://doi.org/10.1136/HEARTJNL-2017-311198

Sharieff, A. A., & Sameer, R. (2023a). Artificial Intelligence Techniques in Bioinformatics: Unravelling Complex Biological Systems. *International Journal of Advanced Research in Science, Communication and Technology*, *3*(1), 269–275. https://doi.org/10.48175/IJARSCT-14033

Sharieff, A. A., & Sameer, R. (2023b). Artificial Intelligence Techniques in Bioinformatics: Unravelling Complex Biological Systems. *International Journal of Advanced Research in Science, Communication and Technology*, *3*(1), 269–275. https://doi.org/10.48175/IJARSCT-14033

Sharma, M., Deswal, S., Gupta, U., Tabassum, M., & Lawal, I. A. (2023). Soft Computing Techniques in Connected Healthcare Systems. In M. Sharma, S. Deswal, U. Gupta, M. Tabassum, & I. A. Lawal (Eds.), *Soft Computing Techniques in Connected Healthcare Systems*. CRC Press. https://doi.org/10.1201/9781003405368

Shastry, K. A., & Sanjay, H. A. (2020). Machine Learning for Bioinformatics. In K. G. Srinivasa, G. M. Siddesh, & S. R. Manisekhar (Eds.), *Statistical Modelling and Machine Learning Principles for Bioinformatics Techniques, Tools, and Applications* (pp. 25–39). Springer. https://doi.org/10.1007/978-981-15-2445-5_3

Shinde, S., Yadav, S., & Somvanshi, A. (2022). Epidemic Outbreak Prediction Using Machine Learning Model. *2022 5th International Conference on Advances in Science and Technology (ICAST)*, 127–132. https://doi.org/10.1109/ICAST55766.2022.10039594

Silva, M., Pratas, D., & Pinho, A. J. (2020). Efficient DNA Sequence Compression with Neural Networks. *GigaScience, 9*(11), 1–15. https://doi.org/10.1093/GIGASCIENCE/GIAA119

Sisk, B. A., Antes, A. L., Burrous, S., & Dubois, J. M. (2020). Parental Attitudes toward Artificial Intelligence-Driven Precision Medicine Technologies in Pediatric Healthcare. *Children, 7*(9), 145. https://doi.org/10.3390/CHILDREN7090145

Spaulding, W., & Deogun, J. (2011). A Pathway to Personalization of Integrated Treatment: Informatics and Decision Science in Psychiatric Rehabilitation. *Schizophrenia Bulletin, 37*(SUPPL. 2), 129–137. https://doi.org/10.1093/SCHBUL/SBR080

Srinivasu, P. N., Shafi, J., Krishna, T. B., Sujatha, C. N., Praveen, S. P., & Ijaz, M. F. (2022). Using Recurrent Neural Networks for Predicting Type-2 Diabetes from Genomic and Tabular Data. *Diagnostics, 12*(12), 3067. https://doi.org/10.3390/DIAGNOSTICS12123067

Stein, B. Van, Raponi, E., Sadeghi, Z., Bouman, N., Van Ham, R. C. H. J., & Back, T. (2022). A Comparison of Global Sensitivity Analysis Methods for Explainable AI With an Application in Genomic Prediction. *IEEE Access, 10*, 103364–103381. https://doi.org/10.1109/ACCESS.2022.3210175

Supriya, M., & Chattu, V. K. (2021). A Review of Artificial Intelligence, Big Data, and Blockchain Technology Applications in Medicine and Global Health. *Big Data on Cognitive Computation, 5*(3), 41. https://doi.org/10.3390/BDCC5030041

Syrowatka, A., Kuznetsova, M., Alsubai, A., Beckman, A. L., Bain, P. A., Craig, K. J. T., Hu, J., Jackson, G. P., Rhee, K., & Bates, D. W. (2021). Leveraging Artificial Intelligence for Pandemic Preparedness and Response: A Scoping Review to Identify Key Use Cases. *NPJ Digital Medicine, 4*(1), 96. https://doi.org/10.1038/S41746-021-00459-8

Tabassum, M., Perumal, S., Afrouzi, H. N., Abdul Kashem, S. Bin, & Hassan, W. (2021). Review on Using Artificial Intelligence Related Deep Learning Techniques in Gaming and Recent Networks. In V. Chaudhary, M. Sharma, P. Sharma, & D. Agarwal (Eds.), *Deep Learning in Gaming and Animations* (pp. 65–90). CRC Press. https://doi.org/10.1201/9781003231530-4

Tabassum, M., Perumal, S., Mohanan, S., Suresh, P., Cheriyan, S., & Hassan, W. (2021). *IoT, IR 4.0, and AI Technology Usability and Future Trend Demands* (pp. 109–144). https://doi.org/10.4018/978-1-7998-4610-9.ch006

Takei, T., Yokoyama, K., Yusa, N., Nakamura, S., Ogawa, M., Kondoh, K., Kobayashi, M., Kobayashi, A., Ito, M., Shimizu, E., Yamamoto, M., Kasajima, R., Yuji, K., Yamaguchi, R., Furukawa, Y., Imoto, S., Miyano, S., & Tojo, A. (2018). Artificial Intelligence Guided Precision Medicine Approach to Hematological Disease. *Blood, 132*(Supplement 1), 2254–2254. https://doi.org/10.1182/BLOOD-2018-99-117941

Terranova, N., Venkatakrishnan, K., & Benincosa, L. J. (2021). Application of Machine Learning in Translational Medicine: Current Status and Future Opportunities. *AAPS Journal, 23*(4), 74. https://doi.org/10.1208/S12248-021-00593-X

Thapen, N., Simmie, D., Hankin, C., & Gillard, J. (2016). DEFENDER: Detecting and Forecasting Epidemics Using Novel Data-Analytics for Enhanced Response. *PLoS ONE, 11*(5), e0155417. https://doi.org/10.1371/JOURNAL.PONE.0155417

Volkova, S., Ayton, E., Porterfield, K., & Corley, C. D. (2017). Forecasting Influenza-Like Illness Dynamics for Military Populations Using Neural Networks and Social Media. *PLoS ONE, 12*(12), e0188941. https://doi.org/10.1371/JOURNAL.PONE.0188941

Vuori, H. (1977). Privacy, Confidentiality and Automated Health Information Systems. *Journal of Medical Ethics, 3*(4), 174–178. https://doi.org/10.1136/JME.3.4.174

Wang, H., Wang, T., Zhao, X., Wu, H., You, M., Sun, Z., & Mao, F. (2020). AI-Driver: An Ensemble Method for Identifying Driver Mutations in Personal Cancer Genomes. *NAR Genomics and Bioinformatics, 2*(4), lqaa084. https://doi.org/10.1093/NARGAB/LQAA084

Wankhede, Mrs. D. S., Sadawarte, R. R., Mulla, M. I., & Jadhav, S. R. (2022). An Analysis of Methods for Forecasting Epidemic Disease Outbreaks using Information from Social Media. *International Journal of Recent Technology and Engineering (IJRTE), 11*(2), 128–137. https://doi.org/10.35940/IJRTE.B7160.0711222

Wei, M. Y. K., Zhang, J., Schmidt, R., Miller, A. S., & Yeung, J. M. C. (2023). Artificial intelligence (AI) in the management of colorectal cancer: on the horizon? *ANZ Journal of Surgery, 93*(9), 2052–2053. https://doi.org/10.1111/ANS.18504

Welivita, A., Perera, I., Meedeniya, D., Wickramarachchi, A., & Mallawaarachchi, V. (2018). Managing Complex Workflows in Bioinformatics: An Interactive Toolkit With GPU Acceleration. *IEEE Transactions on NanoBioscience, 17*(3), 199–208. https://doi.org/10.1109/TNB.2018.2837122

Wiemken, T. L., & Kelley, R. R. (2019). Machine Learning in Epidemiology and Health Outcomes Research. *Annual Review of Public Health, 41*, 21–36. https://doi.org/10.1146/ANNUREV-PUBLHEALTH-040119-094437

Wiens, J., & Shenoy, E. S. (2018). Machine Learning for Healthcare: On the Verge of a Major Shift in Healthcare Epidemiology. *Clinical Infectious Diseases, 66*(1), 149–153. https://doi.org/10.1093/CID/CIX731

Winter, N. R., & Hahn, T. (2020). [Big Data, AI and Machine Learning for Precision Psychiatry: How are they changing the clinical practice?]. *Fortschritte Der Neurologie-Psychiatrie, 88*(12), 786–793. https://doi.org/10.1055/A-1234-6247

Xu, Y., Zhu, Y., & Chen, Z. (2021). Research on Lung Medical Image based on Convolution Neural Network Algorithm. *Journal of Physics: Conference Series, 1802*(3), 32111. https://doi.org/10.1088/1742-6596/1802/3/032111

Yang, Y., Sun, H., Zhang, Y., Zhang, T., Gong, J., Wei, Y., Duan, Y. G., Shu, M., Yang, Y., Wu, D., & Yu, D. (2021). Dimensionality Reduction by UMAP Reinforces Sample Heterogeneity Analysis in Bulk Transcriptomic Data. *Cell Reports, 36*(4), 109442https://doi.org/10.1016/J.CELREP.2021.109442

Yu, K. H., Beam, A. L., & Kohane, I. S. (2018). Artificial intelligence in healthcare. *Nature Biomedical Engineering 2018 2:10, 2*(10), 719–731. https://doi.org/10.1038/s41551-018-0305-z

Zhang, X. M., Liang, L., Liu, L., & Tang, M. J. (2021). Graph Neural Networks and Their Current Applications in Bioinformatics. *Frontiers in Genetics, 12* Article 690049. https://doi.org/10.3389/FGENE.2021.690049/PDF

10 Revolutionizing Drug Development

The Role of AI in Modern Pharmaceutical Research

Samridhi Agarwal and Amit Kumar Dutta

10.1 INTRODUCTION

The advancements in drug discovery have significantly transformed the field of medical practices, turning formerly fatal illnesses into routine therapeutic procedures that can be treated easily. One contributing factor for this progress in medicine and medical practices is the improvement in the techniques for developing and testing new medications. Typically, a novel drug does not replicate the chemical structure of an existing one, thus requiring the identification of a fresh molecule. One of the most recent advancements in this field is the utilization of artificial intelligence (AI), which has contributed significantly to the process of discovering and development new drugs (Katzung et al., 2019; Singh, et al., 2023). AI refers to scenarios where machines can mimic human cognitive processes such as learning and analysis which enables it to solve problems. This form of intelligence is often termed as machine learning (ML). AI is also referred to as an "intelligent agent," which typically refers to any agent or device that can sense and comprehend its environment and subsequently take suitable actions to optimize the chance of attaining its goals (Rong, et al., 2020). ML, a popular type of AI, can be defined as the capacity of computers to learn by using different algorithms in order to extract features from data. ML algorithms enhance performance by automating the process of model creation, enabling the identification of patterns or providing decision support based on the analyzed data (Tekkeşin, 2019).

There are various steps in the process of drug discovery and development, including: Identifying the disease and its required medical needs, selecting a suitable molecular target for drug action, and validating it, creating in vitro assays using high-throughput screening libraries to discover initial hits against the target which are then optimized to develop lead compounds with sufficient potency and selectivity under in vitro condition, as well as its effectiveness in animal-based models. Further, the entire process requires preclinical trials and clinical trials on model organisms and then further trials on humans with FDA approval and postmarket surveillance reporting (Paul et al., 2010). The process of discovering a drug is a very complex process that is a very time consuming and laborious. AI technologies, particularly natural language processing and ML, offer opportunities to expedite and amplify the drug discovery process by enabling more precise and efficient analysis of large datasets. Deep learning (DL) has demonstrated to be effective in accurately predicting and identifying the effectiveness of medicinal compounds, suggesting a significant advancement in this domain (Xu et al., 2021). From the identification of a new chemical entity to prescribing it for use in market the entire process takes around 10-15 years with a minimum expenditure of USD 2.6 billion (Sarkar et al., 2023).

Numerous AI tools have been developed over the past few years, which have entirely changed the drug discovery process. Some of these tools include Alphafold, DeepChem, DeepBar, AtomNet, etc.

Today, computational drug design is used not only in academia but also in several pharmaceutical firms for drug lead discovery. Various docking programs have been created to see the arrangement of structure of molecules in 3D space. The analysis of docking scores (i.e., value) predicts the binding affinity between molecules after docking is performed using computer-aided software designed for drug development and discovery. This approach uses structure-based digital screening to identify the conformity, position, and location inside the framework of the subject molecule (Bhagat, et al., 2021). Another important and accepted computer-aided method especially in medicinal chemistry for lead optimization and discovery of drug is QSAR, also known as quantitative structure-activity relationship modeling (Wang et al., 2015). It is especially useful when the 3D structures of specific drug targets are unavailable. Since their initial introduction (Mishra et al., 2024), QSAR technologies have attracted increased attention. QSAR is primarily a ligand-based drug design strategy for developing mathematical frameworks with the goal of identifying relationships with statistical significance between chemical makeup and biological or hazardous features utilizing regression and classification techniques (Neves et al., 2018).

In recent years, many AI applications for drug development have been discovered. The pharmaceutical and medicinal industries have recognized the effective application of AI-based tools, particularly in processing biological data. To date, AI techniques have been used in drug discovery methodologies such as molecule target prediction, availability prediction, and de novo drug design. AI has also been used in predicting drug toxicity and drug bioactivity and also used in development of different antibiotics (Chen et al., 2023). With the swift incorporation of AI in healthcare, particularly during 2016 and 2017, numerous pharmaceutical companies have started investing in AI and formed joint associations with AI-based firms. Their main goal is to develop enhanced healthcare tools, including advancements in creating biomarkers and diagnostic products as well as help in the recognition of possible drug targets and the designing of new drugs. A number of significant pharmaceutical corporations, including Pfizer, Roche, Bayer, and others, have also started working with IT firms to create AI-based drug creation techniques (Swan et al., 2015).

Due the widespread use of AI techniques in healthcare ethical dilemmas regarding the use of AI in this sector have arisen, requiring the need to maintain ethical standards and protect patient data (Mirbabaie et al., 2022). Although the use of AI in clinical settings can enhance the healthcare sector, ethical problems need to be discussed more (Gerke et al., 2020).

This chapter covers the drug development process using AI; different algorithms used for clinical trials and FDA approval process; machine learning and ML-based software; current AI-based software and tools for discovering and developing drugs; AI-based databases available for designing drugs; applications of AI for target identification; de no vo drug design; drug repositioning; and treating different diseases like cancer, helping in pain relief, and providing anesthesia. It also discusses the significance of collaboration between AI scientists and pharmaceutical experts, including a list of businesses who are partnering to produce enhanced resources for drug development. It also deals with the raising ethical concerns of using AI technology in drug development

10.2 PROPOSED METHOD

This chapter covers the drug development process using AI; different algorithms used for clinical trials and FDA approval process; machine learning and ML-based software; current AI-based software and tools for discovering and developing drugs; AI-based databases available for designing drugs; applications of AI for target identification; de no vo drug design; drug repositioning; and treating different diseases like cancer, helping in pain relief, and providing anesthesia. It also discusses the significance of collaboration between AI scientists and pharmaceutical experts, including a list of businesses who are partnering to produce enhanced resources for drug development. It also deals with the raising ethical concerns of using AI technology in drug development.

10.3 RESULT AND DISCUSSION

10.3.1 Overview of Drug Discovery and Development

Drug discovery is a process initiated when there are no treatments for a disease or when current treatments are ineffective or highly toxic. At the earliest phase, a core hypothesis must be formulated, proposing that activating or inhibiting a specific target (like an enzyme, receptor, or ion channel) can lead to therapeutic outcomes for a disease. This mainly involves identifying and validating the target. For the chosen target, extensive assays are conducted to recognize hits and then further leads are developed for potential drug applicants. This process includes hit discovery, the hit-to-lead phase, and lead optimization. The drug applicants further undergo preclinical trials and clinical studies for testing. If the trials are successful, the drug applicant can be marketed as a medical product to cure a particular disease. Today, to speed up this process various AI tools are used (Deng et al., 2022). The process of drug discovery is shown in Figure 10.1.

10.3.1.1 Target Identification and Validation Using AI

In drug development, target identification is the process that identifies molecules, usually proteins, that can change or alter the state of a disease if its activity is changed or adjusted accordingly. In order to recognize possible drug targets that are expected to be involved in the disease progress, different ML algorithms can examine a variety of

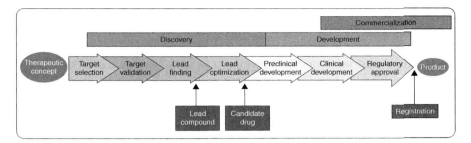

FIGURE 10.1 Process of Drug Discovery.

data sources that includes finding protein-protein interaction networks, information regarding the genome and proteome, and gene expression patterns (Sliwoski et al., 2014). The initial step in target identification is to determine a causal connection between the target and the disease. After the connection is made, target validation is done, in which the function of the selected target in disease development is presented using physiologically significant laboratory and in vivo studies (Failli et al., 2019). Different ML techniques can categorize proteins as either drug targets or non-targets for a particular condition, including some cancers like lung, ovarian, and pancreatic using features like interactions between proteins, level of expression of genes, DNA copy numbers, and presence of mutations (Jeon et al., 2014). The main source of information on the basis of link between targets and diseases is present in the literature. To identify targets and build databases for target identification, pertinent target-disease pairings can be extracted from the literature using text mining and natural language processing (NLP) techniques (Khan et al., 2020, Malik et al., 2023).

10.3.1.2 Hit Identification and Lead Optimization Using AI

The term "hit identification" describes the process of identifying molecules or fragments that can successfully bind to a biological target that can vary from micro- to millimolar. Through the advancement of HTS methods, especially through automated and miniaturized techniques, along with improvements in chemical library design, these methods have significantly improved hit identification by enabling the visualization and gathering of high-quality compounds. Fragment-based methods have become a viable option in recent years for identifying loosely associated fragments among small and varied archive of chemicals (Hoffer et al., 2018). After hit identification is done the next step is to optimize molecule by modifying the molecular structure. The efficiency of these processes can be improved by using data-driven generative and predictive models provided by AI and ML algorithms. These algorithms help in screening of potential compounds in chemical space by using quick and precise prediction models to cut down on expensive computational and experimental efforts (Yoo et al., 2023). The aim of the optimization phase typically involves improving the selectivity, affinity, and physical-chemical properties of substances with the help of tests and in silico methods (Bleicher et al., 2003). The structure-activity relationship (SAR) is one method for categorizing the chemical region surrounding a hit using similar commercially accessible chemicals (Wang et al., 2015).

10.3.1.3 Preclinical and Clinical Trials Using AI

The preclinical trial is an essential step in determining the potential response of a drug under development. ML techniques that are feature- or similarity-based can be used to assess the effectiveness of drug-target interactions by looking at binding affinity or free energy of binding, as well as to anticipate how a medication will affect certain cells. The similarity approaches work on assumptions that similar drugs have similar responses on similar targets (Sachdev & Gupta, 2019). To investigate mechanical models, assess toxicity, and create in vivo predictive models, researchers may utilize data from animal-based pharmacodynamic, pharmacokinetic, and toxicological analysis, exploratory in vitro and in vivo machine-driven studies, multi-organ chip and organ-on-chip systems, and cell assay platforms (Shroff et al., 2022).

AI-based techniques can also assist in anticipating the outcomes of clinical studies well in advance of the actual trials, thereby reducing the risk of toxic effects on patients (Harrer et al., 2019). These methods provide a complete set of tools to speed up the development of innovative medicinal therapies, making the process quicker, more efficient, and patient-centric. These tools encompass patient engagement, real-time adaptation, predictive modeling, and ethical conduct. They bridge the distance between established techniques and new demands, resulting in a more efficient and adaptable approach to healthcare research and development (Niazi, 2022). The clinical trial algorithms utilizing AI/ML is shown in Table 10.1.

10.3.1.4 FDA Approval and Postmarket Monitoring

Regulatory bodies like the U.S. Food and Drug Administration (FDA) and the European Medicines Agency (EMA) are working to address its regulation and execution of the role of AI in medicine. The FDA has been at the forefront of adopting AI/ML-based medical techniques and has established a specific outline for AI/ML algorithms (Food and Drug Administration, 2019). The parent company must submit the medical devices or software to the FDA for review prior to its lawful release in the market. The FDA has created three levels of clearance for AI/ML-based algorithms meant for medical use, which includes premarket approval, the de novo pathway, and 510(k). For each of these paths to be approved, there are requirementsthat must be fulfilled (Benjamens et al., 2020). The list of AI/ML-based medical technologies that are accepted by FDA is shown in Table 10.2.

Pharmacovigilance (PV) is the study and methods used to understand, assess, detect, and prevent harmful events or other drug-related problems (including medication mistakes and issues regarding quality of product). Postmarketing safety monitoring, also known as PV, includes the reporting of adverse events associated with drugs used during the postapproval period. In AI it promotes patient safety during clinical trials by assisting in the early identification and reporting of adverse medication responses (Bate & Hobbiger, 2021).

10.3.2 Machine Learning

ML, a popular type of AI, can be defined as the capacity of the computers to learn by using different algorithms in order to extract features from data. ML techniques

TABLE 10.1
Clinical trial algorithms utilizing AI/ML

Reference	Description	Link
Siemens Healthineers	Offers DL applications and AI-based radiology solutions.	https://www.siemens-healthineers.com/en-us
Deep 6 AI	Specializes in employing AI and natural language analysis to recruit patients for clinical trials.	https://deep6.a
Google Health	Focuses on using AI to solve healthcare issues, such as illness detection, radiology, and predictive modelling	https://health.google
GE Healthcare	Offers a range of AI-powered medical imaging and monitoring systems.	https://www.gehealthcare.com
Owkin	Focuses on creating ML methods for treatment analysis and predictive modelling in medical research.	https://www.owkin.com/
Philips	Provides AI-powered solutions for healthcare informatics, observing patients, and diagnostics.	https://www.usa.philips.com/healthcare/solutions/ai-in-healthcare
Quartic.ai	Provides AI-based pharmaceutical research and manufacturing solutions, such as analytics for developing and discovering drugs	https://quartic.ai/
DataRobot	Offers enterprise AI solutions for risk assessment, predictive modelling, and other data-driven insights in the healthcare industry.	https://www.datarobot.com/
BlackThorn Therapeutics	Employs a data-driven strategy for neurological health, using AI- and ML-based techniques to design clinical trials and choose patients.	https://www.datarobot.com/
Bioclinica	Provides specialised equipment and services for clinical trials, such as AI-powered risk-based monitoring, patient recruiting, and medical imaging.	https://www.bioclinica.com/

Source: Niazi (2022).

provide improved performance by automating model that is formed to create different patterns or provide decision assistance based on the data that has been analyzed (Tekkeşin, 2019). ML algorithms enhance performance by automating the process of model creation to identify patterns or provide decision support based on the analyzed data (Tripathi et al., 2021). Various models have been developed using ML algorithms that can efficiently predict the chemical, physical, and biological models of substances in drug discovery. Random forest, SVM, and others are some of the algorithms used for discovering drugs (Patel et al., 2020).

TABLE 10.2
List of AI/ML-based medical technologies accepted by the FDA

Name of apparatus or algorithm	Parent company's name	Description	FDA approval type	Type of algorithm used
OsteoDetect	Imagen Technologies, Inc.	Diagnosis of radiocarpal joint fracture via X-ray	De novo pathway	Deep learning
DreaMed	DreaMed Diabetes, Ltd	Management of diabetes type 1	De novo pathway	AI
FerriSmart Analysis System	Resonance Health Analysis Service Pty Ltd	Measures concentration of liver iron	510(k) premarket notification	AI
Deep Learning Image Reconstruction	GE Medical Systems, LLC.	Computed Tomography scan image reformation	510(k) premarket notification	DL
Critical Care Suite	GE Medical Systems, LLC	Pneumothorax evaluation on a chest X-ray	510(k) premarket notification	AI algorithms
EchoMD Automated Ejection Fraction	Bay Labs, Inc.	Investigation of an echocardiogram	510(k) premarket notification	ML
Icobrain	Icometrix NV	Interpretation of brain MRI	510(k) premarket notification	ML and DL
Arterys MICA	Arterys Inc.	Uses CT and MRI to diagnose lung and liver cancer	510(k) premarket notification	AI

Source: Benjamens et al. (2020).

10.3.2.1 Random Forest

RF is a popular approach for managing huge datasets with various characteristics. It can efficiently remove outliers and can classify datasets based on their relative features. It is frequently used for huge inputs and variables, and is generally designed for more variables and inputs with accessibility obtained from multiple data sources (Breiman & Cutler, 2003). In drug development RFs are primarily utilized for feature selection, classification, or for feature selection. This approach can also be used to enhance ligand-protein affinity prediction through computer-based visualization by giving molecular descriptors for enzymes like kinase ligands and nuclear hormone receptors selected based on training data (Cano et al., 2017).

10.3.2.3 Support Vector Machines

SVMs are supervised ML algorithms used in drug discovery to classify classes of compounds using a feature selector, which helps to find how compounds might interact with biological targets by deriving a hyperplane or affine subspace. They utilize class similarities to produce an infinite number of hyperplanes. In the case of linear data, SVMs train by distinguishing classes that are composed of compounds based on selecting features and mapping them into a chemical feature space (Heikamp & Bajorath, 2014). To aid in drug development, substances can be ranked based on their likelihood of being active in computational visualization. SVM may be applied in several ways to identify active and inactive substances. To alter the process, the algorithm may be trained using several feature selectors, including 2D fingerprints and target proteins (Wassermann et al, 2011).

10.3.3 CURRENT AI-BASED SOFTWARE

AI-based software has increased the potential for developing new drugs by analyzing large data, predicting the drug activity, establishing protein-ligand relationship, etc. Over the past few years many types of software have been developed that include AlphaFold, Torch drug, DeepPurpose, DeepTox, AtomNet, and many more.

10.3.3.1 AlphaFold

The AlphaFold network helps to forecast the 3D arrangement of all heavy atoms in a protein using the main amino acid sequence and aligned sequences from related proteins as input source. It significantly increases structure prediction accuracy by adding training processes and unique neural network topologies that take into account the physical, evolutionary, and geometric limitations of protein structures (Jumper et al., 2021).

10.3.3.2 TorchDrug

TorchDrug evaluates a number of major drug discovery tasks, including molecular property forecasting, pretrained molecular depiction, optimization, retrosynthesis prediction, de novo molecular design, and biomedical knowledge graph reasoning. These tasks use cutting-edge approaches like knowledge graph reasoning, graph-based ML approaches, deep generative model-based approaches, and augmentation learning. TorchDrug has a hierarchical frontier that allows customization by both novices and specialists in this field (Zhu et al., 2022).

10.3.3.3 DeepPurpose

This is a library for programming and predicting proteins and different chemical compounds. DeepPurpose enables quick prototyping using a computational structure that includes over 50 seven protein encoders, DL models, and eight compound encoders. The DeepPurpose library includes a unified encoder-decoder structure, giving it a unique flexibility (Huang et al., 2020).

10.3.3.4 DeepTox

This is a multi-learning neural network that predicts the toxicity of any compound. DeepTox normalizes the chemical representations of the substances and further computes a vast number of chemical descriptors, which are fed into ML algorithms. The next step is to train models, assess them, and merge the best ones into ensembles and finally predict the toxicity of novel chemicals (Mayr et al., 2016).

10.3.3.5 AtomNet

AtomNet is the very first structure construction-based, convolutional neural network developed to forecast the biological activity of tiny compounds for drug development. It uses convolutional notions of feature localization (Sharma et al., 2023) and multi-layer composition to represent bioactivity and chemical connections. AtomNet's use of local convolutional filters on structural target data successfully predicts the novel active substances for targets with no previously identified modulators (Wallach et al., 2015). List of AI-based softwares is shown in Table 10.3.

10.3.4 Current Databases for Drug Development

The drug-target interaction data that is produced either experimentally or computationally should be merged and made freely available for predicting interactions. Some

TABLE 10.3
List of AI-based software

Reference	Description	Link
AlphaFold	Model for 3-D prediction of proteins	https://github.com/deepmind/alphafold/
TorchDrug	Pytorch-based adaptable structure for models based on drug discovery.	https://torchdrug.ai/
DeepPurpose	Library for protein function prediction and interaction between drug, target and proteins.	https://github.com/kexinhuang12345/DeepPurpose
DeepTox	Predicts adverse effects of chemical substances	http://www.bioinf.jku.at/research/DeepTox/
DeepScreen	Predicts high performance drug-target interaction	https://github.com/cansyl/DEEPScreen
Chemical VAE	An auto-encoder-based foundation for creating novel compounds.	https://github.com/aspuru-guzik-group/chemical_vae/
DeepBlind	Tool for analyzing attachment between protein and DNA/RNA	https://github.com/MedChaabane/DeepBind-with-PyTorch

Source: Qureshi et al. (2023).

of the currently available databases include PDBBind, UniProt, PDB, ADME, ZINC, DrugBank, Atom3D, MoleculeNet, GDSC, SIDER, etc.

10.3.4.1 UniProt

The UniProt Knowledgebase provides information on order of protein with functional details of proteins. It aims to provide a broad, top-notch and openly accessible database of protein sequences to users. The UniProt databases serve to facilitate biological and biomedical research by offering a comprehensive collection of known protein sequences. Each entry is accompanied by experimentally validated or computationally predicted functional data regarding the protein (UniProt, 2021; Gupta et al., 2024).

10.3.4.2 Protein Data Bank

The PDB is a worldwide storehouse of empirically determined structure of biological macromolecules and their complexes in 3D space, which was initially developed in 1971 as the first open access source in the field of biological science. Currently it stores data from different experiments, accompanying metadata, and 3D-atomic-level structural models that are created using three widely recognized methods like electron microscopy (3DEM), crystallography, and nuclear magnetic resonance spectroscopy (NMR) (Burley et al., 2017). According to the data of 2021, PDB has grown and now contains over 180,000 structures of nucleic acids and proteins (Burley et al., 2022).

10.3.4.3 ZINC

ZINC is a research tool designed for scientists seeking chemical compounds relevant to their biological targets. It stores the information. The ZINC database comprises about 20 million commercially accessible compounds that are already demonstrated in physiologically appropriate representations. The compounds can be obtained in standard, ready-to-dock versions or combinations. Data can be found by biological activity, structure with physical property, name, etc. Small custom-designed subsets that can be docked, downloaded, edited, updated, shared, and sent to a vendor for purchase can also be generated (Irwin et al., 2012).

10.3.4.4 SIDER

This is a tool that incorporates data on targets, medications, and undesirable effects of a drug to provide a full understanding of the biological mechanism of actions of pharmaceuticals and the adverse responses they generate (Kuhn et al., 2016). The list of databases for drug development is shown in Table 10.4.

10.4 APPLICATION OF AI IN DRUG DISCOVERY AND DEVELOPMENT

The application of AI in drug discovery and development has increased significantly in recent years, enabling its use in many steps of the entire process.

TABLE 10.4
List of databases for drug development

Reference	Description	Link
PDBBind	The Protein Data Bank has a large library of binding preferences for protein-ligand complexes.	http://www.pdbbind.org.cn/
canSar	Cancer experimental Research and Development of Drugs Database.	https://cansarblack.icr.ac.uk/
TTD	A targeted therapeutic database.	https://moleculenet.org/
MoleculeNet	A collection of datasets for molecular ML.	https://moleculenet.org/
ChEMBL	A massive biological activity database for drug research.	https://www.ebi.ac.uk/chembl/
STITCH	A comprehensive database of chemical-protein interactions.	http://stitch.embl.de/
PDB	Database for 3D structure of proteins, DNA and RNA	https://www.rcsb.org/
ZINC	Vital visualization of compounds database	https://zinc.docking.org/

Source: Qureshi et al. (2023).

10.4.1 PREDICTING DRUG TOXICITY

The prediction of harmful effects caused by a drug is one of the most important steps in drug discovery. This process requires animal testing and is time consuming and costly (Basile et al., 2019). However, computational-based AI methods (Gupta & Gupta, 2023) have eased this process and enabled scientists to understand the harmful effects of a drug. One such AI tool for predicting toxicity of drug is DeepTox (Mayr et al., 2016). Another example could be of terbinafine toxicity, which can be better understood by a metabolite TBF-A. The mechanism in which TBF-A was formed in the liver was unknown. A student from Na Le Dang at Washington University in St. Louis used a ML/AI algorithm to determine the potential metabolic paths that terbinafine may follow when biotransformed by the liver. With the appropriate inputs to the algorithm provided, the student was able to determine that the breakdown of terbinafine to TBF-A was a two-step process (Barnette et al., 2019).

10.4.2 DRUG REPOSITIONING

Drug repositioning, which is also known as repurposing, refers to fresh indications for current drugs and is an alternate option to de novo drug development. Over the years complicated biology and pharmacology have hampered drug repurposing or repositioning efforts, but smart computational algorithms have provided a method that recognizes potential drug indicators by merging large-scale mixed data that includes transcriptomic, genomic, bioactivity, phenotypic, and chemical data from 100 accepted drugs. Several custom AI/ML models (Gupta et al., 2024) have been

Revolutionizing Drug Development

proposed for finding unique indications for medication (Vatansever et al, 2021). In 2020, MIT researchers reported on a deep learning method for developing antibiotics. They trained the deep GCN model using chemical characteristics and identified Halicin as an antibacterial compound from the Drug-Repurposing Hub. It has successfully demonstrated wide spectrum efficacy against strains resistant to drugs in mice. This was the first use of AI/ML-based techniques in the history of medical science to completely discover new kinds of antibiotics from scratch with no preexisting human assumptions (Stokes et al., 2020).

10.4.3 DE NOVO DRUG DESIGN

De novo drug design is a process focused on discovering new bioactive molecules that can address multiple optimization criteria, such as selectivity, activity, and important physicochemical properties, including ADMET (Absorption, Distribution, Metabolism, Excretion, and Toxicity) characteristics (Blaschke et al., 2020). Although ligand-based and structure-based drug design approaches have improved the identification of small-molecule drug applicants they still require understanding of a biological target's active site or the pharmacophores of an active binder, which limits their use in modern drug discovery and thus requires new approaches. In recent years, AI technology has expedited drug development by opening up new options for de novo design (Mouchlis et al., 2021) such as the encoder-decoder-based model ChemVAE (Gómez-Bombarelli et al., 2018), GAN-based model GraphINVENT (Mercado et al., 2021), and the RNN-based model MolRNN (Li et al., 2018). Validation and screening of de novo drug design is shown in Figure 10.2.

10.4.4 TREATMENT OF CANCER

Cancers coming from a single source of tissue and clinical categorization show significant genetic and phenotypic heterogeneity among individuals (Gerdes et al., 2021). In execution, this variability leads to patients having varying reactions to therapy. To solve this problem, the field of personalized medicine tries to create measurable biomarkers that relate to the efficacy of medical therapy in humans, allowing

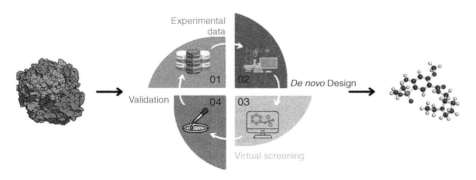

FIGURE 10.2 Validation and Screening of De Novo Drug Design.

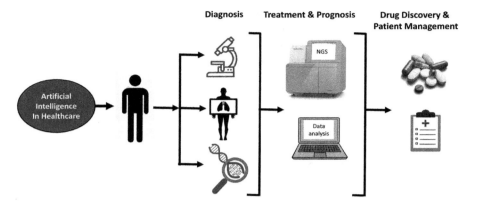

FIGURE 10.3 Artificial Intelligence in Healthcare (https://ars.els-cdn.com/content/image/1-s2.0-S200103702030372X-ga1.jpg).

clinicians to anticipate patient reactions to particular medicines (Sawyers, 2008; Gupta et al., 2021). Different targeted cancer therapies like protein biomarkers have been used for decades. Examples include human epidermal growth factor receptor 2 and estrogen receptor, whose activity forecasts the responses of breast cancer patients to trastuzumab and tamoxifen. The development of effective and inexpensive next-generation sequencing technology enables the discovery of genetic markers that forecast how people respond to numerous specialized drugs (Myers, 2016). The various roles of artificial intelligence in healthcare is shown in Figure 10.3.

10.4.5 Applications in Anesthesia

In recent years, there has been an increasing application of AI in administering anesthesia to patients. AI technologies are being utilized in the field of anesthesia management to enhance patient safety and optimize dosing (Hemmerling et al., 2011). To provide the correct amount of drug dosage these robots use complex AI/ML-based data, which are generally based on patient data like blood pressure, heart rate, etc., with pharmacokinetic features of drug. The function of pharmacological robots and even more sophisticated autonomous systems like cognitive robots, which are capable of recognizing key clinical states that require human supervision in the anesthesia sector, has been thoroughly reviewed (Zaouter et al., 2020).

10.4.6 Pain Treatment

AI systems in pain treatment are used to determine degree of pain, analgesic needs in clinical settings, personalized medicinal decision support in analgesic treatment, efficacy of analgesics, and in pharmaceutical overusage of medicines (Vatansever et al., 2021). Researchers have also used ML techniques to find new genes and pathways linked to acute and chronic pain (Chidambaran et al., 2020) along with predicting inhibitors of therapeutic targets for pain (Kong et al., 2020). To facilitate this, researchers developed a pain-domain-specific chemogenomics data library for forecasting novel multi-target analgesics and medication combinations for pain treatment. The database comprises currently available analgesics, pain-related targets with 3D structures, as well as reported compounds for these target proteins (Feng et al., 2020).

10.5 COLLABORATION BETWEEN AI AND PHARMACEUTICAL INDUSTRY

For the development of novel and effective therapies for diverse disorders collaboration between AI researchers and pharmaceutical scientists is essential to construct strong data and ML models to forecast possible drug candidates and thus accelerate the drug discovery process. This collaboration can enhance clinical trial accuracy, efficiency, affordability, and accessibility and help in the process of preparing personalized medications for patients (Blanco-Gonzalez et al, 2023). In recent years many pharmaceutical industries have started investing and creating joint ventures with AI-based industries in order to develop tools and speed up the process of drug development. This includes improving diagnostics tests, developing biomarkers, identifying therapeutic targets, and developing novel medications (Swan et al., 2015). AI-powered drug discovery startups received over $1 billion in investment in 2018 and $1.5 billion in 2019 (Freedma, 2019).

DeepMind Technologies, a Google-based company, has partnered with the Royal Free London NHS Foundation Trust to help treat acute renal damage (Powles & Hodson, 2017). Cyclica, a computation biophysics and ML business, collaborated with Bayer's G4A project in 2018 to build AI-driven prediction models for off-target effects.

In 2017, GSK and Excientia began developing a medicinal agent for chronic obstructive pulmonary disease using excientia's AI-based automated drug discovery platform. The Oxford-based AI pharma-tech business utilized its AI technology to identify preclinical prospects for Roche. In 2019, Roche collaborated with Sensyne Health to create clinical trials using AI technology. Collaboration of Asian and HK companies with big pharma is shown in Table 10.5 and Figure 10.4.

10.6 ETHICAL ISSUES RELATED TO THE USE OF AI IN DRUG DEVELOPMENT

The use of AI in healthcare especially in the pharmaceutical industry is increasing rapidly but so are the ethical issues, which must be addressed. Ethical considerations such as justice, autonomy preservation, explainability, and patient privacy are frequently debated questions (Murphy et al., 2021). Currently there are no practical methods or frameworks to verify whether methods used by AI adhere to justice, damage prevention, human autonomy, and explainability principles. Creating and carrying out reliable AI in healthcare involves recognizing stakeholders, analyzing their ethical concerns, and capturing their requirements and preferences. There is a lack of information on stakeholder perspectives on these topics, with a few studies addressing just a small number of healthcare practitioners and patients (Karimianet al., 2022). The topic of whether AI "works for within current legal classifications or

TABLE 10.5
List of collaborations in AI and pharmaceutical companies

Company's name	Application of AI	Partnership	Platform Advanced/Lead Agents for Clinical Studies
IBM Watson Health Cambridge, MA 02142, USA	Provides an outline for evaluating clinical and health data.	Novartis	Using current time patient monitoring to enhance breast cancer treatment outcomes.
Atomwise San Francisco, CA 94103, USA	A framework for AI-powered structural modelling.	Lilly	Uses agent BBT-401 in Phase 2 of clinical trial.
Numerate San Francisco, CA 94107, USA	AI-based medication discovery for cancer and gastrointestinal specialties.	Takeda	Clinical testing for Ryanodine receptor 2 in Phase 1by using agent S48168.
Benevolent AI London, UK	AI-powered Judgement Augmented Cognition System (JACS) which identifies and helps in advancement of potential clinical lead medicines for neurodegenerative diseases.	Janssen	Collaboration can lead to the advancement of new medicinal molecules.
Exscientia Oxford, UK	A proposal for AI-powered discovery drug and refining lead.	Sanofi	Advanced Centaur Chemist™ Scheme for AI-enabled Drug Discovery, Use agent DSP-1181 in Phase I clinical testing for obsessive-compulsive disorder.
XtalPi Shenzhen, Guangdong, China	Uses both QM and ML approaches for target validation and identification.	Pfizer	Refining and preparation of crystallized drug candidates for early screening process.
BioXcel therapeutics New Haven, CT, USA	Uses AI mechanisms for drug discovery process	Pfizer	BXCL501 lead agent is now in Phase 3 clinical development, while BXCL701 a drug agent, is being evaluated in Phase 2.
Benevolent AI London, UK	Use AI agents which can treat chronic kidney diseases	AstraZeneca	Effective drug candidate for treating chronic kidney disease was investigated in Phase 2b trial.

Source: Sarkar et al. (2023).

Revolutionizing Drug Development

FIGURE 10.4 Collaboration of Asian and HK Companies with Big Pharma (https://media.licdn.com/dms/image/D4D12AQHDg9HOSqtwmQ/article-inline_image-shrink_1000_1488/0/1691914199963?e=1720051200&v=beta&t=26xpn_WPbxRKDkx6momT8DXlF6xdNy3Ln2p5dm8iDIM).

within another group with unique characteristics and effects should be established" is constantly open for debate. Although utilizing AI in clinical settings has enormous potential to improve healthcare, there are existing ethical considerations that must be addressed (Gerke et al., 2020). Overall the ethical application of AI in the drug industry necessitates cautious assessment and ensuring that AI systems are trained on numerous and accurate information, routinely checking and inspecting AI systems for bias, and putting in place robust data privacy and security standards. Addressing these concerns will allow the pharmaceutical business to employ AI responsibly and ethically (Blanco-Gonzalez et al., 2023).

10.7 CONCLUSION

The use of AI/ML technologies in the field of drug discovery and development have significantly increased in the past few years. This technology has enabled scientists to create low-cost drugs using intelligent and adaptable tools such as Alphafold, DeepChem, DeepBar, AtomNet, etc. (Qureshi, et al., 2023). The database includes three-dimensional structures of molecules, proteins, and nucleic acid sequences and functional information, providing insights into the biological roles of proteins, which aids in target identification for drug discovery. AI can be used to predict toxicity of a drug, determine adverse side effects, find potential pathways, in drug repositioning, to find cures for diseases, and in pain management. Over the years scientists and AI researchers have pooled their ideas and resources for designing better tools for the drug development process. Companies like Pfizer, Novartis, Roche, and many more have collaborated with different AI-based companies. While there have been many successes, ethical concerns still need to be addressed. As the use of AI in healthcare is increasing the ethical concern is also increasing. This is because there are no established practical frameworks to verify if AI methodologies comply with principles like justice, harm prevention, human autonomy, and explainability.

REFERENCES

Barnette, Dustyn A., Mary A. Davis, Noah Flynn, Anirudh S. Pidugu, S. Joshua Swamidass, and Grover P. Miller. "Comprehensive kinetic and modeling analyses revealed CYP2C9 and 3A4 determine terbinafine metabolic clearance and bioactivation." *Biochemical Pharmacology* 170 (2019): 113661.

Basile, Anna O., Alexandre Yahi, and Nicholas P. Tatonetti. "Artificial intelligence for drug toxicity and safety." *Trends in Pharmacological Sciences* 40, no. 9 (2019): 624–635.

Bate, Andrew, and Steve F. Hobbiger. "Artificial intelligence, real-world automation and the safety of medicines." *Drug Safety* 44, no. 2 (2021): 125–132.

Benjamens, Stan, Pranavsingh Dhunnoo, and Bertalan Meskó. "The state of artificial intelligence-based FDA-approved medical devices and algorithms: an online database." *NPJ Digital Medicine* 3, no. 1 (2020): 118.

Bhagat, Rani T., Santosh R. Butle, Deepak S. Khobragade, Sagar B. Wankhede, Chandani C. Prasad, Divyani S. Mahure, and Ashwini V. Armarkar. "Molecular docking in drug discovery." *Journal of Pharmaceutical Research International* 33, no. 30B (2021): 46–58.

Blaschke, Thomas, Josep Arús-Pous, Hongming Chen, Christian Margreitter, Christian Tyrchan, Ola Engkvist, Kostas Papadopoulos, and Atanas Patronov. "REINVENT 2.0: an AI tool for de novo drug design." *Journal of Chemical Information and Modeling* 60, no. 12 (2020): 5918–5922.

Blanco-Gonzalez, Alexandre, Alfonso Cabezon, Alejandro Seco-Gonzalez, Daniel Conde-Torres, Paula Antelo-Riveiro, Angel Pineiro, and Rebeca Garcia-Fandino. "The role of ai in drug discovery: challenges, opportunities, and strategies." *Pharmaceuticals* 16, no. 6 (2023): 891.

Bleicher, Konrad H., Hans-Joachim Böhm, Klaus Müller, and Alexander I. Alanine. "Hit and lead generation: beyond high-throughput screening." *Nature Reviews Drug Discovery* 2, no. 5 (2003): 369–378.

Breiman, Leo. 2002. 'Manual on Setting up, Using, and Understanding Random Forests v3. 1'. *Statistics Department University of California Berkeley, CA, USA* 1 (58): 3–42.

Burley, Stephen K., Charmi Bhikadiya, Chunxiao Bi, Sebastian Bittrich, Li Chen, Gregg V. Crichlow, Jose M. Duarte et al. "RCSB Protein Data Bank: Celebrating 50 years of the PDB with new tools for understanding and visualizing biological macromolecules in 3D." *Protein Science* 31, no. 1 (2022): 187–208.

Burley, Stephen K., Helen M. Berman, Gerard J. Kleywegt, John L. Markley, Haruki Nakamura, and Sameer Velankar. 2017. 'Protein Data Bank (PDB): The Single Global Macromolecular Structure Archive'. In *Protein Crystallography*, edited by Alexander Wlodawer, Zbigniew Dauter, and Mariusz Jaskolski, 1607:627–41. Methods in Molecular Biology. New York, NY: Springer New York. https://doi.org/10.1007/978-1-4939-7000-1_26.

Cano, Gaspar, Jose Garcia-Rodriguez, Alberto Garcia-Garcia, Horacio Perez-Sanchez, Jón Atli Benediktsson, Anil Thapa, and Alastair Barr. "Automatic selection of molecular descriptors using random forest: Application to drug discovery." *Expert Systems with Applications* 72 (2017): 151–159.

Chen, Wei, Xuesong Liu, Sanyin Zhang, and Shilin Chen. "Artificial intelligence for drug discovery: Resources, methods, and applications." *Molecular Therapy-Nucleic Acids* 31 (2023): 691–702.

Chidambaran, Vidya, Maria Ashton, Lisa J. Martin, and Anil G. Jegga. "Systems biology-based approaches to summarize and identify novel genes and pathways associated with acute and chronic postsurgical pain." *Journal of Clinical Anesthesia* 62 (2020): 109738.

Deng, Jianyuan, Zhibo Yang, Iwao Ojima, Dimitris Samaras, and Fusheng Wang. "Artificial intelligence in drug discovery: applications and techniques." *Briefings in Bioinformatics* 23, no. 1 (2022): bbab430.

Failli, Mario, Jussi Paananen, and Vittorio Fortino. "Prioritizing target-disease associations with novel safety and efficacy scoring methods." *Scientific Reports* 9, no. 1 (2019): 9852.

Feng, Zhiwei, Maozi Chen, Mingzhe Shen, Tianjian Liang, Hui Chen, and Xiang-Qun Xie. "Pain-CKB, a Pain-domain-specific chemogenomics knowledgebase for target identification and systems pharmacology research." *Journal of Chemical Information and Modeling* 60, no. 10 (2020): 4429–4435.

Food and Drug Administration. "Proposed regulatory framework for modifications to artificial intelligence/machine learning (AI/ML)-based software as a medical device (SaMD)." (2019).

Freedman, David H. "Hunting for new drugs with AI." *Nature* 576, no. 7787 (2019): S49–S53.

Malik, Kamal, Harsh Sadawarti, Moolchand Sharma, Umesh Gupta, and Prayag Tiwari, eds. *Computational Techniques in Neuroscience*. CRC Press, 2023.

Mishra, Shambhavi, Tanveer Ahmed, Mohd. Abuzar Sayeed, and Umesh Gupta. 2023. 'Artificial Neural Network Model for Automated Medical Diagnosis'. In *Soft Computing Techniques in Connected Healthcare Systems*, by Moolchand Sharma, Suman Deswal, Umesh Gupta, Mujahid Tabassum, and Isah Lawal, 1st ed., 34–54. Boca Raton: CRC Press. https://doi.org/10.1201/9781003405368-3.

Gerdes, Henry, Pedro Casado, Arran Dokal, Maruan Hijazi, Nosheen Akhtar, Ruth Osuntola, Vinothini Rajeeve et al. "Drug ranking using machine learning systematically predicts the efficacy of anti-cancer drugs." *Nature Communications* 12, no. 1 (2021): 1850.

Gerke, Sara, Timo Minssen, and Glenn Cohen. "Ethical and legal challenges of artificial intelligence-driven healthcare." In *Artificial Intelligence in Healthcare*, pp. 295–336. Academic Press, 2020.

Harrer, Stefan, Pratik Shah, Bhavna Antony, and Jianying Hu. "Artificial intelligence for clinical trial design." *Trends in Pharmacological Sciences* 40, no. 8 (2019): 577–591.

Hemmerling, Thomas M., Riccardo Taddei, Mohamad Wehbe, Joshua Morse, Shantale Cyr, and Cedrick Zaouter. "Robotic anesthesia–A vision for the future of anesthesia." *Translational Medicine@ UniSa* 1 (2011): 1.

Hoffer, Laurent, Yuliia V. Voitovich, Brigitt Raux, Kendall Carrasco, Christophe Muller, Aleksey Y. Fedorov, Carine Derviaux et al. "Integrated strategy for lead optimization based on fragment growing: the diversity-oriented-target-focused-synthesis approach." *Journal of Medicinal Chemistry* 61, no. 13 (2018): 5719–5732.

Gómez-Bombarelli, Rafael, Jennifer N. Wei, David Duvenaud, José Miguel Hernández-Lobato, Benjamín Sánchez-Lengeling, Dennis Sheberla, Jorge Aguilera-Iparraguirre, Timothy D. Hirzel, Ryan P. Adams, and Alán Aspuru-Guzik. "Automatic chemical design using a data-driven continuous representation of molecules." *ACS Central Science* 4, no. 2 (2018): 268–276.

Gupta, Deepak, Ambika Choudhury, Umesh Gupta, Priyanka Singh, and Mukesh Prasad. "Computational Approach to Clinical Diagnosis of Diabetes Disease: A Comparative Study." In *Multimedia Tools and Applications* (2021): 1–26.

Gupta U, Gupta D. "Least squares structural twin bounded support vector machine on class scatter." *Applied Intelligence* 53, no. 12 (2023): 15321–51.

Gupta, Umesh, Ayushman Pranav, Anvi Kohli, Sukanta Ghosh, and Divya Singh. "The Contribution of Artificial Intelligence to Drug Discovery: Current Progress and Prospects for the Future." In *Microbial Data Intelligence and Computational Techniques for Sustainable Computing* (2024): 1–23.

Heikamp, Kathrin, and Jürgen Bajorath. "Support vector machines for drug discovery." *Expert Opinion on Drug Discovery* 9, no. 1 (2014): 93–104.

Huang, Kexin, Tianfan Fu, Lucas M. Glass, Marinka Zitnik, Cao Xiao, and Jimeng Sun. "DeepPurpose: a deep learning library for drug–target interaction prediction." *Bioinformatics* 36, no. 22–23 (2020): 5545–5547.

Irwin, John J., Teague Sterling, Michael M. Mysinger, Erin S. Bolstad, and Ryan G. Coleman. "ZINC: a free tool to discover chemistry for biology." *Journal of Chemical Information and Modeling* 52, no. 7 (2012): 1757–1768.

Jeon, Jouhyun, Satra Nim, Joan Teyra, Alessandro Datti, Jeffrey L. Wrana, Sachdev S. Sidhu, Jason Moffat, and Philip M. Kim. "A systematic approach to identify novel cancer drug targets using machine learning, inhibitor design and high-throughput screening." *Genome Medicine* 6 (2014): 1–18.

Jumper, John, Richard Evans, Alexander Pritzel, Tim Green, Michael Figurnov, Olaf Ronneberger, Kathryn Tunyasuvunakool et al. "Highly accurate protein structure prediction with AlphaFold." *Nature* 596, no. 7873 (2021): 583–589.

Karimian, Golnar, Elena Petelos, and Silvia MAA Evers. "The ethical issues of the application of artificial intelligence in healthcare: a systematic scoping review." *AI and Ethics* 2, no. 4 (2022): 539–551.

Katzung, Bertram G., Marieke Kruidering-Hall, and Anthony J. Trevor. *Katzung & Trevor's Pharmacology Examination & Board Review*. McGraw-Hill Education, 2019.

Kempt, Hendrik, and Saskia K. Nagel. "Responsibility, second opinions and peer-disagreement: ethical and epistemological challenges of using AI in clinical diagnostic contexts." *Journal of Medical Ethics* 48, no. 4 (2022): 222–229.

Khan, Junaed Younus, Md Tawkat Islam Khondaker, Iram Tazim Hoque, Hamada RH Al-Absi, Mohammad Saifur Rahman, Reto Guler, Tanvir Alam, and M. Sohel Rahman. "Toward preparing a knowledge base to explore potential drugs and biomedical entities related to COVID-19: automated computational approach." *JMIR Medical Informatics* 8, no. 11 (2020): e21648.

Kong, Weikaixin, Xinyu Tu, Weiran Huang, Yang Yang, Zhengwei Xie, and Zhuo Huang. "Prediction and optimization of NaV1. 7 sodium channel inhibitors based on machine learning and simulated annealing." *Journal of Chemical Information and Modeling* 60, no. 6 (2020): 2739–2753.

Kuhn, Michael, Ivica Letunic, Lars Juhl Jensen, and Peer Bork. "The SIDER database of drugs and side effects." *Nucleic Acids Research* 44, no. D1 (2016): D1075–D1079.

Li, Yibo, Liangren Zhang, and Zhenming Liu. "Multi-objective de novo drug design with conditional graph generative model." *Journal of Cheminformatics* 10 (2018): 1–24.

Mayr, Andreas, Günter Klambauer, Thomas Unterthiner, and Sepp Hochreiter. "DeepTox: toxicity prediction using deep learning." *Frontiers in Environmental Science* 3 (2016): 80.

Mercado, Rocío, Tobias Rastemo, Edvard Lindelöf, Günter Klambauer, Ola Engkvist, Hongming Chen, and Esben Jannik Bjerrum. "Graph networks for molecular design." *Machine Learning: Science and Technology* 2, no. 2 (2021): 025023

Mirbabaie, Milad, Lennart Hofeditz, Nicholas RJ Frick, and Stefan Stieglitz. "Artificial intelligence in hospitals: providing a status quo of ethical considerations in academia to guide future research." *AI & Society* 37, no. 4 (2022): 1361–1382.

Mouchlis, Varnavas D., Antreas Afantitis, Angela Serra, Michele Fratello, Anastasios G. Papadiamantis, Vassilis Aidinis, Iseult Lynch, Dario Greco, and Georgia Melagraki. "Advances in de novo drug design: from conventional to machine learning methods." *International Journal of Molecular Sciences* 22, no. 4 (2021): 1676.

Murphy, Kathleen, Erica Di Ruggiero, Ross Upshur, Donald J. Willison, Neha Malhotra, Jia Ce Cai, Nakul Malhotra, Vincci Lui, and Jennifer Gibson. "Artificial intelligence for good health: a scoping review of the ethics literature." *BMC Medical Ethics* 22 (2021): 1–17.

Myers, Meagan B. 2016. 'Targeted Therapies with Companion Diagnostics in the Management of Breast Cancer: Current Perspectives'. *Pharmacogenomics and Personalized Medicine* 9 (January):7–16. https://doi.org/10.2147/PGPM.S56055.

Neves, Bruno J., Rodolpho C. Braga, Cleber C. Melo-Filho, José Teófilo Moreira-Filho, Eugene N. Muratov, and Carolina Horta Andrade. "QSAR-based virtual screening: advances and applications in drug discovery." *Frontiers in Pharmacology* 9 (2018): 1275.

Niazi, Sarfaraz K. "The coming of age of ai/ml in drug discovery, development, clinical testing, and manufacturing: The FDA perspectives." *Drug Design, Development and Therapy* (2023): 2691–2725.

Patel, Lauv, Tripti Shukla, Xiuzhen Huang, David W. Ussery, and Shanzhi Wang. "Machine learning methods in drug discovery." *Molecules* 25, no. 22 (2020): 5277.

Paul, Steven M., Daniel S. Mytelka, Christopher T. Dunwiddie, Charles C. Persinger, Bernard H. Munos, Stacy R. Lindborg, and Aaron L. Schacht. "How to improve R&D productivity: the pharmaceutical industry's grand challenge'. *Nature Reviews. Drug Discovery* 9, no. 3 (March 2010): 203–14.

Powles, Julia, and Hal Hodson. "Google DeepMind and healthcare in an age of algorithms." *Health and Technology* 7, no. 4 (2017): 351–367.

Qureshi, Rizwan, Muhammad Irfan, Taimoor Muzaffar Gondal, Sheheryar Khan, Jia Wu, Muhammad Usman Hadi, John Heymach, Xiuning Le, Hong Yan, and Tanvir Alam. 2023. 'AI in Drug Discovery and Its Clinical Relevance'. *Heliyon* 9 (7): e17575. https://doi.org/10.1016/j.heliyon.2023.e17575.

Rong, Guoguang, Arnaldo Mendez, Elie Bou Assi, Bo Zhao, and Mohamad Sawan. "Artificial intelligence in healthcare: review and prediction case studies." *Engineering* 6, no. 3 (2020): 291–301.

Sachdev, Kanica, and Manoj Kumar Gupta. "A comprehensive review of feature based methods for drug target interaction prediction." *Journal of Biomedical Informatics* 93 (2019): 103159.

Sarkar, Chayna, Biswadeep Das, Vikram Singh Rawat, Julie Birdie Wahlang, Arvind Nongpiur, Iadarilang Tiewsoh, Nari M. Lyngdoh, Debasmita Das, Manjunath Bidarolli, and Hannah Theresa Sony. "Artificial intelligence and machine learning technology driven modern drug discovery and development." *International Journal of Molecular Sciences* 24, no. 3 (2023): 2026.

Sawyers, Charles L. "The cancer biomarker problem." *Nature* 452, no. 7187 (2008): 548–552.

Sharma, Moolchand, Suman Deswal, Umesh Gupta, Mujahid Tabassum, and Isah Lawal, eds. *Soft Computing Techniques in Connected Healthcare Systems*. CRC Press, 2023.

Shroff, Tanvi, Kehinde Aina, Christian Maass, Madalena Cipriano, Joeri Lambrecht, Frank Tacke, Alexander Mosig, and Peter Loskill. "Studying metabolism with multi-organ chips: new tools for disease modelling, pharmacokinetics and pharmacodynamics." *Open Biology* 12, no. 3 (2022): 210333.

Singh, Divya, Deepti Singh, Manju, and Umesh Gupta. 2023. 'Smart Healthcare: A Breakthrough in the Growth of Technologies'. In *Artificial Intelligence-Based Healthcare Systems*, edited by Manju, Sandeep Kumar, and Sardar M. N. Islam, 73–85. The Springer Series in Applied Machine Learning. Cham: Springer Nature Switzerland. https://doi.org/10.1007/978-3-031-41925-6_5.

Sliwoski, Gregory, Sandeepkumar Kothiwale, Jens Meiler, and Edward W. Lowe. "Computational methods in drug discovery." *Pharmacological Reviews* 66, no. 1 (2014): 334–395.

Stokes, Jonathan M., Kevin Yang, Kyle Swanson, Wengong Jin, Andres Cubillos-Ruiz, Nina M. Donghia, Craig R. MacNair et al. "A deep learning approach to antibiotic discovery." *Cell* 180, no. 4 (2020): 688–702.

Swan, Anna L., Dov J. Stekel, Charlie Hodgman, David Allaway, Mohammed H. Alqahtani, Ali Mobasheri, and Jaume Bacardit. "A machine learning heuristic to identify biologically relevant and minimal biomarker panels from omics data." *BMC Genomics* 16 (2015): 1–12.

Tripathi, Manish Kumar, Abhigyan Nath, Tej P. Singh, A. S. Ethayathulla, and Punit Kaur. "Evolving scenario of big data and Artificial Intelligence (AI) in drug discovery." *Molecular Diversity* 25 (2021): 1439–1460.

Tekkeşin, Ahmet İlker. "Artificial intelligence in healthcare: past, present and future." *Anatolian Journal of Cardiology* 22, no. Suppl 2 (2019): 8–9.

"UniProt: the universal protein knowledgebase in 2021." *Nucleic Acids Research* 49, no. D1 (2021): D480–D489.

Vatansever, Sezen, Avner Schlessinger, Daniel Wacker, H. Ümit Kaniskan, Jian Jin, Ming-Ming Zhou, and Bin Zhang. "Artificial intelligence and machine learning-aided drug discovery in central nervous system diseases: State-of-the-arts and future directions." *Medicinal Research Reviews* 41, no. 3 (2021): 1427–1473.

Wallach, Izhar, Michael Dzamba, and Abraham Heifets. "AtomNet: a deep convolutional neural network for bioactivity prediction in structure-based drug discovery." *arXiv preprint arXiv:1510.02855* (2015).

Wassermann, Anne Mai, Hanna Geppert, and Jürgen Bajorath. 2011. 'Application of Support Vector Machine-Based Ranking Strategies to Search for Target-Selective Compounds'. In *Chemoinformatics and Computational Chemical Biology*, edited by Jürgen Bajorath, 672:517–30. Methods in Molecular Biology. Totowa, NJ: Humana Press. https://doi.org/10.1007/978-1-60761-839-3_21.

Wang, Tao, Mian-Bin Wu, Jian-Ping Lin, and Li-Rong Yang. "Quantitative structure–activity relationship: promising advances in drug discovery platforms." *Expert Opinion on Drug Discovery* 10, no. 12 (2015): 1283–1300.

Xu, Yongjun, Xin Liu, Xin Cao, Changping Huang, Enke Liu, Sen Qian, Xingchen Liu et al. "Artificial intelligence: a powerful paradigm for scientific research." *Innovation* 2, no. 4 (2021).

Yoo, Jiho, Tae Yong Kim, InSuk Joung, and Sang Ok Song. "Industrializing AI/ML during the end-to-end drug discovery process." *Current Opinion in Structural Biology* 79 (2023): 102528.

Zaouter, Cédrick, Alexandre Joosten, Joseph Rinehart, Michel MRF Struys, and Thomas M. Hemmerling. "Autonomous systems in anesthesia: where do we stand in 2020? A narrative review." *Anesthesia & Analgesia* 130, no. 5 (2020): 1120–1132.

Zhu, Zhaocheng, Chence Shi, Zuobai Zhang, Shengchao Liu, Minghao Xu, Xinyu Yuan, Yangtian Zhang et al. "Torchdrug: a powerful and flexible machine learning platform for drug discovery." *arXiv preprint arXiv:2202.08320* (2022).

11 Impact of AI on Healthcare from Diagnostics to Drug Discovery

Smrita Singh and Ashutosh Singh Chauhan

11.1 INTRODUCTION

11.1.1 Scope and Applications of Bioinformatics

Bioinformatics is an interdisciplinary field that combines biology and information technology. It leverages computational and analytical tools to interpret biological data. Following are scopes of bioinformatics:

1. Understanding Genes: Bioinformatics helps us understand gene functions.
2. Cell Organization and Function: It analyzes cellular processes.
3. Drug Target Analysis: Identifying potential drug targets.
4. Disease Characteristics: Examining disease-related features.
5. Biological Database Management: Developing tools for managing biological data.

Following are Applications of Bioinformatics:

1. Medicine: Bioinformatics aids in drug discovery, personalized medicine, and understanding disease mechanisms.
2. Genomics and Proteomics: Analyzing genetic and protein data.
3. Drug Development: Predicting drug side effects and designing more effective drugs.
4. Agriculture: Understanding crop patterns, pest control, and crop management.
5. Environmental Conservation: Studying biodiversity and ecosystems.

11.1.2 AI in Bioinformatics

AI has made significant strides in various scientific and engineering domains, including computational systems biology. Traditionally, computational biology tackles biological problems across different scales, from individual organisms to cellular components, DNA, and external ecosystems, using mathematical models and

Artificial Intelligence Impact

statistical analysis. Bioinformatics and Genomics, closely related to computational systems biology, benefit from AI's versatile toolbox. Unlike conventional methods, AI enables more efficient problem-solving. Machine learning, particularly neural networks and deep learning, dominates the current AI landscape in bioinformatics and genomics. However, AI's applications extend beyond these techniques, encompassing molecular structure analysis, gene sequence matching, protein interactions, and more.

11.2 AI TYPES RELEVANT TO HEALTHCARE

AI encompasses a range of technologies rather than a single entity. While their specific applications and functions vary widely, many of these technologies hold significant relevance for the healthcare sector (see Figure 11.1).

11.2.1 Machine Learning

Arthur Samuel (1959) defined machine learning as the branch of study that allows computers to learn without explicit programming. Consequently, ML can be defined as the branch of computer science that studies how to create self-programming computers (Edwards, 2018). ML involves four steps for data prediction as follows:

1. Training: Historical data is used to train a ML model.
2. Testing and Prediction: The trained model is then tested on new data to make predictions.
3. Validation: A portion of the available historical data (not used during training) is reserved for validation. The model's performance is evaluated using this validation data.

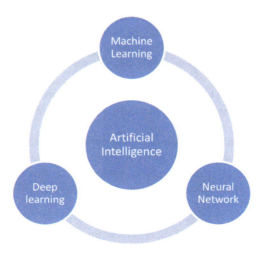

FIGURE 11.1 AI Technologies.

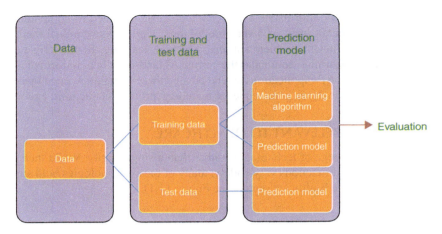

FIGURE 11.2 Machine learning process.

4. Performance Measure: Accuracy is commonly used to assess the model's performance. It represents the ratio of correctly predicted features to the total features to be predicted.

To put it succinctly, past data trains the model, which is then evaluated using unseen data to ensure accurate predictions (see Figure 11.2).

11.2.1.1 Machine Learning Methods

There are two types of ML: supervised and unsupervised learning.

1. Supervised learning: Supervised ML is a technique where an algorithm learns from labeled data to make predictions or decisions. In this approach:
   a) **Training**: The model is trained on a labeled dataset, which includes both input features and corresponding output labels.
   b) **Prediction**: Using this training, the model predicts output values for new data.
   c) **Validation**: The predicted output is compared with the actual output, allowing for error identification and model refinement (Edward, 2018).
2. Unsupervised ML algorithms: Unsupervised learning occurs without a supervisor to guide or correct. It is used when unclassified data is available for training. The system explores data to infer rules and describe hidden structures from unlabeled datasets (Edward, 2018; Gupta et al., 2018 & Quinlan, 2014).

Precision medicine, which predicts which treatment regimens are likely to be effective for a patient based on a variety of patient features and the treatment environment, is the most popular use of classical ML in the healthcare industry.

Artificial Intelligence Impact

11.2.2 Deep Learning

The productive AI fields have been shaped by deep learning. Its applications span a wide range of areas, including sequence and text prediction, computational biology, machine vision, voice, and signal processing (Zhang et al., 2016; Esteva et al., 2017; Ching et al., 2018). Artificial neural networks, deep structured learning, and hierarchical learning are a few examples of deep learning implementation models. These models typically use a class of structured networks to infer quantitative properties between responses and causes within a set of data (Ditzler et al., 2015; Liang et al., 2015; Xu J. et al., 2016; Giorgi and Bader, 2018).

11.2.2.1 Applications of Deep Learning in Bioinformatics

Deep learning has made significant strides in the field of bioinformatics, leveraging its capabilities to handle large-scale data, including in the following:

1. **Omics Research**:
   - Deep learning architectures, such as **Deep Neural Networks (DNNs)**, **Convolutional Neural Networks (CNNs)**, and **Recurrent Neural Networks (RNNs)**, are used for analyzing various omics data. These include genomics, transcriptomics, proteomics, and metabolomics.
   - DNNs can learn complex patterns from high-dimensional data, aiding in tasks like gene expression prediction, variant calling, and functional annotation.
2. **Biomedical Image Processing**:
   - CNNs excel in biomedical image analysis. They can automatically extract features from medical images, such as X-rays, MRIs, and histopathology slides.
   - Applications include tumor detection, cell segmentation, and identifying disease-related patterns in images.
3. **Biomedical Signal Processing**:
   - RNNs are useful for processing sequential data, such as time-series signals from wearable devices or electrocardiograms (ECGs).
   - They can predict disease progression, detect anomalies, and assist in personalized medicine.
4. **Protein Classification**:
   - Deep learning models can classify proteins based on their sequences or structures.
   - This aids in understanding protein function, predicting protein-protein interactions, and drug discovery.
5. **Antiviral Peptide Prediction**:
   - Deep learning techniques are employed to predict antiviral peptides from protein sequences.
   - These peptides can potentially inhibit viral infections.
6. **Antibiotic Resistance Identification**:
   - Deep learning models analyze genomic data to identify antibiotic resistance genes.
   - This information guides antibiotic treatment decisions.

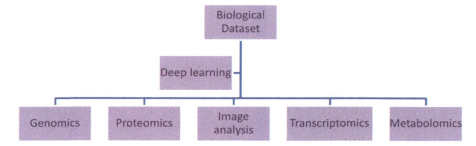

FIGURE 11.3 Deep learning in bioinformatics research.

7. **Computer-Aided Drug Design**:
   - Deep learning assists in virtual screening of compounds for drug discovery.
   - It predicts binding affinities, toxicity, and pharmacokinetic properties.
8. **Drug Formulation Optimization**:
   - Deep learning can optimize drug formulations by predicting solubility, stability, and bioavailability.

Deep learning plays a crucial role in bioinformatics, addressing challenges posed by big data and complex biological systems (see Figure 11.3 and Tables 11.1 and 11.2).

11.3 APPLICATION OF ARTIFICIAL NEURAL NETWORKS

There are various applications of ANN in the field of information, medicine, economic, control, transport and psychology (Wu et al., 2018).

1. Information processing: The complex issues that modern information processing must address are numerous. ANNs can mimic or replace human thought processes. They can also be used to automatically diagnose issues and solve problems that conventional approaches are unable to address. ANN systems are frequently employed in military system electronic equipment applications because of their high fault tolerance, robustness, and self-organization. They can function optimally even in the event of significant damage to the connection. Intelligent devices, automated fault diagnosis and alarm systems, autonomous control guidance systems, automatic tracking and monitoring instrumentation systems are some examples of the intelligent information systems that are currently in use.
2. Pattern recognition: Processing and analyzing different forms of information that characterize objects or occurrences allows for the procedure for characterizing, recognizing, grouping, and analysingthem. This process is known as pattern recognition. The technology uses Shennong's information theory and Bayesian probability theory as its theoretical foundation, bringing information

TABLE 11.1
Summary of Bioinformatics database

S.No	Bioinformatics Database	Features
1	GenBank	Vast collection of DNA and RNA sequences, annotated with source organism, gene information, etc. Integrated with other NCBI tools.
2	RefSeq	Curated subset of GenBank, aiming for non-redundant, well-annotated sequences for each gene/transcript.
3	Ensembl	Genome browser with gene annotation, variation data, comparative genomics. Strong on vertebrate genomes
4	UCSC Genome Browser	Visual exploration of genomes, customizable tracks for various data types (genes, regulation, variation).
5	UniProt	Central protein sequence and function database, combining Swiss-Prot (manually curated) and TrEMBL (automatically annotated).
6	Protein Data Bank	3D structures of proteins and other macromolecules, determined experimentally.
7	InterPro	Protein family and domain signatures, predicting function based on sequence patterns.
8	KEGG	Metabolic pathways, signaling networks, disease associations. Strong on linking genes to biological processes.
9	Reactome	Curated pathways, emphasizing human biology but with cross-species comparisons.
10	BioGRID	Protein-protein interaction database, collecting data from various sources.
11	Gene Ontology	Controlled vocabulary for describing gene product functions.
12	dbSNP	Database of single nucleotide polymorphisms and other variations.
13	ClinVar	Links genetic variants to their clinical significance (disease associations).
14	ArrayExpress/GEO	Repositories for microarray and high-throughput sequencing data.

processing closer to the logical thought process seen in the human brain. Statistical pattern recognition and structural pattern recognition are now the two fundamental techniques for pattern recognition.

3. Biological signal detection and analysis: The continuous waveform data output by the majority of medical testing apparatus serves as the foundation for medical analysis Neural networks are primarily used in biomedical signal detection and processing for the following purposes: analysis of EEG signals, extraction of auditory evoked potential signals, identification of EMG and

TABLE 11.2
Deep learning applications in bioinformatics research fields

S.No	Deep learning techniques	Applications
1	Convolutional Neural Networks	• Image analysis (microscopy, medical imaging) • Genomic sequence analysis (motif finding, regulatory element prediction)
2	Recurrent Neural Networks	• DNA/RNA sequence analysis (gene prediction, splicing analysis) • Protein structure prediction (secondary structure, contact maps) • Time-series data (gene expression over time)
3	Long Short-Term Memory Networks	• Similar to RNNs, but often preferred for longer sequences or complex relationships.
4	Autoencoders	• Gene expression analysis (identifying patterns, clustering) • Drug discovery (finding similar molecules)
5	Generative Adversarial Networks	• Drug discovery (designing novel molecules) • Generating synthetic biological data for simulations
6	Graph Neural Networks	• Drug-target interaction prediction • Pathway analysis
7	Transformers	• Protein language modeling (predicting structure, function from sequence) • Linking genes to literature

gastrointestinal signals, compression of ECG signals, and recognition of medical images.

4. Medical Expert system: The classic expert system builds a knowledge base, uses logical reasoning to diagnose medical conditions, and stores the expertise experience and knowledge as dataset in computer. In real-world applications, however, growing database sizes will cause a "explosion" of information and a "bottleneck" in knowledge access, which will lead to low efficiency. Because it addresses the issues raised by expert systems, enhances their capacity for knowledge inference, self-organization, and self-learning, and opens up new avenues for research, neural networks based on nonlinear parallel processing are extensively utilized by medical professionals, The system has been created and utilized extensively. In the field of anesthesia, critical medicine, and related research, there is ongoing interest in analyzing and predicting physiological variables using clinical data. However, there is still insufficient evidence regarding the relationship between phenomena, signal processing, automatic interference signal detection, and the prediction of clinical conditions. Artificial neural network technology can be applied to address these challenges (see Figures 11.4 and 11.5).

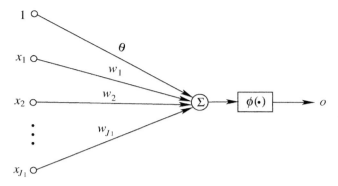

FIGURE 11.4 Artificial neural network (ANN) structure with input unit. (Adopted from Du, K.-L.; Leung, C.-S.; Mow, W.H.; Swamy, M.N.S. Perceptron: Learning, Generalization, Model Selection, Fault Tolerance, and Role in the Deep Learning Era. *Mathematics* **2022**, *10*, 4730. https://doi.org/10.3390/math10244730.)

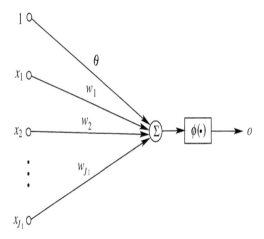

FIGURE 11.5 Single-layer perceptron with input layer and output layer. (Adopted from Du, K.-L.; Leung, C.-S.; Mow, W.H.; Swamy, M.N.S. Perceptron: Learning, Generalization, Model Selection, Fault Tolerance, and Role in the Deep Learning Era. *Mathematics* **2022**, *10*, 4730. https://doi.org/10.3390/math10244730.)

11.3.1 Role of AI in Autonomous Molecular Design

Recently, there has been a growing adoption of domain-aware AI to speed up molecular design for a variety of applications, such as drug discovery and design. Recent advances in disciplines such as computer infrastructures, software engineering, and high-end hardware development make it possible to create buildable and explicable

AI chemical discovery systems. This could improve a design hypothesis through more intelligent searches of chemical space, feedback analysis, and data integration that can act as a basis for the introduction of end-to-end automation for compound discovery and optimization. A number of cutting-edge MLarchitectures are primarily and separately employed for small molecule property prediction, high-throughput synthesis, and screening, iteratively finding and refining therapeutic compounds. Nevertheless, many technical, conceptual, scalability, and end-to-end error quantification challenges are also raised by these deep learning and MLtechniques.

11.3.1.1 Components of Computational Autonomous Molecular Design

The computational autonomous molecular design (CAMD) consists of the following steps:

1. Effective instruments for creating and extracting data
2. Effective data representation methods
3. Predictive MLmodels with physics-based constraints
4. Molecular generation tools informed by learned knowledge

A computational workflow for lead molecule discovery should ideally be able to learn from its mistakes and adapt its functionality through active learning when the targeted functionality or the chemical environment changes. When all the parts cooperate with one another, giving input and enhancing the model's performance as we go through the steps, this can be accomplished (see Figure 11.6).

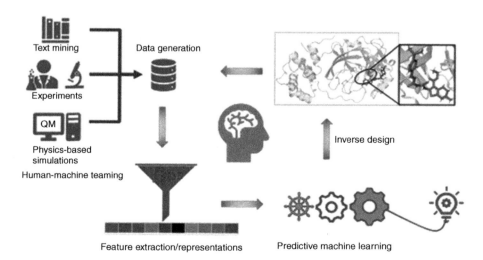

FIGURE 11.6 Autonomous computational molecule discovery flow diagram. (Adopted from Joshi, R.P.; Kumar, N.Artificial Intelligence for Autonomous Molecular Design: A Perspective. Molecules **2021**, 26, 6761. https://doi.org/10.3390/molecules26226761.)

1. **Effective instruments for creating and extracting data:** Density functional theory (DFT) is a popular option for data creation in CAMD, mostly because to its reasonable accuracy and efficiency. Typically, 3D structures are fed into DFT in order to forecast the desired attributes. The more pertinent structural and property data are extracted from the data generated by DFT simulations through processing, and these can be employed as targets for MLmodels. The obtained data can be applied in two distinct ways: first, by utilizing a direct supervised MLapproach to calculate the properties of new molecules, and second, by employing inverse design to create new molecules possessing the necessary characteristics of the target. Databases and other auxiliary components can be integrated with CAMD to store and display data.
2. **Robust data representation techniques:** For MLmodels to perform accurately, molecules must be robustly represented (Huang & Lilienfeld, 2016). A perfect molecule representation would be one that is distinct, invariant to various symmetry operations, invertible, and effectively captures the physics, stereochemistry, and structural motif. Certain objectives can be met by utilizing the physical, chemical, and structural features (Chen et al., 2019), all of which are rarely thoroughly recorded, making it difficult to collect this information. Several different strategies that are effective for particular issues have been used to address this over time (Elton, 2019; Bjerrum, 2017; Gilmer et al., 2017; Hamilton et al., 2017; Kearnes et al., 2016; Wu et al., 2018). Various molecular representation formulations have been used for predictive and generative modeling (see Figure 11.7).
3. **Physics-Informed Machine Learning:** An intriguing model that connects MLand physical rules is called Physics-Informed Machine Learning (PIML). It improves performance on tasks involving a physical process by utilizing empirical data and physically accessible prior knowledge. PIML combines domain knowledge and physical rules with data-driven models. It aims to solve complex problems in science and engineering by combining empirical data and mathematical physics models. By incorporating physical priors into MLmodels, PIML guides the learning process toward physically plausible solutions. PIML involves two key phases:

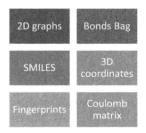

FIGURE 11.7 Molecular representation in every formulation that has been employed in research on generative and predictive modeling.

1. Representation Learning: This phase focuses on learning meaningful representations of molecules (or other systems) from data.
2. Property Prediction: Once the representations are learned, they are used for accurate property prediction (e.g., predicting molecular properties).
Graph neural networks (GNNs) are a popular approach within PIML. GNNs learn both the representation and prediction simultaneously, benefiting from both data and physical knowledge. It improves accuracy and efficiency, even in uncertain and high-dimensional contexts. PIML can be applied to various domains, including drug discovery, molecular modeling, and more. Researchers are actively exploring how to encode different forms of physical prior knowledge into model architectures, optimizers, and inference algorithms.
3. **Molecular generation tools informed by learned knowledge**: Here are some tools and approaches for generative molecule design using AI:
   REINVENT 4: An open-source generative AI system called REINVENT 4 was created especially for the design of tiny molecules. It uses transformer designs and recurrent neural networks (RNNs) to power the synthesis of molecules. De novo design (building molecules from the ground up), R-group substitution, library and linker design, scaffold hopping, and molecular optimisation are among its uses. REINVENT 4 is a command-line tool available on GitHub under the Apache 2.0 license. It provides reference implementations for common AI-based molecule generation algorithms, fostering transparency, collaboration, and education.
   Unified Framework for Predicting Properties and Generating Molecules: Researchers from MIT and the MIT-IBM Watson AI Lab developed a unified framework. It simultaneously predicts molecular properties and generates new molecules more efficiently than traditional deep-learning approaches.It efficiently predicts properties and generating novel molecules.
   Generative Models: State-of-the-art generative models include Recurrent neural networks (RNNs),Variational autoencoders (VAEs), Adversarial autoencoders (AAEs), Generative adversarial networks (GANs), It is used for De novo drug design and molecular optimization.
   Learning Molecular Grammar: A ML system learns the "language" of molecules using a small, domain-specific dataset. It constructs viable molecules and predicts their properties based on this learned grammar. These tools contribute to advancing molecular design and drug discovery through AI-driven approaches.
5. **Challenges of Artificial intelligence in Healthcare Sector**: The challenges of achieving explainable AI in biomedical data science are multifaceted.
   1. **AI Method Customization**: Most state-of-the-art AI techniques are not inherently designed for biomedical data. Adapting these methods to the unique characteristics of biological and clinical data poses a challenge (Han & Liu, 2022).
   2. **Nonlinear Data**: Biomedical data often exhibits high nonlinearity due to complex interactions between genes, proteins, and other molecular components. Traditional linear models may struggle to capture these intricate relationships effectively.

3. **Problem-Solving Complexity**: Biomedical problems can be highly complex, involving multiple variables, noisy data, and intricate dependencies. Developing AI solutions that address such complexity while maintaining interpretability is a significant challenge.
4. **Learning Bias**: AI models can inadvertently learn biases present in the training data. Ensuring fairness, transparency, and unbiased decision-making is crucial, especially in healthcare applications.

Achieving explainable AI in bioinformatics requires tailored methods, handling nonlinear data, addressing problem complexity, and mitigating learning bias (Han & Liu, 2022).

11.4 FUTURE PROSPECTS OF ARTIFICIAL INTELLIGENCE

AI has a bright future ahead of it in a lot of different fields. AI will keep improving drug development, personalized care, and diagnostics to revolutionize healthcare. Early disease detection can be achieved using predictive models, and routine chores can be automated by robots with AI. AI can customize medicines by analyzing genetics, medical history, and patient data. Predictive models assist in identifying people who are at-risk and in suggesting protective actions. AI algorithms improve pathology, radiology, and other imaging-related diagnostic accuracy. They are faster than human experts at detecting abnormalities, tumors, and fractures. AI predicts molecular interactions, finds possible drug candidates, and optimizes chemical structures to speed up the drug development process. This lowers expenses and simplifies the procedure. Chatbots and virtual nurses driven by AI offer round-the-clock assistance, respond to medical questions, and help with appointment scheduling. They lessen administrative load and enhance patient participation. AI algorithms forecast illness outbreaks, patient readmissions, and therapeutic results. This makes resource allocation and proactive actions possible. AI-guided surgical robots improve accuracy, minimize invasiveness, and shorten recovery times. To get better results, surgeons work in tandem with these intelligent systems. Real-time health data is gathered by sensors and wearable technology. AI examines this data to track chronic disorders and identify early indicators of disease. It is critical to address privacy, prejudice, and transparency as AI becomes more and more integrated into healthcare. It is crucial to strike the correct balance between automation and human knowledge. In conclusion, AI will transform healthcare by enhancing patient outcomes, treatment, and diagnosis. A healthier future will be shaped by responsible AI development and engagement with healthcare experts.

REFERENCES

Bjerrum, E.J. (2017). SMILES Enumeration as Data Augmentation for Neural Network Modeling of Molecules.arXiv:1703.07076.
Chen, C., Ye, W., Zuo, Y., Zheng, C.,and Ong, S.P.(2019). Graph networks as a universal machine learning framework for molecules and crystals. *Chem. Mater.* 31, 3564–3572.

Ching, T., Himmelstein, D. S., Beaulieu-Jones, B. K., Kalinin, A. A., Do, B. T.,Way, G. P., et al. (2018). Opportunities and obstacles for deep learning in biology and medicine. *J. R. Soc. Interface* 15, 20170387. doi: 10.1098/rsif.2017.0387.

Ditzler, G., Polikar, R., Member, S., Rosen, G., and Member, S. (2015). Multi-layer and recursive neural networks for metagenomic classification. *IEEE. Trans. Nanobiosci.* 14, 608. doi: 10.1109/TNB.2015.2461219.

Du, K.-L., Leung, C.-S., Mow, W.H.,and Swamy, M.N.S. (2022). Perceptron: Learning, generalization, model selection, fault tolerance, and role in the deep learning era. *Mathematics* 10, 4730. https://doi.org/10.3390/math10244730.

Elton, D.C., Boukouvalas, Z., Fuge, M.D.,and Chung, P.W.(2019). Deep learning for molecular design—A review of the state of the art. *Mol. Syst. Des. Eng.* 4, 828–849.

Esteva, A., Kuprel, B., Novoa, R. A., Ko, J., Swetter, S. M., Blau, H. M., and Thrun, S. (2017). Dermatologist-level classification of skin cancer with deep neural networks. *Nature* 542, 115–118. doi: 10.1038/nature21056.

Gauthier, J., Vincent, A. T., Charette, S. J., and Derome, N. (November, 2019). A brief history of bioinformatics.*Brief.Bioinf.* 20(6), 1981–1996.https://doi.org/10.1093/bib/bby063

Gilmer, J., Schoenholz, S.S., Riley, P.F., Vinyals, O.,and Dahl, G.E.(2017). Neural Message Passing for Quantum Chemistry. arXiv:1704.01212.

Giorgi, J. M., and Bader, G. D. (2018). Transfer learning for biomedical named entity recognition with neural networks. *Bioinformatics* 34, 4087–4094.doi: 10.1093/bioinformatics/bty449.

Gupta, D., Julka, A., Jain, S., Aggarwal, T., Khanna, A., Arunkumar, N., and DeAlbuquerque, V.H.C. (2018). Optimized cuttlefish algorithm for diagnosis of Parkinson's disease.*Cognit. Syst. Res.* 52, 36e48.

Hamilton, W.L., Ying, R.,and Leskovec, J.(2017). Representation Learning on Graphs: Methods and Applications. arXiv:1709.05584.

Han,H., and Liu,X. (January 20, 2022). The challenges of explainable AI in biomedical data science. *BMC Bioinformatics*.22(Suppl 12), 443. doi: 10.1186/s12859-021-04368-1. PMID: 35057748; PMCID: PMC8772040.

Hinton, G. E., and Salakhutdinov, R. R. (2006). Reducing the dimensionality of data with neural networks. *Science* 313, 504–507. doi: 10.1126/science.1127647.

Hogeweg, P. (1978). Simulating the growth of cellular forms. *Simulation* 31(3), 90e96.

Huang, B., and von Lilienfeld, O.A.(2016). Communication: Understanding molecular representations in machine learning: The role of uniqueness and target similarity. *J. Chem. Phys.* 145, 161102.

Hunt, L.T. (1983). Margaret O. Dayhoff 1925e1983. *DNA Cell Biol.* 2(2), 97e98.

Joshi, R.P., and Kumar, N.(2021). Artificial intelligence for autonomous molecular design: A perspective. *Molecules* 26, 6761.

Kearnes, S., McCloskey, K., Berndl, M., Pande, V.,and Riley, P.(2016). Molecular graph convolutions: Moving beyond fingerprints. *J. Comput. Aided Mol. Des.* 30, 595–608.

LeCun, Y., Bengio, Y., and Hinton, G. (2015). Deep learning. *Nature* 521, 436. doi: 10.1038/nature14539.

Lesko, L.J. (2012). Drug research and translational bioinformatics. *Clin. Pharmacol. Ther.* 91(6), 960e962.

Liang, M., Li, Z., Chen, T., and Zeng, J. (2015). Integrative data analysis of multiplatform cancer data with a multimodal deep learning approach. *IEEE/ACM Trans. Comput. Biol. Bioinf.* 12, 928–937. doi: 10.1109/TCBB.2014.2377729.

Mamoshina, P., Vieira, A., Putin, E., and Zhavoronkov, A. (2016). Applications of deep learning in biomedicine. *Mol. Pharmaceut.* 13, 1445–1454. doi: 10.1021/acs.molpharmaceut.5b00982.

Mnih, V., Kavukcuoglu, K., Silver,D., Rusu, A. A.,Veness, J., Bellemare,M.G., et al.(2015). Human-level control through deep reinforcement learning. *Nature* 518, 529–533. doi: 10.1038/nature14236.

Nussinov, R. (2015). Advancements and challenges in computational biology. *PLoSComput. Biol.* 11, e1004053. doi: 10.1371/journal.pcbi.1004053.

Ouzounis, C.A. (2012). Rise and demise of bioinformatics? Promise and progress. *PLoSComput. Biol.* 8(4), 1e5.

Pevsner, J.(2009). *Pairwise Sequence Alignment. Bioinformatics and Functional Genomics*, second ed.Wiley-Blackwell, ISBN 978-0-470-08585-1,pp. 58e68.

Quinlan, R.(2014). *C4.5: Programs for Machine Learning*, Morgan Kaufmann Publishers, San Mateo, CA.

Ranganathan, S., Menon, R., and Gasser, R. B. (2009). Advanced in silico analysis of expressed sequence tag (EST) data for parasitic nematodes of major socio-economic importance — Fundamental insights toward biotechnological outcomes.*Biotechnol Adv.* 27(4), 439–448, ISSN 0734-9750.https://doi.org/10.1016/j.biotechadv.2009.03.005.

Rashedi, K.A., Ismail, M.T., Al Wadi, S., Serroukh, A., Alshammari, T.S., and Jaber, J.J.(2024). Multi-layer perceptron-based classification with application to outlier detection in Saudi Arabia Stock Returns. *J. Risk Financial Manag.*17, 69. https://doi.org/10.3390/jrfm17020069.

Sarkar, I.N., Butte, A.J., et al.(2011). Translational bioinformatics: linking knowledge across biological and clinical realms. *J. Am. Med. Inf. Assoc.* 18(4), 345e357.

Wu, Z., Ramsundar, B., Feinberg, E., Gomes, J., Geniesse, C., Pappu, A.S., Leswing, K., Pande, V.(2018). MoleculeNet: A benchmark for molecular machine learning. *Chem. Sci.* 9, 513–530.

Xu, T., Zhang, H., Huang, X., Zhang, S., and Metaxas, D. N. (2016). Multimodal Deep Learning for Cervical Dysplasia Diagnosis.In *International Conference on Medical Image Computing and Computer-Assisted Intervention (Boston, MA)*,115–123.

Zhang, S., Zhou, J., Hu, H., Gong, H., Chen, L., Cheng, C., and Zeng, J. (2016). A deep learning framework for modeling structural features of RNA-binding protein targets. *Nucleic Acids Res.* 44, e32. doi: 10.1093/nar/gkv1025.

12 Biomedical Imaging Techniques in AI Applications

T. Kalpana, R. Thamilselvan, and T. M. Saravanan

12.1 INTRODUCTION

Recent research has concentrated on creating non-invasive molecular imaging technologies and automated image processing tools. These developments are critical for improving the precision of imaging at the cellular and molecular levels, which is vital for precise diagnosis and therapy planning [1]. The definitive diagnosis is typically obtained following a clinical examination and interpretation of imaging modalities such as magnetic resonance imaging (MRI) or computed tomography (CT), followed by pathological testing [2]. MRI, CT methods generate 3D images from 3D using digital geometry [3]. In addition to RECIST-based approaches, multimodal imaging technologies with superior artificial intelligence (AI) should be developed to assess cancer-targeted therapy fully and reliably [4]. The research explores the application of deep learning to biological microscopic images, emphasizing advancements in image classification, object recognition, and image segmentation. Deep learning has outperformed traditional machine learning approaches, making it more useful in medical and biological research [5]. The AlphaFold 3 model utilizes an enhanced diffusion-based architecture to predict complex structures, including proteins, nucleic acids, small molecules, ions, and mutated residues [6]. This chapter focuses on generating high-resolution images from low-resolution counterparts to improve image quality [7]. Genetic and environmental risk factors for neurodegenerative illnesses are discussed, along with the connection between the microbiota-gut-brain axis components and the fecal microbiota transplantation (FMT) therapy technique [8].

12.2 VARIOUS METHODS OF BIOIMAGING

Bioimaging refers to the techniques and processes used to create visual representations of biological processes and structures, which are typically at the cellular or molecular level.

Bioimaging methods cover a variety of techniques that are employed to observe biological structures, activities, and functions at different levels, ranging from molecules to tissues.

The bioimaging methods are Light Microscopy, Electron Microscopy, X-ray Imaging, MRI (Magnetic Resonance Imaging), Ultrasound Imaging, Nuclear

Biomedical Imaging Techniques in AI Applications 243

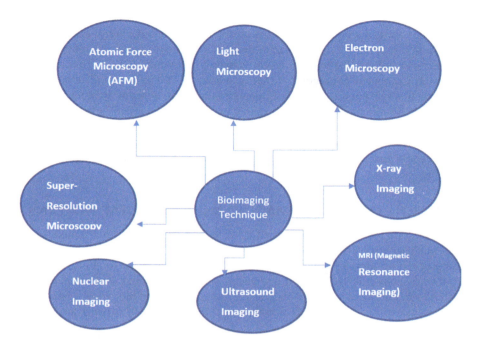

FIGURE 12.1 Types of bio-medical imaging.

Imaging, Super-resolution Microscopy, and Atomic Force Microscopy (AFM) (see Figure 12.1).

Atomic Force Microscopy (AFM) uses a pointed probe to monitor surface forces and capture high-resolution photographs of biological samples at nanoscale. Light Microscopy integrates bright-field microscopy, which involves direct illumination of specimens. Fluorescence microscopy labels certain molecules with fluorescent dyes. Focused Microscopy uses a pinhole and laser scanning to improve resolution and decrease out-of-focus blur. Multiphoton microscopy uses longer wavelength light to provide deeper imaging of tissues.

Electron Microscopy (TEM) is a technique that uses electrons passed through a thin object to produce high-resolution images. Scanning Electron Microscopy (SEM) creates comprehensive 3D images by scanning a concentrated electron beam across the specimen's surface. X-ray imaging incorporates the X-ray microscopy to capture high-resolution images of biological material. MicroCT (micro-computed tomography) creates 3D images by merging numerous x-ray images acquired from various angles. MRI (Magnetic Resonance Imaging) is used for detailed imaging of organs and tissues.

The Functional MRI (fMRI) measures brain activity by detecting variations in blood flow.

The Diffusion MRI maps the diffusion of water molecules in biological tissues, providing information on tissue structure. By Ultrasound imaging the Medical ultrasound imaging, which uses high-frequency sound waves to visualize soft tissues within the body, is commonly utilized in obstetrics and cardiology. PET (Positron Emission Tomography) detects gamma rays released by a positron-emitting radioactive tracer put into the body to reveal metabolic activities. SPECT (Single Photon Emission Computed Tomography) that is similar to PET but uses different tracers and detects single photon emissions. Super-resolution microscopy techniques include Structured Illumination Microscopy (SIM), which uses patterned light to overcome the diffraction limit and achieve resolutions beyond conventional light microscopy, and Stimulated Emission Depletion (STED) microscopy, which employs laser beams to selectively deactivate fluorescence outside the focal spot, enabling super-resolution imaging.

12.3 ROLE OF AI IN BIOIMAGING AND MOLECULAR PATTERNS

AI is being used to evaluate single-cell RNA sequencing data for understanding cellular heterogeneity, cell type classification, and lineage tracking. This is critical for comprehending complex biological processes on the single-cell level. Integrating data from several omics levels (genomics, transcriptomics, proteomics, and metabolomics) with AI can provide a more comprehensive understanding of biological systems and disease causes. Techniques such as multi-omics data fusion and network analysis are intensively investigated. AI techniques such as network inference, dynamical modeling, and pathway analysis are used to model complicated biological systems to comprehend how molecules interact within cells and animals. This aids in forecasting system behavior and responses to perturbations. AI approaches including feature selection, classification algorithms, and deep learning models are being used to identify illness biomarkers.

Biomarkers are essential for early detection, prognosis, and individualized treatment options. Biomedical text mining uses natural language processing (NLP) and AI to extract information from large amounts of biomedical literature, clinical notes, and electronic health records. This helps with knowledge discovery, literature curation, and evidence-based medicine. AI algorithms are being developed to precisely anticipate protein structures and generate novel proteins with specific functionalities. This has potential uses in drug creation, enzyme engineering, and synthetic biology. AI is used to integrate genomic and microbiome data with dietary information in order to deliver individualized nutrition recommendations. This strategy seeks to enhance health outcomes by customizing meals to individuals' genetic and metabolic characteristics. Using AI to improve medical imaging analysis for automated illness identification, organ or tumor segmentation, and patient outcome prediction based on imaging data enhances both diagnosis accuracy and treatment planning.

12.4 AUGMENTATION TECHNIQUES IN BIOIMAGING

In bioinformatics and AI, augmentation approaches improve model performance and resilience by enhancing datasets or features. Data augmentation entails creating

Biomedical Imaging Techniques in AI Applications 245

synthetic data from existing data samples using changes that preserve the underlying biological information.

Common augmentation/transformation techniques include:

- Image augmentation altering biological images (e.g., microscopy, medical scans) to increase training data diversity through rotation, flipping, scaling, cropping, noise addition, or contrast/brightness adjustments.
- Sequence augmentation modifying nucleotide or amino acid sequences while maintaining biologically relevant patterns and structures. Time-series augmentation is done for real time data that deals with time relationship which removes noise on data and can generate synthetic data.
- Features are also need to be identified from the collected data. By creating new feature or modifying the existing feature we can enhance the model for prediction by listening molecular factors. Derived Features generate additional features from raw data, such as statistical summaries (mean, variance), frequency domain features (FFT coefficients), or domain-specific features (protein structure descriptors). Embedding techniques, such as those using biological ontologies (e.g., Gene Ontology) or word embeddings (e.g., vectorized protein sequences), represent biological entities in a continuous space. These approaches help capture the relationships and similarities between entities, enabling more effective analysis and interpretation of complex biological data.
- Transfer Learning and Pre-training encapsulated the pre-trained models or datasets to enhance learning efficiency and generalization. Transfer Learning involves fine-tuning pre-trained models (e.g., language models trained on biomedical text) for specific bioinformatics tasks like protein-protein interaction prediction and drug-target interaction prediction. These models have already acquired key aspects like edges, textures, and forms, which can help with medical pictures. Pretraining involves training models on huge biological datasets (e.g., gene expression, protein sequences) to learn generic properties that can be applied to downstream tasks. Pretrained Models for Cell Type Annotation: Use models pretrained on large single-cell RNA-seq datasets to automatically annotate cell types in new datasets. Use transfer learning to predict the functional characteristics of microbial communities using 16S rRNA or metagenomic sequencing data.
- Adversarial Augmentation specify the GANs or related approaches to generate realistic synthetic data to enhance the original dataset. This strategy is especially effective when data collection is expensive or limited. Microbiome Classification classifies the Pre-train algorithms on huge microbiome datasets to learn microbial community representations. Refine these models for tasks like disease association research or environmental impact evaluations. Cross-validation and Ensemble Techniques enhancing training data with approaches such as k-fold cross-validation or ensemble learning, which train several models on diverse subsets of data or representations to increase overall prediction accuracy and robustness.

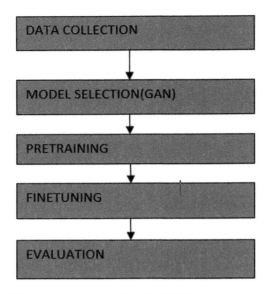

FIGURE 12.2 Workflow of proposed model using GAN.

12.5 WORKFLOW OF BIOIMAGING USING GAN ARCHITECTURE AND APPLICATIONS

- Data Collection: Create a huge dataset for pretraining, such as genetic sequences or protein structures (see Figure 12.2).
- Pretraining: Train a deep learning model (e.g., a convolutional neural network or a transformer) using the huge dataset.
- Fine-Tuning: Fine-tune the pretrained model using a smaller, task-specific dataset. BioBERT: A pretrained biomedical language model that can be fine-tuned for various bioinformatics tasks.
- Drug discovery: Tokenizing the drug, compound, and interaction texts. Identifying and marking entity pairs (e.g., drugs and compounds) within the text. Labeling the data according to interaction types (e.g., "activates," "inhibits," "binds to," or "no interaction"). Based on this, virtual screening is done on images to identify new compounds on drugs.
- Evaluation: Evaluate the model's performance on the specified job, and iterate as necessary.

12.6 APPLICATIONS

- Genomic Sequence Analysis and Protein Structure Prediction: Understanding the biological functions encoded by genes and proteins
- Medical Imaging and Drug Discovery: Medical imaging involves the use of various technologies to create visual representations of the interior of a body for clinical analysis, diagnosis, and treatment planning

Biomedical Imaging Techniques in AI Applications

- AlphaFold and Beyond: Utilize pretrained models like AlphaFold for predicting protein structures. Fine-tune these models on specific protein families or adapt them to predict protein-protein interactions. Transfer Learning for Function Prediction: Pre-train models on huge protein sequence databases before fine-tuning them to predict specific protein functions, enzyme activities, or binding locations.
- Molecular Representation Learning: Molecular representation learning involves encoding this information in ways that AI models can process. These representations are then used to train models for various drug discovery tasks, such as predicting drug-target interactions, pharmacokinetics, and toxicity.
- **Deep Generative Models**: Variational autoencoders (VAEs) and generative adversarial networks (GANs) can generate new molecular structures based on learned representations, aiding in the discovery of novel drug candidates.
- Disease Diagnosis: Fine-tune pretrained models on specific datasets for diagnosing diseases from medical images, such as detecting tumors, identifying pathological features in histopathology images, or diagnosing retinal diseases from eye scans.

12.7 GAN ARCHITECTURE

Generative Adversarial Networks (GANs) have shown great potential in a variety of applications, including bioinformatics. GANs are made up of two neural networks, the generator and the discriminator, which compete against one another (see Figure 12.3). The generator strives to generate realistic data, whereas the discriminator seeks to discern between genuine and generated data. This adversarial approach teaches GANs to generate high-quality synthetic data. Adversarial training in GANs is consistent with the chapter, focus on AI applications in drug development and biomedical imaging by providing a potent tool for producing high-quality synthetic data. GANs are used in drug development to generate novel molecular structures, and in biomedical imaging to provide realistic medical images for training diagnostic models. This method overcomes data scarcity, improves model accuracy, and speeds the creation of new treatments, thus contributing to the evolution of AI-driven healthcare solutions.

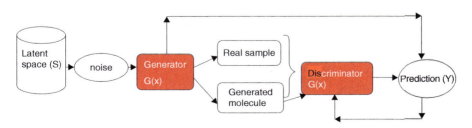

FIGURE 12.3 Prediction of molecular structure using GAN.

12.7.1 Prediction Flow of GAN Architecture Based on Molecular Structure

With bioinformatics, GANs may produce synthetic genomic sequences and single-cell RNA sequencing (scRNA-seq) data that closely resemble the statistical features of real biological data. This generated data is invaluable for supplementing restricted datasets and enabling for more robust training of ML models. In genomics, synthetic sequences aid in the research of genetic variation and disease causes without relying entirely on genuine samples, which can be limited or costly to collect. Similarly, in single-cell RNA-seq, synthetic data can help with the investigation of cellular heterogeneity and the identification of cell kinds and states by giving more high-quality training examples. This method quickens genomics and transcriptomics research by overcoming data restrictions.

Data gathering for bioinformatics tasks requires actual, high-quality data relevant to the issue at hand. Protein structure analysis entails gathering information such as amino acid sequences, 3D protein conformations, and functional annotations from sources such as the Protein Data Bank (PDB) or UniProt. This information forms the basis for a variety of tasks, including protein structure prediction, drug-binding site identification, and functional analysis. Accurate and comprehensive protein structure data is essential for understanding biological processes and applications like medication discovery and enzyme engineering. In the evaluation phase, the quality of generated data is assessed through a variety of metrics. For visual datasets, such as medical images, metrics like **Fréchet Inception Distance (FID)** and **inception score** are used to gauge similarity and diversity. In bioinformatics, it's crucial to apply domain-specific metrics: for example, **Root Mean Square Deviation (RMSD)** is used to evaluate the accuracy of protein structures, while **sequence alignment scores** help determine the accuracy of genetic sequences. These metrics ensure the generated data is accurate and applicable for its intended uses.

12.7.2 Types of GAN and Applications

GANs provide high-quality synthetic data, which is critical for drug development and bioinformatics research. This function solves the prevalent issue of data scarcity, especially in sectors where getting real-world data is costly or time-consuming.

12.7.2.1 Conditional GANs (cGANs)

> Goal: Use an amino acid sequence to infer a protein's three-dimensional structure.
> Method: Convert protein sequences into corresponding protein structures by training a

12.7.2.2 CycleGAN

Understanding the mapping between the 1D sequence data and the 3D structure data can help you accomplish this. CycleGANs are useful for image-to-image translation tasks in bioimaging. They can transform images from one domain to another without paired examples. For instance, translating histopathology images from one tissue type to another for diagnostic purposes. This can help comprehend how proteins work based on their architectures, which is important for researching diseases and developing new drugs

12.7.2.3 Attention-based GANs

These GAN variants incorporate attention mechanisms to focus on relevant regions of the image, enhancing the generation process. In bioimaging, they can help generate images with more anatomical or pathological relevance by focusing on specific regions of interest.

12.7.2.4 Stacked GANs (StackGAN)

StackGANs generate high-resolution images progressively, starting from low-resolution representations. In bioimaging, this approach can be used to generate detailed and realistic high-resolution images of cellular structures or tissue morphology.

12.7.2.5 3D GANs

In bioimaging, especially in fields like medical imaging (e.g., MRI, CT scans), 3D GANs generate and manipulate volumetric data. They can synthesize realistic 3D representations of organs or tissues for diagnostic purposes or drug discovery.

12.7.2.6 Super-resolution GANs (SRGANs)

SRGANs enhance the resolution of low-quality bioimaging data, such as low-resolution microscopy images. They generate high-resolution images that preserve biological details, aiding in more accurate analysis and interpretation.

12.7.2.7 Multi-modal GANs

These GANs handle multi-modal data, such as combining different imaging modalities (e.g., MRI and PET scans) or combining imaging with genomic data. They can generate integrated representations that provide a comprehensive view of biological systems.

In bioimaging, GANs are advancing techniques for generating synthetic data, enhancing image quality, translating between imaging modalities, and aiding in data augmentation for training robust models. These applications demonstrate the versatility and potential of GANs to contribute significantly to biomedical research and healthcare applications.

12.8 BIOGAN: A SPECIALIZED LIBRARY FOR APPLYING GANS TO BIOLOGICAL DATA

The "biogan" a specialized library for applying GANs to biological data. GAN is an advanced technology that, based on various classifications of GANs and molecular structure augmentation, enables the prediction of biological data using gene information. This field is at the intersection of biology and advanced technology. Similar to biogenetics or biotechnology, it can play a crucial role in predicting diseases through several key mechanisms.

12.8.1 Genome Sequencing and Analysis

Personal Genomics: By sequencing an individual's genome, biotechnology can detect genetic variants associated with an increased risk of certain diseases. Consumers can already learn about their genetic predispositions through businesses such as 23andMe and AncestryDNA.

Polygenic Risk Scores: These scores use data from numerous genetic variants to determine an individual's risk of complicated diseases such as diabetes, heart disease, and certain malignancies. GANs have shown significant promise in various applications within bioimaging, leveraging their ability to generate realistic synthetic data and enhance existing datasets. Here are some types and applications of GANs specifically in bioimaging.

Blood and Tissue Biomarkers: Biogan technologies allow for the detection of certain biomarkers in blood or tissue samples that can suggest the early start of diseases like cancer or Alzheimer's. Elevated protein levels, for example, or the presence of certain genetic abnormalities, can act as early warning indications.

Metabolomics and Proteomics: Advanced techniques in metabolomics (the study of metabolites) and proteomics (the study of proteins) can detect alterations in biological molecules that occur prior to disease onset.Biogan in Identifying Molecular Structures.

12.8.2 X-ray Crystallography & Cryo-electron Microscopy (Cryo-EM)

Technique: X-ray crystallography is the process of crystallizing a molecule and then diffracting X-rays through it to generate a pattern. This pattern can be used to determine the molecule's three-dimensional structure at an atomic level.

Applications: Widely used to determine the structures of proteins, DNA, and tiny chemical molecules, which is important for medication design and understanding biological mechanisms.

Cryo-EM is a technique in which a sample is flash frozen and then imaged using an electron microscope. Advances in this approach have enabled researchers to see molecular structures at near-atomic resolution without the necessity for crystallization.

Applications: Specifically beneficial for researching big complexes such as viruses, ribosomes, and membrane proteins, providing insights into their function and relationships.

12.8.3 Nuclear Magnetic Resonance (NMR) Spectroscopy and Mass Spectrometry (MS)

NMR spectroscopy employs the magnetic characteristics of atomic nuclei to determine the physical and chemical properties of atoms and molecules. Using a magnetic

Biomedical Imaging Techniques in AI Applications 251

field and radiofrequency pulses, researchers may obtain precise information about molecule structure, dynamics, and interactions. Applications: Ideal for examining tiny proteins and nucleic acids in solution, revealing how these molecules act in their native state.

Mass spectrometry uses the mass-to-charge ratio of ions to identify and quantify compounds. It can be coupled with techniques such as liquid chromatography (LC-MS) to analyze complex mixtures. Applications include estimating molecular weight, structural characterization, and investigating post-translational changes in proteins.

12.8.4 COMPUTATIONAL MODELING AND BIOINFORMATICS AND INDIVIDUAL MOLECULE METHODS

Single-molecular and multimolecular simulations are used to predict molecular structures and understand molecular interactions. Using FAM (Feature-Activity Modeling) methods, these simulations analyze how individual molecules react, which can then be used to predict disease patterns. By modeling the behavior of molecules in various biological contexts, these simulations help identify key molecular mechanisms and potential targets for disease treatment.

Technique: Computational approaches, such as molecular dynamics simulations and homology modeling, use known structures to anticipate molecule structure. Bioinformatics tools use sequence data to predict structural features. Applications include predicting protein structure from amino acid sequences, analyzing molecular interactions, and developing novel compounds for medicinal use.

Method: Researchers can handle and view individual molecules using methods like optical tweezers and atomic force microscopy (AFM), which shed light on the structure and mechanical characteristics of the molecules.Applications: Used to examine molecular interactions, analyze the behavior of biomolecules under tension, and comprehend the mechanical characteristics of nucleic acids and proteins.

12.8.5 DRUG DESIGN

The use of structural biology methods has proven essential in the development of medications that specifically target particular proteins. For instance, protease inhibitors used in HIV treatment were developed as a result of the structure of the HIV protease enzyme being discovered by X-ray crystallography.

Understanding Disease Mechanisms: The architecture of the amyloid fibrils implicated in Alzheimer's disease have been studied with the help of cryo-EM, which has helped to shed light on how these formations arise and how they might be addressed therapeutically.

12.9 PREDICTION OF A SINGLE MOLECULE AND MULTI-MOLECULE

1. Quantum Mechanical Techniques
   Energy levels, charge distribution, and reactivity are just a few of the electronic characteristics that can be predicted using density functional theory, or DFT
2. **Prediction of a Multi-Molecule**
   Host-Guest Interactions: Research the stability and binding characteristics of compounds that form complexes with one another. Predict the characteristics and crystal structure of multi-component systems by co-crystallization (see Figure 12.4).

12.10 EVALUATION

RMSD: This is defined as the square root of the average of the squared differences between corresponding atomic coordinates of two structures. Consider N points with coordinates A_i and B_i. To reduce the RMSD, the structures must be aligned or stacked

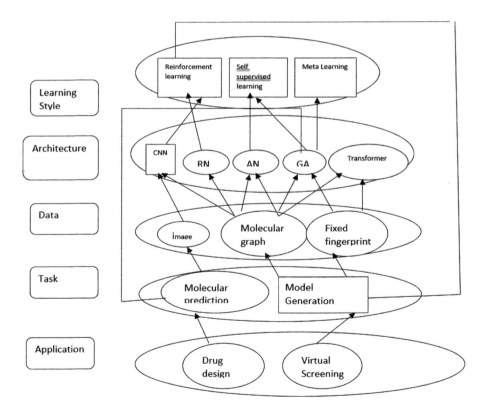

FIGURE 12.4 Prediction of a single molecule and multi-molecule.

Biomedical Imaging Techniques in AI Applications

before the RMSD is calculated. Typically, quaternion-based techniques or algorithms like Kabsch are used to determine the ideal translation and rotation that minimizes the RMSD. Accurate analysis and interpretation of complicated biological data are made easier with the use of RMSD.

$$\text{RMSD} = \sqrt{\frac{1}{N} Ni = 1|Ai - Bi|2} \qquad (12.1)$$

Quaternion-based approaches, such as the Kabsch algorithm, are commonly used in bioinformatics to align proteins and other macromolecules. The Kabsch algorithm, in particular, is used to determine the ideal rotation that reduces the root-mean-square deviation (RMSD) between two sets of points (e.g., atoms in protein structures). Here's a quick description of the Kabsch algorithm and its relationship to quaternions.

Kabsch Algorithm

1. **Compute the Centroids:** Calculate the centroids of the two-point sets P and Q:

$$C_P = \frac{1}{n}\sum_{i=1}^{n} P_i, \quad C_Q = \frac{1}{n}\sum_{i=1}^{n} Q_i \qquad (12.2)$$

P_i and Q_i are the coordinates of P and Q, respectively.

2. **Center the Points**: Subtract the centroids from the point sets

$$P_i' = P_i - CP, \quad Q_i' = Q_i - CQ \qquad (12.3)$$

3. **Compute the Covariance Matrix(H) and Rotation Matrix (R)**

Quaternion-Based Techniques (Q)

Quaternions can express rotations in 3D space and are widely utilized due to their numerical stability and efficiency. Converting the rotation matrix (R) to a quaternion representation provides computing benefits

$$\text{Quaternion Representation } Q = W + Xi + Yj + ZK \qquad (12.4)$$

W,X,Y,Z are real numbers I,J,K are quaternion units.

Rotation Using Quaternions where Q = (w,v) and v = (x,y,z) and then vector P is represented as **P = qPq*** (12.5)

Converting Rotation Matrix to Quaternion

The rotation matrix R from the Kabsch algorithm can be converted to a quaternion using the following steps:

Compute the trace of the matrix

$$Tr(R) = R_{00} + R_{11} + R_{22} \tag{12.6}$$

Compute the quaternion components

$$W = (1 + R_{00} + R_{11} + R_{22}/2)^2 \tag{12.7}$$

The Kabsch algorithm and quaternions are used to align protein structures, minimizing RMSD. To make protein alignment.

$$RMSD = \sqrt{\frac{1}{N} Ni = 1 |Ai - Bi| 2} \tag{12.8}$$

F1-Score: Gather protein sequences from databases (like UniProt) in order to prepare the data. Use techniques such as embedding or one-hot encoding to encode sequences. Utilize AI algorithms to deduce protein activities from sequence and structural data, such as deep neural networks, sequence alignment tools (like BLAST), or structure-based techniques (like homology modeling). To assess the precision of protein function predictions and confirm the specificity and sensitivity of functional annotations, calculate the F1 score.

Understand the Confusion Matrix by using TP,TN,FP,FN.

Calculate Precision and Recall
Precision(P): Ratio of true positive predictions to the total predicted positive instances.

$$Precision(P) = TP/TP + FP \tag{12.9}$$

Recall(R): Ratio of true positive predictions to the total actual positive instances.

$$Recall(R) = TP/TP + FN \tag{12.10}$$

F1-Score(F1): To determine the Similarity Measure

$$F1 = 2 * \frac{P * R}{P + R} \tag{12.11}$$

The degree of similarity or resemblance between two biological sequences, such as protein, RNA, or DNA sequences, is referred to as sequence similarity. It is the proportion of identical places in matched sequences. Sequence similarity includes conservative substitutions (such as amino acid substitutions in proteins) in addition to identical places.

12.11 CONCLUSION

Bioimaging techniques help to predict disease and identify the disease pattern with molecular structure at single and multi-molecular level by enhancing the accuracy, speed, and efficiency of bioimaging. By identifying both single and multi-molecule structures using various methods such as GANs, this study provides predictions, highlighting potential gaps in our understanding of molecular interactions. AI allows for the fast and accurate examination of massive volumes of medical pictures. AI models can detect early indicators of diseases that the human eye may miss, allowing for more timely intervention. By combining images from many modalities and utilizing the advantages of each imaging method, GANs can provide more thorough diagnostic insights. GANs can be employed to improve the resolution of molecular images, which can be acquired by atomic force microscopy (AFM) or cryo-electron microscopy (cryo-EM). This technique is known as high-resolution molecular imaging. Combined, these technologies improve imaging resolution, predict molecular structures, simulate drug interactions, and enable personalized medicine, which leads to more accurate diagnosis and successful therapies.

REFERENCES

1. Yadav, P., Mandal, C. C. "Bioimaging: Usefulness in Modern Day Research". In: Mukherjee, T.K., Malik, P., Mukherjee, S. (eds) *Practical Approach to Mammalian Cell and Organ Culture*. Springer, Singapore. 2023.
2. Irmak, E. "Multi-classification of brain tumor MRI images using deep convolutional neural network with fully optimized framework". *Iran J Sci Technol Trans Electr Eng* 45: 1015–1036. 2021.
3. Singh, V. K., Kolekar, M. H. "Deep learning empowered COVID-19 diagnosis using chest CT scan images for collaborative edge-cloud computing platform". *Multimed Tools Appl* 81:3–30. 2022.
4. Bai, J. W., Qiu, S. Q., Zhang, G. J. "Molecular and functional imaging in cancer-targeted therapy: current applications and future directions". *Sig Transduct Target Ther* 8: 89. 2023.
5. Wang, H., Shang, S., Long, L., Hu, R., Wu, Y., Chen, N., Zhang, S., Cong, F., Lin, S. "Biological image analysis using deep learning-based methods: Literature review". *Dig Med* 4(4):157–165. October 2018.
6. Abramson, J., Adler, J., Dunger, J., Evans, R., Green, T., Pritzel, A., Ronneberger, O., Willmore, L., Ballard, A. J., Bambrick, J., Bodenstein, S. W. "Accurate structure prediction of biomolecular interactions with AlphaFold 3". *Nature* 630: 493–500. 2024.
7. Yang, W., Zhang, X., Tian, Y., Wang, W., Xue, J. -H., Liao, Q. "Deep learning for single image super-resolution: A brief review". *IEEE Trans Multimedia* 21(12): 3106–3121. December 2019.
8. Wang, H., Yang, F., Zhang, S., Xin, R., Sun, Y. "Genetic and environmental factors in Alzheimer's and Parkinson's diseases and promising therapeutic intervention via fecal microbiota transplantation". *NPJ Parkinsons Dis* 7: 70. 2021.

13 Ethical Implications of AI in CRISPR

Responsible Genome Editing

Umesh Gupta, Ayushman Pranav, Rajesh Kumar Modi, and Ankit Dubey

13.1 INTRODUCTION TO CRISPR AND AI IN GENOME EDITING

13.1.1 Brief Overview of CRISPR Technology

CRISPR/Cas9, otherwise known as CRISPR, is an innovative genome-editing technology that has revolutionized molecular biology. CRISPR stands for the Clustered Regularly Interspaced Short Palindromic Repeats and the (CRISPR) – associated protein 9. Scientists can easily use this method to modify DNAs of living things like people. This procedure involves an RNA molecule joining with the Cas9 protein that can "zero in on" specific type of DNA sequences. Then they act as though the small knives and then cut the DNA where it should be completed or done [1–5]. The resulting incision will affect the cells' DNA through its own repair mechanisms such as the adding or the deleting specific type of genetic material. The possibilities of this CRISPR technology are limitless. It has been extensively used in a variety of areas ranging from basic research to potential medical cures (see Figure 13.1). For example, GMOs have been created by employing CRISPR to investigate how genes function while determining if gene therapy could be used to treat some inherited diseases. On the other hand, major issues with this method are its precision and fastness leading to high probability of manipulation of human germline.

13.1.2 Introduction to Artificial Intelligence in CRISPR

The combination of artificial intelligence (AI) with CRISPR technology represents a new frontier of genome editing. Precision and efficiency of CRISPR methods take on a new lease of life thanks to AI algorithms. In particular, algorithms based on AI can analyze large datasets including genomic information and predict the outcomes of CRISRP-mediated gene edits with high precision levels [9]. This partnership has far reached implications in terms of diseases study and therapyCRISPR has been successfully used in the treatment of disorders caused by specific genetic mutations,

Ethical Implications of AI in CRISPR

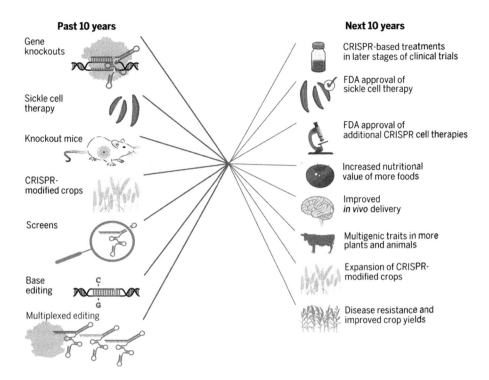

FIGURE 13.1 CRISPR: past, present, and future [6].

including in cancer research, making it the most precise and efficient gene-editing tool compared to previous methods. Its ability to target and modify genes with high accuracy has revolutionized the potential for treating genetic diseases and advancing personalized medicine. CRISPR/Cas9 is used to target specific genomic sites for knock-out or knock-in approaches, enabling cancer modeling, drug target identification, drug resistance studies, and biomarker development, including exploration of noncoding genome regions [10–14].

However, this is a promising technology that has some ethical and social implications that cannot be ignored. Thus, continuous research is necessary to evaluate both the advantages and disadvantages of CRISPR in various applications and develop best practices for its use.

The accuracy of the CRISPR system can be optimized by AI. For example, using AI algorithms researchers have been managed to analyze the genomic data that predicts potential off-target effects, thus enhancing specificity during the gene editing procedures such as those connected within cancer treatment where the accuracy is critical [15, 16]. In summary, the integration of AI and the CRISPR technology offers many exciting opportunities in precision medicine and targeted therapies. However, it also presents challenges that require careful consideration and ethical oversight to ensure responsible and beneficial use [3, 16].

13.1.3 THE INTERSECTION OF MATHEMATICS AND GENOME EDITING ETHICS

Math and genome editing ethics overlap in a difficult field that also has many aspects. The overlap between math and genome editing ethics involves using mathematical models to address complex ethical issues surrounding CRISPR technology, such as its limits and equitable access. They include how far CRISPR should be deployed and assuring equitable access to CRISPR applications. The development of regulatory frameworks for human germline genome editing that includes germline modification and international norms against misuse are needed. These moral considerations have been quantified through mathematical explanations. By formulating such types of models, policymakers can then evaluate whether the certain interventions like genome-editing may be more harmful than being beneficial. For example, mathematical models can be employed to determine whether germline gene therapy will alter the population over generations [17–19].

Furthermore, math analysis can assist in the formulation of moral benchmarks. Thus, ethical principles and the potential harms are measured by the mathematical models to inform evidence-based policies that promote human health and progress while minimizing the severe risks. Mathematics, on the ethical side of genome editing, meets with the ethics of the genome editing. It is important that the evidence based on the decision-making and the responsible governance are emphasized as CRISPR technology becomes even more advanced. This points to the need for collaboration between the scientists, ethicists, politicians, and mathematicians in addressing the ethical issues around gene editing and its socially suitable and acceptable application [1–3, 20–24].

13.2 ETHICAL FOUNDATIONS

13.2.1 THE PRINCIPLE OF AUTONOMY IN GENOME EDITING

AI-assisted genome editing refers to the use of AI in genetic modifications, where ethical considerations emphasize the importance of self-determination, ensuring individuals have autonomy over their genetic choices. AI autonomy refers to enabling AI systems to rely on formulas and data to independently make their own decisions. AI-enabled CRISPR-Cas9 genome editors autonomously design and execute precise gene-editing strategies without external control, raising ethical concerns about unchecked genetic manipulation. But this math autonomy carries serious ethical problems [25–27].

1. **Autonomy in Mathematical Terms**: These AI algorithms are a self-contained processor that can edit genes by examining genetic data and proposing possible changes. These algorithms are based on ML, and therefore, they rely on scientific principles to refine over time their general accuracy. This mathematical self-management at the same time improves performance while it becomes more complex. There is thus an urgent need to ensure transparency and accountability in these algorithms in order to prevent unintended consequences or biases resulting from editing decisions.

Ethical Implications of AI in CRISPR 259

There may be ethical problems as math might enhance eugenic motivations related to choosing for traits like intellect and physical beauty. This creates moral quandaries particularly when coupled with respect to autonomy, which necessitates a highly robust regulatory framework that upholds human rights [1, 28–30].

2. **Ethical Implications of Autonomous AI Systems**: The ethical implications of autonomous AI systems in gene editing include concerns about informed consent, privacy of genetic data, and equitable access to AI technologies. Informed consent may be sought voluntarily by patients and research subjects if they understand how AI interacts with their genes during editing. Privacy becomes an issue when algorithms have access to genetic data that is not secure.

Additionally, CRISPR technologies enabled by AI must ensure equal accessibility, hence avoiding health disparities and genetic enhancement for example where children can be designed to have preferred features. The debate about how genome-editing tools consisting of AI abused like creating offspring with certain properties raises the need for global guidelines and moral framework guiding ethical development and use of these genome-editing tools based on Artificial Intelligence [2, 31–35].

13.2.2 Beneficence and Non-Maleficence in AI-Augmented CRISPR

The integration of CRISPR technology with AI helps address ethical concerns by focusing on principles such as non-maleficence (avoiding harm) and beneficence (promoting good). AI can assist in ensuring that CRISPR interventions are precise and minimize unintended consequences, while also maximizing the potential benefits for patients, thus fostering a more ethical approach to gene editing.

1. **Models for Assessing Benefits and Risks:** – The principle of maximizing benefits calls for some interventions like fixing inherited diseases or improving the state of health. The types of algorithms used to assess benefits and risks in situations like genomic editing are often predictive models and machine learning algorithms. These algorithms analyze large amounts of data to forecast the outcomes of potential interventions, such as fixing inherited diseases. For example, by analyzing genomic data, they can predict the effects of a genomic edit before it's even attempted or completed, helping to maximize the benefits while minimizing potential risks [36].

However, mathematical models must also consider weaknesses like off-targeting or adverse side effects. Therefore, it is important to ensure that in their development, the focus is on patient safety and minimizing uncontrolled injuries, thereby then contravening the principle of 'do no harm' in ethical frameworks [37].

2. **Quantifying Potential Harm through Equations:** Using the equations and the algorithms to measure the risks inherent in CRISPR interventions driven

by AI. These mathematical tools can estimate how likely it is that genes will be manipulated accidentally or other negative outcomes may result. The principles of nonmaleficence must be followed by these formulas in order for them to reduce harm.

AI-backed CRISPR systems, while offering significant advantages, also come with certain risks, though fewer compared to earlier methods. When developing mathematical models for these systems, ethical considerations must carefully balance the potential benefits—such as targeted gene editing and disease prevention—with the risks, including unintended genetic changes or misuse of the technology. Striking this balance ensures responsible use of AI in gene editing, aiming for positive outcomes while minimizing harm. AI's ethical ramifications combined with CRISPR technology entail autonomy, beneficence, and non-maleficence, which are convoluted to say the least. Most of the ethical concerns deal with mathematical algorithms that control autonomous decision-making and enable determinants on which genome editing is judged as positive or negative. Ethical concerns must be addressed in order for AI-powered CRISPR to be a powerful scientific research tool and medicine according to principles based on ethics that avoid its abuse [1–2, 39–40].

13.3 ACCOUNTABILITY AND TRANSPARENCY

Thus, responsible behavior and openness in every aspect of AI-controlled gene-editing is about morality. Such aspects are fundamental in ensuring responsible and ethical use of technologies like CRISPR-Cas9, particularly when combined with AI. These can be explained through the concepts of explainability, transparency, and accountability involving their mathematical and AI aspects [15, 28–29].

13.3.1 EXPLAINABILITY AND TRANSPARENCY IN AI-DRIVEN GENOME EDITING

13.3.1.1 Mathematical Models for Explaining AI Decisions

Several mathematical models have been developed to articulate AI decisions in genome editing. These models are designed to explain how AI processes and evaluates genetic data, ensuring transparency and accuracy in decision-making. The key models include:

13.3.1.1.1 Algorithmic Transparency

Mathematical equations can represent algorithms that control the functioning of AI systems. This reduces complicated decision-making processes to a series of steps outlined in clear language. For instance, the system's process of identifying target genes and subsequently designing genetic modifications applicable to CRISPR/Cas9 can be described by means of equations [3, 33–38].

13.3.1.1.2 Probability and Confidence Scores

These models use an AI-driven decision-making based upon the given probability scores, which means that the researchers could have some idea on how the certain genetic can be edited. For instance, such formulas can then be used to predict successful gene-editing through off-target effects and failure modes [3].

13.3.1.1.3 Error Analysis

AI prediction accuracy can be tested using mathematical techniques like statistical analysis. To clarify, scientists could employ equations in order to identify the boundaries of error for genome editing results related to the unintended consequences. Some of these equations are influenced by variables such as seriousness of genetic illness or its possible effects on health [39–40].

13.3.1.1.4 Ethical Safeguards

Furthermore, mathematical modeling plays a crucial role in ethical decision-making, especially in the context of AI-based genome editing. These models help establish boundaries and guidelines that ensure the technology is used in ways that align with ethical principles. By incorporating ethical constraints into the formulas, mathematical modeling helps ensure that AI-driven genome editing remains within safe and morally acceptable limits.

13.3.1.2 Quantifying Transparency Levels

To measure transparency in AI-driven genome editing, metrics need to be formulated. Such metrics help to assess the extent to which AI systems can make their logic clear. For example, a metric for transparentness might look at how much information an AI algorithm gives about data sources or biases and criteria it uses for making decisions. Thus, quantifying transparency will help us understand if these technologies are reliable and moral [4]. In simple terms, by quantifying how transparent an AI system is—through examining its data sources, biases, and decision-making criteria—we can assess its reliability and ensure it is developed responsibly [40].

13.3.2 Accountability Frameworks in the Context of Mathematical Ethics

13.3.2.1 Mathematical Equations for Attributing Responsibility

It is through these equations that accountability structures can be formulated:

13.3.2.1.1 Attribution Models

Equations help clarify responsibility for outcomes in AI-based genome editing by distributing accountability based on the contributions and decisions of scientists, doctors, and policymakers. Another key thing that models know includes; engagement of scientists, doctors as well as policy makers and so on. Equations assist in distributing responsibilities based on contributions made and choices reached by respective parties [20–22].

13.3.2.1.2 Impact Assessment

The societal effects of genome editing decisions may be evaluated by means of mathematical equations. This helps in counting potential gains and losses from

AI-supported genome editing so that the correct decision can be made to do editing [22–23].

13.3.2.1.3 Regulatory Compliance

Equations, for instance, are what guide regulatory frameworks. They are used to guide regulatory frameworks and provide a quantitative basis for defining genome editing applications. For AI-based gene therapy to be ethical enough, obligatory math ethics remain important [22–23].

13.3.2.1.4 International Collaboration

The mathematical modeling backs up global cooperation in conversations on genomes ethics. Equations facilitate unification of legislations that touch on moral issues across many nations under this discipline. It is critical that the alliance refrains from misusing and instead promotes right usage of AI-based gene therapy [23–24].

13.3.2.2 Role of AI in Ensuring Accountability

AI itself plays a crucial role in ensuring that genome editing is not only accountable but also responsible. AI algorithms are used to track and audit genetic editing procedures with continuous evaluation of their adherence to ethical standards and regulatory frameworks. For example, AI can prescribe specific protocols that ensure informed consent is obtained from patients and provide immediate feedback on the ethical implications of proposed genetic changes. This helps ensure that individuals fully understand the potential consequences of genome editing, fostering transparency and promoting ethical practices in genetic interventions. To boot, the same technology can be used to detect scientific studies that have violated allowable behaviors and therefore put in place a mechanism for giving alerts. Thus, applying AI in such a proactive manner enables the observance of supreme ethical norms with regard to accountability for genome editing according to societal values and ethical principles [36].

Additionally, this kind of technology may be utilized to find out where scientific research has crossed the line of acceptable behavior so that action can be taken using alarm systems.

13.4 BIAS AND FAIRNESS

13.4.1 Addressing Bias in AI algorithms for CRISPR

The implications are extensive ethics and practicality when bias is built into CRISPR's machine learning algorithms. To enable responsible use of gene therapies/genome manipulation for this purpose we need to locate and eliminate bias from such algorithms. This section will explain how bias in CRISPR's machine learning algorithms can be identified and removed using mathematical formulas and and the ways to mitigate it [29].

13.4.1.1 Mathematical Formulations of Bias

There can be different kinds of bias in AI algorithms such as gender, racial or socioeconomic bias. Mathematically, bias can also be described as a systematic deviation of algorithm predictions from the ground truth or desired outcomes. Statistical metrics express this in terms like:

- **Bias Score**: A quantification of the extent to which algorithm outputs are biased. The approach is concerned with comparing the distribution of predicted results for various groups (e.g., male and female).
- **Confusion Matrix**: A confusion matrix is a tool that shows the true positives, true negatives, false positives, and false negatives from an algorithm's predictions, helping to assess its accuracy and potential bias.

13.4.1.2 Equations for Detecting and Mitigating Bias

To avoid bias in AI algorithms used in CRISPR numerous equations and techniques could be used:

- **Bias Detection Equation**: This equation determines biases by comparing prediction probabilities between different groups. It can also be expressed as:

$$\text{Bias} = |P(\text{outcome}|\text{group A}) - P(\text{outcome}|\text{group B})| \qquad (i)$$

where P(outcome|group A) represents the probability of a certain outcome for the group A.

- **Bias Mitigation Equations**: Various methods, such as the re-weighting or the re-sampling data, can be used to mitigate the biases. These equations aim to adjust to the dataset or the algorithm parameters to reduce bias.
- **Fairness Constraints**: For several groups, it is possible to design algorithms with fairness constraints that guarantee fair results. This may take the form of mathematical language and be incorporated into an optimization process.

13.4.2 ENSURING EQUITABLE ACCESS TO AI-ENHANCED CRISPR TECHNOLOGY

To prevent the healthcare and genetic research disparities, equitable access to AI-driven CRISPR technology must be ensured. Thus, this entails mathematical models for resource allocation and ethical implications.

13.4.2.1 Mathematical Models for Resource Allocation

Equitable access can be ensured through development of mathematical models that allocate resources such as funding, research opportunities, and clinical applications of CRISPR technology. These models can include considerations like:

- **Need-Based Allocation**: Allocating resources based on the severity of genetic disorders and the potential impact of CRISPR-based treatments.
- **Geographical Allocation**: Ensuring that research and applications are distributed geographically to avoid concentration in specific regions.
- **Affordability Models**: Developing pricing models that make CRISPR treatments accessible to diverse socioeconomic groups.

13.4.2.2 Ethical Implications of Equitable Distribution

Equitably distributing of the AI enhanced CRISPR technology also raises crucial ethical questions:

- **Prioritization of Vulnerable Populations**: Ethical considerations may necessitate prioritizing vulnerable populations to address health disparities.
- **Informed Consent**: Ensuring that individuals have informed consent regarding CRISPR treatments, especially if they are part of research studies.
- **Global Cooperation**: Ethical frameworks may require international cooperation to ensure that the benefits of CRISPR technology are accessible to all nations.

Finally, making sure that AI-powered CRISPR is available without the variations and then creating a bias-free algo for the CRISPR-AI is a necessary step for the unlocking of the gene editing's potential while adhering to moral values. This includes math formulas and ethics put in place to ensure equitable use of the CRISPR technology in healthcare and research [32].

13.5 PRIVACY AND DATA SECURITY

With new technological advancements for gene editing such as CRISPR-Cas9, and the integration of AI in this field, there have been concerns about privacy and data security. In this section, we investigate the ethical problems concerning privacy of genetic data and how AI can help secure genomic data [33].

13.5.1 Ethical Concerns Related to Genetic Data Privacy

13.5.1.1 Equations for Assessing Privacy Risks

Calculations involving mathematics could also be used to find out the possible risks of data confidentiality breaches like unauthorized access and de-anonymization probabilities. Such computations consider genetic information complexity, existence of external sources having information and how anonymous methods are effective. The quantification of these risks enables researchers and policymakers to decide better on their data sharing and access control mechanisms. At the same time, it is an ethical issue that may require a trade-off between necessary research work and personal protection with informed consent [23].

Ethical Implications of AI in CRISPR 265

13.5.1.2 Mathematical Representations of Genetic Data

Genetic information can be mathematically represented in several ways:

- **Genetic Sequences:** Genetic information is often written as a letter sequence where each letter represents different nucleotide bases such as A (adenine), C (cytosine), G (guanine), or T (thymine). For instance, one can represent it using a string where every character stands for a single nucleotide base; for example, "ATCGTACG."
- **Matrices:** Genetic information can be transformed into matrices whereby rows stand for alternative genes or sequences while columns symbolize various positions within such sequences. Every square in this new matrix gives nucleotide details about that place.
- **Graphs:** Graphs may be used to represent genetic interactions and relations between genes. In these diagrams, nodes are the genes while edges denote their interactions or similarities.

Mathematical notations play a crucial part in the analysis and manipulation of genetic data to facilitate efficient algorithms that can properly understand it [35].

13.5.1.3 Equations for Assessing Privacy Risks

To measure privacy risks associated with genomic data, mathematical modeling and equations that can quantify possible vulnerabilities are necessary. Some of these key factors include:

- **K-Anonymity:** This implies that any individual's genetic data would resemble at least k-1 other individuals in a dataset. Consequently, the equation representing k-anonymity is expressed as follows:

$K = f(QI, S)$, where K is the level of k-anonymity, QI denotes quasi-identifiers (e.g., age, gender), and S represents sensitive genetic data.

- **Differential Privacy:** Differential privacy involves quantifying how changing or removing a person's record would change a dataset. This parameter is referred to as privacy loss, which equals ε. In mathematical terms:

$$Pr[Algo(Dataset1) \in Output] \leq e^{\wedge}\varepsilon * Pr[Algo(Dataset2) \in Output] \qquad (ii)$$

where ε controls the privacy level.

- **Entropy:** This measures how uncertain or random genetic entropy is. It can be used to determine information on the leakage in the dataset. The equation of entropy:

$H(X) = -\Sigma P(x) * log2(P(x))$, where $H(X)$ is the entropy of the dataset X and $P(x)$ is the probability of observing value x.

13.5.2 AI Solutions for Securing Genomic Data

13.5.2.1 Mathematical Encryption and Decryption Algorithms

To protect the genomic data, sensitive genetic information is encrypted within the encryption algorithms that make it inaccessible without decryption. Key algorithms used for this purpose include [23]:

Advanced Encryption Standard (AES): AES is a symmetric encryption algorithm that is widely used for the data protection. It uses mathematical operations including methods such as substitution, permutation, and the XOR operations to encrypt and decrypt data.

Homomorphic Encryption: This advanced encryption technique enables computations on encrypted data without decryption. Its reliance on the mathematical structures allows operations like addition and multiplication on the ciphertexts.

RSA Encryption: RSA uses a pair of keys (public and private) for encryption and decryption, and the reference to "very large prime numbers" should specify that these primes are used to generate the keys.

Using encryption methods combined with AI-driven control access and secure storage helps protect genomic data, addressing ethical and privacy concerns commonly associated with genome editing and AI technologies [3].

13.5.2.2 Balancing Security with Accessibility Equations

It is important to find the right level of security and accessibility when it comes to data. Based on how sensitive the genetic information is, mathematical formulae can be used in determining the amount of security that should be accorded. Success in this area will mean that it can still be utilized for proper medical research while avoiding its misuse or other related effects. Ethical issues include making genetic information accessible for all (without the discrimination) and preventing it from being used in a discriminatory manner. From such type of equations, ethical sharing agreements are created [34].

To this end, given that the genomics is becoming more dependent on AI while also pushing forward advances in genetic technologies, the ethical consideration concerning protection and security of genetic data is vital in this industry. To address these concerns, mathematized representations, risk assessment formulas, and safekeeping algorithms plus accessibility equations must be employed. In order to make ethical decisions within this framework individual privacy must be maintained as well as responsible use of data emphasized to enable fair distribution of genomic research and healthcare benefits [35].

13.6 HUMAN ENHANCEMENT AND DESIGNER BABIES

13.6.1 The Mathematical Concept of Enhancement

13.6.1.1 Quantifying Genetic Enhancement through Equations

Genetic enhancement involves purposely changing an individual's genetic structure to improve particular characteristics or abilities. This can also be summed up in mathematical terms using equations that represent genes.

Ethical Implications of AI in CRISPR

One of the fundamental equations used in genetics is the Hardy-Weinberg equilibrium equation:

$$p2 + 2pq + q2 = 1 \quad p2 + 2pq + q2 = 1 \tag{iii}$$

- p represents the frequency of one allele in a population.
- q represents the frequency of the other allele in the same population.

This equation assists in calculating genetic frequencies for various alleles within a population and allows for the prediction of trait distributions.

For genetic enhancement, we can introduce a parameter, EE, that represents the extent of improvement done to a certain characteristic. The modified equation becomes:

$$p2 + 2pq + q2 + E = 1 \quad p2 + 2pq + q2 + E = 1 \tag{iv}$$

Here, EE represents the genetic enhancement applied to the trait in question. Positive values in EE imply enhancement while negative ones indicate decrease of trait expression.

Once the well-being or lives of individuals and society become at stake with respect to how far genetic enhancements can go, ethical aspects must come into play. By adding some parameters related to safety, fairness, and societal impact this equation can even be made more extensive [39].

13.6.1.2 Ethical Limits on Enhancement

Different factors are considered when setting limits on ethical genetic enhancement such as:

- Safety: In terms of mathematics, safety can be expressed as a probability equation. The probability of adverse effects (PadverseP) occurring as a result of genetic enhancements may be approximated by:

$$Padverse = (E) \quad Padverse = f(E) \tag{v}$$

The function $E(f) = f(E)$ is a graph that relates how much enhancement (Ee) there would be to the probability of adverse effects. The aim is to minimize $Padverse$.

- Fairness: The distribution of enhanced traits within a population can be mathematically examined to determine equity in access to genetic enhancement. Gini coefficient (GG) can measure inequality in trait distribution:

$$G = \frac{\sum_{i=1}^{n} \sum_{j=1}^{n} |x_i - x_j|}{2n^2 \bar{x}} \tag{vi}$$

where x_i and x_j represent the enhanced trait values of individuals ii and jj and $\bar{x}$ is the mean trait value. A lower Gini coefficient indicates greater fairness.

- Societal Impact: The consequences of widespread enhancement on social structures, resources, and norms can be captured by mathematical models.

In summary, the mathematics behind genetic enhancement involve equations for measuring degree of enhancement and models that also consider ethical concerns such as safety, fairness, and societal good. A balance must therefore be struck between mathematical quantification and ethical boundaries in discussions on human enhancement and designer babies [1–5].

13.6.2 Ethical Debates Surrounding Designer Babies

13.6.2.1 Mathematical Equations for Defining "Designer" Traits

Mathematicians and geneticists have explored the concept of "designer" traits using many mathematical models. One of the approaches involves quantifying the heritability of specific traits using the equations like the following:

$$\text{Heritability\_}(H^2) = \text{Genetic Variance\_}(Vg) / \text{Total Variance\_}(Vt) \quad \text{(vii)}$$

where:

- **Heritability-(H^2)** represents the proportion of the variance in the particular trait that is attributable to the genetic factors.
- **Genetic Variance-(Vg)** quantifies the genetic contribution to the variation in the trait.
- **Total Variance-(Vt)** represents the overall variability in the trait, including genetic and environmental factors.

These formulas allow us to judge which traits are most influenced by genes; these can be starting points of arguments about altering or determining traits.

But it should be noted that applying mathematical assessments to issues related to designer babies is complicated because it is interlaced with moral considerations. Nonetheless in utilizing numerical measures of traits like height, we ask ourselves whether this kind of trait must be subject to change, what are limits of genetics, and what happens when the natural variation in genes is altered.

13.6.2.2 Society's Role in Setting Ethical Boundaries

It is not a matter of math alone to determine the ethical boundaries of designer babies. The society is at stake in forming this boundary. The ethical debates on designer babies revolve around cultural norms, values, and potential for increasing inequalities.

Society must engage in the discussions about which traits are acceptable to modify, how much control parents should have over their own children's genetics, and the potential consequences of widespread genetic customization. Striking the balance between individual desires for genetically tailored offspring and the broader societal implications requires open dialogue and the ethical guidelines that reflect a consensus among the stakeholders [38–40].

In conclusion, enhancement as a mathematical concept and the moral debate about designer babies are two closely inter-related topics. Mathematical models can help quantify genetic enhancements, but it is crucial to set ethical boundaries to ensure that any modifications, like those for designer babies, are done responsibly and regulated properly.

13.7 GLOBAL GOVERNANCE AND REGULATIONS

13.7.1 INTERNATIONAL PERSPECTIVES ON AI IN CRISPR

There have been many advancements in scientific knowledge due to genome editing development through CRISPR/Cas9 technology, which is promising for furthering science and medicine. Nonetheless just like everything else with great power comes great responsibility and hence this has made the ethical and regulatory aspects of CRISPR crucial worldwide.

13.7.1.1 Mathematical Models for Global Governance

The ethical issues surrounding CRISPR and AI integration involve the need for global governance, which can be guided by mathematical models that quantify risks and benefits to help policymakers evaluate ethical implications and societal impacts, such as safety, informed consent, and the risk of eugenics.

The ethical concerns surrounding CRISPR/Cas9, as outlined in [1], include safety, morality, informed consent, and the risk of eugenics. Mathematical equations can be used to develop a model of the government that comprehensively deals with these issues through the quantification of risks, ethics, and societal impact.

An example is a Risk-Benefit Equation that measures the benefits of CRISPR applications like stopping genetic problems against its possible hazards such as unwarranted genetic alterations brought about by some mutations. Thus, the numerical values can be assigned to these factors so that policy-makers and ethicists can evaluate objectively if some CRISPR applications are proper.

Furthermore, public consent index might incorporate statistical methods aimed at gauging public opinion regarding various forms of CRISPR applications and its eagerness about them so as to make sound decisions concerning this technology [12, 15, 21].

13.7.1.2 Equations for Harmonizing International Regulations

For international cooperation to play a role in integrating regulatory paths affecting CRISPR and AI, there is a need for complete alignment. The ethical issues here are not solely about governance, but rather about creating a global ethical index to evaluate CRISPR/Cas9 technologies, considering factors like the diseases they target, alternatives to gene editing, and the permanence of changes to the human genome.

To do an ethics review of CRISPR/Cas9 technologies, it is vital to evaluate the disease burden they target, whether there are other feasible alternatives besides this method, and if it results in permanent changes to the human genome. As a result, the international community should establish an ethical index globally that takes these factors into account as varying inputs in order to produce uniform ethical values for each application.

Also, the Regulatory Alignment Equation could be used to compare extant laws across different countries, thus exposing both areas of overlap and divergence. In this equation, however, one can incorporate powers of regulations around public engagement degrees as well as enforcement-inspection processes.

Again, these mathematical models can be employed by international organizations and policy-makers as tools for facilitating conversations and dialogues on CRISPR and AI rules. These mathematical models can help international organizations and policymakers discuss and agree on CRISPR and AI regulations, using the models to find common ground on ethical issues and develop consistent governance for genome editing [22–24].

In conclusion, it is therefore important that we all globally understand and appreciate the moral dimensions of CRISPR/Cas9 and AI convergence. The application of mathematical models is essential for measuring and reconciling these factors thereby promoting responsible use of global CRISPR applications. As a result, there is a need to develop regulatory frameworks that are informed by universal humanity's values and ethics.

13.7.2 Ethical Considerations in Cross-Border Genome Editing

13.7.2.1 Mathematical Approaches to Ethical Consensus

Different cultures and regions have complex ethical issues surrounding genome editing. One approach to address these concerns involves formulation of mathematically based consensus. Additionally, mathematical models allow for evaluation and interpretation of possible ramifications resulting from various genomic changes in terms of their safety, effectiveness, or social consequences.

They can be used as policy tools so that policy-makers have sufficient grounds for making judgements on the morality aspect of certain gene-editing applications. Countries may work jointly with experts drawn from different areas such as biology, mathematics, ethics, etc., to come up with globally recognized methods of assessing the ethical aspects about gene editing in future. Such kind of models aim to reconcile innovation and ethical responsibility.

Nevertheless, it is important to recognize that mathematical models are not a remedy for ethical predicaments. They can help in decision-making but should involve wider ethical considerations with the input of the diverse stakeholders including scientists, ethicists, and the members of the public. Furthermore, these types of models must be constantly updated so as to accommodate new genome-editing techniques and associated ethical problems [17–19].

13.7.2.2 Balancing Local Values with Global Standards

Global governance of genome editing is complex because different nations have diverse ethical values and priorities. In some cases, ethics may be founded on individual autonomy and medical innovation whereas in other instances societal well-being or caution may be prioritized.

Dealing with these differences requires international cooperation and negotiations. Such talks are facilitated by the World Health Organization (WHO) as well

Ethical Implications of AI in CRISPR   271

as the United Nations (UN). Through this platform, different countries express their opinions, deliberate on ethical norms as well as strive to reach a consensus about gene editing for responsible applications.

In order to strike a balance between local values and international standards, diplomacy, compromises, and commitment to resolution of common moral problems like safety, equity, and access are called for. In this regard, international conventions that prescribe universal ethical principles but allow individual nations to adjust to them would be significant.

Furthermore, there should be transparency and accountability mechanisms in these countries, so that the agreed upon principles are carried out. Transparency and accountability mechanisms help ensure that ethical principles are followed, allowing for continuous review and improvement of global policies related to modern technologies like genome editing, which leads to sustainable service delivery from diverse perspectives [26, 27].

To sum up, successful tackling of moral issues related to cross-border genome editing requires different perspectives. While mathematical models might provide information on this topic, discussion about ethics in the field is still necessary. Thus, striking a balance among local norms and global values calls for global cooperation and flexible rules. Through careful thoughtfulness as well as collaboration, responsible and ethical practices for genome editing can be nurtured on a global level while taking into the account divergent moral viewpoints of humans on earth.

13.8 CASE STUDIES

13.8.1 REAL-WORLD EXAMPLES OF AI-DRIVEN CRISPR PROJECTS

1. **Enhancing Precision of CRISPR**: The use of AI will target superior accuracy levels in CRISPR gene editing technology. For example, researchers have been using AI to improve how accurately they can change cells for cancer therapy or even make crops like wheat and corn that are high yielding yet resistant to drought.
2. Predicting RNA-targeting CRISPR Activity: RNA-targeting Cas9s are predicted using AI, especially deep learning models, which helps understand and control better the on- and off-target activity of CRISPR tools [8, 9].
3. Development of Predictive Models: Integration of AI with CRISP/Cas9 has resulted in several predictive models estimating effective targeting by CRISP algorithms such as CRISPrater, Azimuth 2.0, and DeepCRISPr, among others [10].

Ethical Analysis using Mathematical Tools

- **Precision and Unintended Consequences**: This also raises questions on ethics with regard to unintended effects, off-target activities and long-term consequences for genetic diversity due to the higher accuracy brought about by AI in CRISPR technology.

- **Data Privacy and Consent**: For instance, large datasets used in AI assisted CRISPR research may raise ethical problems regarding data confidentiality and informed consent particularly when personal or sensitive genetic information is involved.
- **Accessibility and Equity**: There's a concern about the equitable access to the benefits of AI-driven CRISPR technologies, and whether they would be accessible to all, or only to a privileged few.

Lessons Learned from Case Studies

- **Need for Robust Regulatory Frameworks**: Real-world projects reveal a crucial need for robust regulatory frameworks to address the ethical, legal, and social implications of AI-driven CRISPR technologies.
- **Interdisciplinary Collaboration**: Lessons underscore the importance of interdisciplinary collaboration between geneticists, AI experts, ethicists, and policymakers to navigate the complex landscape of AI-driven CRISPR research [10–13].

These summaries offer a preview of some interactions between AI technology, CRISPR and ethics. In addition to demonstrating how AI can be used with CRISPR for sophisticated genetic research and applications through actual projects carried out worldwide, they demonstrate the ethical considerations involved and important lessons learnt pertinent for guiding future enquiries and ensuring accountable growth and deployment of these emerging technologies.

13.9 CONCLUSION

13.9.1 Recap of Ethical Considerations in AI-enhanced CRISPR

The fusion of the AI and CRISPR-Cas9 genome editing tool has led to anew dimensions in biotechnology and medicine. While this advancement comes with many types of benefits, it also gives rise too many ethical-based questions.

AI-enhanced CRISPR raises ethical issues from the number of facts. Firstly, on the subject of ethics in the use of CRISPR itself, especially concerning editing human germ line. The precision and speed at which CRISPR can accomplish genetic manipulation is unrivaled but using it on the human germline has raised safety concerns as well as questions touching on ethics and the possibility of promoting eugenic practices [1]. To this end, strict policies, international talks, and public awareness campaigns are necessary. It is crucial that scientists, ethicists, policymakers, and the public work together to comprehend some of the many social implications tied to CRISPR-Cas9.

Another key ethical concern is about just access to CRISPR applications so all individuals can profit from these transformative technologies. It is quite essential to ensure that there are regulatory frameworks for ethical practices in AI-enhanced gene editing with respect to germline editing in human beings as well as internationally established rules to prevent misuse [23].

13.9.2 CALL TO ACTION FOR RESPONSIBLE GENOME EDITING

This is a powerful AI-driven CRISPR technology, which elicits an urgent need to act. The main idea here is that the genetic changes must be done responsibly. This means that it is impossible to make fixed lines on ethics while this technology is evolving. Therefore, there should be evidence-based regulations by national and supranational parliaments on various uses of CRISPR for human health worth and progress [12].

There is a need for ongoing crucial research on pros and cons related to CRISPR-Cas9, particularly in cancer research and immunotherapy among others. This technology has possibilities of being precision medicine or targeted therapy even though this comes with ethical and social implications. AI plays a vital role in the improvement of CRISPR's accuracy that may be beneficial towards other industries like cancer treatment [13].

To responsibly edit genomes requires interdisciplinary collaboration, transparency, and commitment to the well-being of all individuals. This dialogue should not be restricted within scientific communities alone but should also include wider societies, thus ensuring these technologies are used for common good.

13.9.3 ONGOING ROLE OF MATHEMATICS IN ETHICAL AI-CRISPR DEVELOPMENT

The integration of ethical AI development, mathematical modeling, and CRISPR technology, which together enhance the precision and accuracy of genome editing, guide decision-making, and help policy-makers balance risks and benefits responsibly. It lays the ground for modeling and optimizing CRISPR-Cas9 processes to be more precise and accurate. Designing CRISPR experiments, predicting outcomes, and using mathematical algorithms to identify druggable targets are critical.

Also, the mathematics is significant for the ethical process of decision-making process. This helps in the process of quantifying risks and benefits of particular CRISPR applications, thus enabling rational regulatory decisions by policy-makers. Mathematical modeling can ensure guidelines that foster a responsible genome editing while minimizing damage.

To sum up, the intersection of AI and CRISPR provides the opportunities for scientific progress but raises many and important ethical question too. Responsible genome editing, equitable access, and continuous study constitute some of the navigation landmarks across this territory. Mathematics acts as a useful tool in optimizing AI-enhanced CRISPR technology on the one hand and guiding moral implications on the other.

REFERENCES

[1] Shinwari, Z. K., Tanveer, F., & Khalil, A. T. (2018). Ethical issues regarding CRISPR mediated genome editing. *Current Issues in Molecular Biology*, 260(1), 103–110.

[2] Bhat, A. A., Nisar, S., Mukherjee, S., Saha, N., Yarravarapu, N., Lone, S. N., ... & Haris, M. (2022). Integration of CRISPR/Cas9 with artificial intelligence for improved cancer therapeutics. *Journal of Translational Medicine*, 20(1), 534.

[3] Ethical Challenges of AI Applications. (2021). Available online. Accessed from https://aiindex.stanford.edu/wp-content/uploads/2021/03/2021-AI-Index-Report-_Chapter-5.pdf
[4] Birney, E. (2023). Available online. Accessed from https://www.embl.org/news/lab-matters/human-genome-editing-regulations-risks-&-ethical-considerations/
[5] Wang, J. Y., & Doudna, J. A. (2023). "CRISPR technology: A decade of genome editing is only the beginning." *Science, 379*(6629), eadd8643.
[6] Roach, J. (2018). Available online. Accessed from https://blogs.microsoft.com/ai/crispr-gene-editing/
[7] Communications, N. W. (2023). Available online. Accessed from https://www.nyu.edu/about/news-publications/news/2023/july/ai-crispr-gene-expression.html
[8] University, C. (2023). Available online. Accessed from https://scitechdaily.com/artificial-intelligence-meets-crispr-the-rise-of-precision-rna-targeting-&-gene-modulation
[9] Bhat, A. A., Nisar, S., Mukherjee, S., Saha, N., Yarravarapu, N., Lone, S. N., Masoodi, T., Chauhan, R., Maacha, S., & Haris, M. (2022). Integration of CRISPR/Cas9 with artificial intelligence for improved cancer therapeutics. *Journal of Translational Medicine, 20*, 534. https://doi.org/10.1186/s12967-022-03765-1
[10] Umar Ibrahim, A., Pwavodi, P. C., Ozsoz, M., Al-Turjman, F., Galaya, T., & Agbo, J. J. (2023). Crispr biosensing and Ai driven tools for detection and prediction of Covid-19. *Journal of Experimental & Theoretical Artificial Intelligence, 35*(4), 489–505.
[11] Nayak, A., & Dutta, D. (2023). A comprehensive review on CRISPR and Artificial Intelligence based emerging food packaging technology to ensure "safe food.". *Sustainable Food Technology.*
[12] Khoshandam, M., Soltaninejad, H., Hamidieh, A. A., & Hosseinkhani, S. (2023). CRISPR and Artificial intelligence to improve precision medicine: Future perspectives and potential limitations. *Authorea Preprints.*
[13] Lee, M. (2023). Deep learning in CRISPR-Cas systems: a review of recent studies. *Frontiers in Bioengineering and Biotechnology, 11*, 1226182.
[14] Durán-Vinet, B., Araya-Castro, K., Zaiko, A., Pochon, X., Wood, S. A., Stanton, J. A. L., Jeunen, G. J., Scriver, M., Kardailsky, A., Chao, T. C., Ban, D. K., & Gemmell, N. J. (2023). CRISPR-Cas-based biomonitoring for marine environments: toward CRISPR RNA design optimization via deep learning. *CRISPR Journal, 6*(4), 316–324.
[15] Ai, P., Xue, D., Wang, Y., & Zeng, S. (2023). An efficient improved CRISPR mediated gene function analysis system established in Lycium ruthenicum Murr. *Industrial Crops and Products, 192*, 116142.
[16] Dara, M., Dianatpour, M., Azarpira, N., & Omidifar, N. (2024). Convergence of CRISPR and artificial intelligence: A paradigm shift in biotechnology. *Human Gene*, 201297.
[17] Sun, H. (2024). Navigating the CRISPR-Cas9 frontier: AI-Enabled off-target prediction and sgRNA design for unprecedented precision. *Transactions on Materials, Biotechnology and Life Sciences, 3*, 522–531.
[18] Danter, W. R. (2022). aiCRISPRL: An artificial intelligence platform for stem cell and organoid simulation with extensive gene editing capabilities. *bioRxiv*, 2022-06.
[19] Ahmar, S., Usman, B., Hensel, G., Jung, K. H., & Gruszka, D. (2024). CRISPR enables sustainable cereal production for a greener future. *Trends in Plant Science.*
[20] Zhang, Y., MA, X., & Xie, X. (2022). *CRISPR.* Springer Singapore.
[21] Lior, A. (2019). AI entities as AI agents: Artificial intelligence liability and the AI respondeat superior analogy. *Mitchell Hamline L. Rev., 46*, 1043.
[22] Umar, T. P. (2024). Artificial intelligence-enhanced application of CRISPR-Cas13a for cancer gene therapy: A breakthrough concept. *Experimed, 14*(1), 61-62.

[23] Vladimirovich, V. (2019). Chapter Three Years 1996 to 2003: "Artificial Intelligence" In Title. In *Patents and Artificial Intelligence: Thinking Computers*, p. 29.
[24] Zheng, X., Cui, J., Wang, Y., Zhang, J., & Wang, C. (2021). CRSIPR-AI: a webtool for the efficacy prediction of CRISPR activation and interference. *bioRxiv*, 2021-12.
[25] Yu, D., Zhong, Q., Xiao, Y., Feng, Z., Tang, F., Feng, S., Cai, Y., Gao, Y., Lan, T., Li, M., Yu, F., & Li, Z. (2024). Combination of MRI-based prediction and CRISPR/Cas12a-based detection for IDH genotyping in glioma. *NPJ Precision Oncology*, 8(1), 140.
[26] Cui, Z., Lin, L., Zong, Y., Chen, Y., & Wang, S. (2024). Precision gene editing using deep learning: A case study of the CRISPR-Cas9 editor. *Applied and Computational Engineering*, 64, 134–141.
[27] Ameen, Z. S. I., Ozsoz, M., Mubarak, A. S., Al Turjman, F., & Serte, S. (2021). C-SVR Crispr: Prediction of CRISPR/Cas12 guideRNA activity using deep learning models. *Alexandria Engineering Journal*, 60(4), 3501–3508.
[28] Hou, T., Zeng, W., Yang, M., Chen, W., Ren, L., Ai, J., Wu, J., Liao, Y., Gou, X., Li, Y., Wang, X., & Xu, T. (2020). Development and evaluation of a rapid CRISPR-based diagnostic for COVID-19. *PLoS Pathogens*, 16(8), e1008705.
[29] Weinberg, R., Mann, C. M., Babnigg, G., Forrester, S., Greenwald, S., Larsen, P. E., Owens, S., Gros, M. F., Noirot, P., & Antonopoulos, D. A. CRISPR-Act: AI-guided prediction of a CRISPR kill-switch across physiological contexts.
[30] Maserat, E. (2022). Integration of artificial intelligence and CRISPR/Cas9 system for vaccine design. *Cancer Informatics*, 21, 11769351221140102.
[31] Ai, Y., Liang, D., & Wilusz, J. E. (2022). CRISPR/Cas13 effectors have differing extents of off-target effects that limit their utility in eukaryotic cells. *Nucleic Acids Research*, 50(11), e65–e65.
[32] Urnov, F. D. (2021). Imagine CRISPR cures. *Molecular Therapy*, 29(11), 3103–3106.
[33] Dai, M., Li, X., Zhang, Q., Liang, T., Huang, X., & Fu, Q. (2024). Health research in the era of artificial intelligence: Advances in gene-editing study. *Medicine Plus*, 100027.
[34] Wang, Z., & Cui, W. (2020). CRISPR-Cas system for biomedical diagnostic platforms. *View*, 1(3), 20200008.
[35] Huang, K., Qu, Y., Cousins, H., Johnson, W. A., Yin, D., Shah, M., Zhou, D., Altman, R., Wang, M., & Cong, L. (2024). Crispr-GPT: An LLM agent for automated design of gene-editing experiments. *arXiv preprint arXiv:2404.18021*.
[36] Zhang, W., Liu, K., Zhang, P., Cheng, W., Li, L., Zhang, F., Yu, Z., Li, L., & Zhang, X. (2021). CRISPR-based approaches for efficient and accurate detection of SARS-CoV-2. *Laboratory Medicine*, 52(2), 116–121.
[37] Umar, A. I. (2020). *Application of Artificial Intelligence in Microbiology and Crispr* (Doctoral dissertation, Near East University).
[38] Gupta, U., Pranav, A., Kohli, A., Ghosh, S., & Singh, D. (2024). The Contribution of Artificial Intelligence to Drug Discovery: Current Progress and Prospects for the Future. In *Microbial Data Intelligence and Computational Techniques for Sustainable Computing* (pp. 1–23).
[39] Pandey, P., Patel, J., & Kumar, S. (2022). CRISPER Gene Therapy Recent Trends and Clinical Applications. In *Gene Delivery* (pp. 179–194). CRC Press.
[40] Baliram, P. M., Sharma, M., & Ganpatrao, W. S. (2019). CRISPR/Cas9 Genome Editing and Its Medical Potential. In *Advances In Biotechnology and Bioscience*, (pp. 71–90). Akinik Publication.

Index

A

Abnormality in heart rhythm, detection of
  methodology of, 53–57
  use of machine learning approach for, 52–53
Absorption, distribution, metabolism, excretion, and toxicity (ADMET), 77, 217
Adaptive and learning algorithms, in managing dynamic health data, 32–33
ADMET Predictor software, 165
Administrative workflow optimization, 96
Advanced Encryption Standard (AES), 266
Adversarial autoencoders (AAEs), 238
AI-based software, list of, 214
Aidoc, 100–101
ALDOB gene, 189–190
ALGOPS software, 71
AlphaFold, 72–73, 207, 213, 247
Alpha-thalassemia, 190
Alzheimer's disease, 99, 190, 194
  BACE1 inhibitors for treatment of, 166
Amazon Web Services (AWS), 183
Analytical modeling, 7
Anesthesia, application of AI in administering, 218
Anisotropic Median Filter (AMM), 137
Anomaly detection, 75
Antibiotic resistance, 99
Antibiotic resistance identification, use of deep learning in, 231
Antibiotic-resistant bacteria, 99
Antiretroviral medication, 162
Antiviral medications, development of, 165
Antiviral peptide prediction, use of AI in, 231
Area under the ROC curve (AUC), 126, 134
Arrhythmia (heart disease), 49–50
  causes of, 50
  detection of, 52
  symptoms of, 50
  types of, 50
Arterial fibrillation (AF), 62
Artificial intelligence (AI), 33, 42, 93
  algorithms of, 29
  bioinformatics and, 178–182
  chemical discovery systems, 236
  meaning of, 131–132
  potential effects in surgeries, 14
  predictive capability of, 76
  as tool for early detection and diagnosis of diseases, 6–10
  transparency and interpretability of, 108
  use in
    biomedical industries, 11–12
    drug discovery and development, *see* Drug discovery and development, application of AI in
    image analysis, 44
    medical imaging, 9
    pharmaceutical industry, 11–12
    robot-assisted surgeries, 13–14
Artificial Neural Networks (ANNs), 27, 130, 134, 137, 141, 146, 231
  application in bioinformatics research, 232–239
Atomic force microscopy (AFM), 243, 251, 255
AtomNet, 207, 214
Atrial fibrillation (AFib), 50
Attention-based GANs, 249
Augmented reality (AR), 43
  robotic kidney removal using, 133
  use in urology, 133
Australian Health Practitioner Regulation Agency (AHPRA), 37
Australia, telemedicine licensing in, 37
Auto Dock tool, 77, 100
Automated eligibility verification systems, 107
Automated image processing tools, 242
Autonomous molecular design, role of AI in, 235–239
Autonomy, principle of, 108

B

BACE1 inhibitors, used in treatment of Alzheimer's disease, 166
Bagging, concept of, 56
Basal cell carcinoma, 102
Bayesian networks, 101
Bayesian probability theory, 232
Beta-thalassemia, 190
Big data
  analytics tools, 78, 154
  incorporation into the field of bioinformatics, 183–184
Bioactive compounds, 162
Bioactivity, use of AI in prediction of, 71–72
BioBERT, 188, 246
Biochemical indicators, 147
Biogan, 249–251
Bioinformatics, 108–109, 118
  applications of, 228
    artificial neural networks, 232–239
    deep learning, 231–232

277

artificial intelligence (AI) and, 178–182, 228–229
  future prospects of, 239
big data integration, 183–184
computational modeling and, 251
database, 233
data gathering for, 248
deep learning approaches in
  use of convolutional neural networks for medical imaging, 187
  use of neural networks in genomic data analysis, 187
  use of recurrent neural networks in healthcare data, 187–188
emerging trends and innovations, 194
ethical and regulatory considerations, 194–195
future directions and ethical considerations, 194–195
in healthcare and public health, 190–194
high-throughput sequencing technology, 183
incorporation of AI and ML in the field of, 196
key technological milestones and methodologies, 182–183
machine learning techniques in, 184–186
natural language processing and genomics, 188
reinforcement learning (RL), 186
scope of, 228
supervised learning, 185
unsupervised learning, 185–186
Biological data analysis, 182
Biological signal detection and analysis, 233–234
Biological text mining, use of NLP for, 188
Biomarkers, 147, 182, 219, 244
  blood and tissue, 250
  development of, 257
  role of AI in discovery of, 190
  use in diagnosis and monitoring of diseases, 190
Biomedical imaging
  AlphaFold 3 model for, 242
  applications of, 246–247
  augmentation techniques in, 244–246
  and drug discovery, 246
  evaluation of, 252–254
  methods of, 242–244
  multimodal imaging technologies, 242
  prediction of a single molecule and multi-molecule, 252
  processing of images, 231
  quaternion-based techniques, 253
  role of AI in, 244
  workflow of, 246
  X-ray crystallography, 250
Biomedical industries, use of AI and machine learning in, 11–12

Biomedical knowledge graph reasoning, 213
Biomedical research, 180, 183, 215, 249
Biomedical signal processing, 231
Biomedical text mining, 244
Biometric monitoring wearables, 43–44
Biopharmaceutical industry, 70
"Black box" phenomenon, 40, 82, 108, 180
Blockchain technology, for securing health data sharing, 44
Blood glucose
  prediction techniques, 10
  prognosis, 9
  simulator, 9
Blood tests, 8–9, 52, 101
BlueDot, 192
Body mass index (BMI), 134, 141
Bone mineral density estimation, DEXA scans for, 101
Bootstraping aggregation, concept of, 56
Breast cancer metastases, 102
Breastfeeding, role of telemedicine in supporting, 28–29

C

Calcium oxide (CaO_x), 147
Cancer
  epidemiology, 193
  predictive algorithm, 8
  use of AI in treatment of, 217–218
Cardiology, AI's role in, 103–105
Cardio myopathy, 49
Cas9 protein, 256–257
Catalyst (BIOVIA) software, 163
Chatbots, AI-powered, 108, 179, 239
Chemical indicators, 147
Chemical library design, 209
Cheminformatics
  case study of
    drug-induced liver injury, 170
    Gleevec (Imatinib), 168
    raltegravir, 164
    substitutes for chloroquine, 166
  growth of computational techniques in, 161
  machine learning, 168–170
  molecular docking, 164–165
  molecular dynamics, 166–168
  mutations, 168
  pharmacophore modeling, 163
  Quantitative Structure Activity Relationship (QSAR), 165–166
  virtual screening, 166
ChemOffice, 165
ChemVAE model, 217
Chloroquine, 166

Index

Chronic Disease (CD) diagnosis, use of ML models in, 100
Chronic illness management, AI models in, 191
Chronic myeloid leukemia, 162, 168
Chronic Obstructive Pulmonary Disease (COPD), 101
Clinical decision support system (CDSS), 8, 33, 106, 131
Clinical trials, application of AI in, 78–79
Cloud computing, 183–184
Clustered Regularly Interspaced Short Palindromic Repeats (CRISPR) technology, 259
 addressing bias in AI algorithms for, 262
 advantages and disadvantages of, 257
 application of AI in, 256–257
 beneficence and non-maleficence in AI-augmented, 259–260
 brief overview of, 256
 case studies, 271–272
 equations for detecting and mitigating bias, 263
 ethical considerations in AI-enhanced, 272
 ethical implications of equitable distribution, 264
 ethical issues surrounding, 258
 global governance and regulations, 269–271
 intersection of mathematics and genome editing ethics, 258
 machine learning algorithms, 262
 mathematical formulations of bias, 263
 mathematical models for resource allocation, 263–264
 models for assessing benefits and risks, 259
 past, present, and future, 257
 precision and efficiency of, 256
 privacy and data security
  AI solutions for, 266
  ethical concerns related to, 264–265
 quantifying potential harm through equations, 259–260
 Regulatory Alignment Equation, 270
 risk-benefit equation, 269
 role of mathematics in ethical AI-CRISPR development, 273
Clustering of compounds, 75
Cognitive robotics, 14
Communication with doctors, role of telemedicine in, 28–29
Compound screening, use of AI in, 76–77
Computational autonomous molecular design (CAMD), 236–239
Computational biology, 228, 231
Computational intelligence tools, 27
Computational power, 184

Computed tomography (CT), 9, 131, 141, 242
 stone composition identification using, 150
Computer-aided detection system, 136
Computer-aided drug design, use of deep learning in, 232
Computer-aided software, 207
Computer vision, 10
Conditional GANs (cGANs), 248
Congenital heart diseases, 49
Congestive heart failure
 degrees of heart block, 49
 factors responsible for, 49–50
 symptoms of, 50
Convolutional neural networks (CNNs), 75, 100, 101, 131, 181, 191, 214, 231
 for medical imaging, 187
Coulomb matrices, 68
COVID-19 disease, 20, 23, 69, 179, 192
 identification and diagnosis of, 9
 RT-PCR test for, 9
Cryo-electron microscopy (Cryo-EM), 99, 250, 255
CycleGAN, 248–249

D

DANAOS expert systems, 103
Databases, for drug development, 214–215
Data collection
 preprocessing of data, 7, 30
 and validation, 7
Data fusion and integration, 31–32
Data management, 188
Data mining, 79, 162
Data preprocessing, for advanced computing analysis, 7, 30
Data privacy, 107, 109, 181, 183–184, 196, 221, 264, 272
Data processing, 7, 132, 178, 182, 190
Data reduction, 7
Data security, 36, 106, 108, 264
Datasets, mining of, 78
Data storage systems, 194
Data transformation, 7
Data visualization tools, 95
DBDermo-Mips, 103
Decision-making, 21, 29, 82, 132, 188, 270
 algorithmic processes, 108
 collaborative, 22
 ethical ramifications of AI decision-making in healthcare, 180
 in healthcare settings, 31
 integration of genomic data into process of, 44
 and rule-based reasoning systems, 25
 usage of soft computing in medical for, 25

Decision support systems (DSSs), 14
   AI-driven, 191
   clinical decision support system (CDSS), 131
   development of, 34
   for healthcare professionals, 33–34
   for prenatal care, 33
Decision trees, 56–57, 185
DeepBar, 207, 222
DeepChem, 170, 207, 222
DeepLabV3, 135
Deep learning (DL), 10, 66, 76, 93, 115, 132, 162, 182, 207, 242
   algorithms, 151
   applications in bioinformatics, 231–232
   models for, 75–76, 82
   techniques for drug target prediction and identification, 97
   use of
      convolutional neural networks for medical imaging, 187
      neural networks in genomic data analysis, 187
      recurrent neural networks in healthcare data, 187–188
DeepMind Technologies, 219
Deep neural networks (DNNs), 71–72, 142, 187, 231
DeepPurpose, 213
Deep structured learning, 231
DeepTox, 214, 216
DeepVariant, 188
De novo molecular design, 213, 217
   validation and screening of, 217
Density functional theory (DFT), 237, 252
Dermatology, AI's role in, 103
DermTech, 103
Designer babies
   ethical debates on, 268–269
   genetic customization of, 268
   mathematical equations for defining "designer" traits, 268
   society's role in setting ethical boundaries of, 268–269
"Designer" traits, concept of, 268
Diabetes detection, use of AI in, 9
Digital Imaging and Dispatches in Medicine (DICOM), 37–38
Dimensionality reduction, 7, 9, 75, 185–186
Discovery Studio software, 163, 165
Disease-associated network, 98
Disease classification, 185, 191
Disease diagnosis and prediction
   overview of, 130–131
   techniques, types of, 104
   use of AI in, 100–101
   aims of, 131–132
   forecasting managerial results, 139–143
   optimizing the surgical process, 144–152
   urology, 133–139
DNA sequences, 108, 181
   use of AI in, 188–189
   "zero in on" specific type of, 256
Docking simulations, 100
DOCK tool, 77, 100
Dragon software, 166
Drug design
   computer-aided, 67
   structure-based, 97
Drug discovery and development, application of AI in, 11, 66, 68–69, 96–97, 215–219
   advancements in, 206
   algorithms for, 68, 170
   applications in anesthesia, 218
   challenges and ethical considerations in
      data quality and bias, 82
      ethical impacts, 83
      interpretability and transparency, 82
      perspectives, 82
   changing landscape, 84
   clinical trials, 78–79
      algorithms utilizing AI/ML, 211
   compound screening, 76–77
   computer-aided software, 207
   conventional *versus* AI-driven process, 80–81
   current AI-based software for developing new drugs, 213–214
   databases for, 214–215
   docking simulations, 100
   drug repositioning, 216–217
   enhanced predictive modeling, 97–98
   ethical issues related to the use of AI in, 219–221
   FDA approval and postmarket monitoring, 210
   fragment-based, 74
   future perspectives, 83–84
   high-throughput screening (HTS), 98–99
   hit identification and lead optimization, 209
   innovative technologies, 83–84
   lead optimization, 77
   machine learning algorithms, 210–211
   modern methods of, 73–74
   objectives of, 67
   overview of, 75–76, 208–210
   pharmaceutical industry, *see* Pharmaceutical industry
   postmarketing safety monitoring, 210
   preclinical and clinical trials using, 210
   preclinical testing, 78
   for predicting drug toxicity, 216

Index

for prediction of physicochemical
properties, 70–71
process of, 209
proposed method, 208
Random Forest approach for managing, 212–213
research gaps, 84
structure-based drug design (SBDD), 99–100
Support Vector Machines (SVMs), 213
target identification and validation, 76, 208–209
for treatment of cancer, 217–218
Drug formulation optimization, use of deep learning in, 232
Drug-induced liver injury, 170
Drug molecules, use of AI in designing
prediction of the target protein structure, 72–73
Drug repositioning, use of AI in, 216–217
Drug-Repurposing Hub, 217
Drug resistance, 168
Drug screening, application of AI in, 69–70
Drug-target binding affinity (DTBA), 71
Drug-target identification, 98, 257
Drug-target interaction prediction, 245, 247
Drug target protein, identification of, 98
Drug toxicity, use of AI in prediction of, 216

E

Early disease detection, exosome analysis for, 44
Echocardiogram (echo), 52, 103
Ehrlich, Paul, 163
Electrocardiograms (ECGs), 52, 61, 231
kinds of, 52–53
Electroencephalography, 132
Electromyography, 132
Electronic health information, 37
Electronic health record (EHR) systems, 5, 40, 78, 92, 94, 106, 147, 179
integration of AI with, 190
Electronic medical records (EMRs), 101
EMBL database, 183
Environmental Protection Agency (EPA), 71
Enzyme engineering, 244, 248
Epidemiology, applications in, 193–194
Estimation Programme Interface (EPI), 71
Ethical and legal considerations, in usage of telemedicine
privacy and data security, 36
reliability and accuracy, 36
European Medicines Agency (EMA), 210
Event monitor, 52–53
Extracorporeal Shockwave Lithotripsy (ESWL), 140–142

enhancement of, 144
for treatment of kidney stones, 144
Extreme Learning Machines (ELM), 70

F

Fast Healthcare Interoperability (FHIR), 37
Feature-Activity Modeling (FAM), 251
Fecal microbiota transplantation (FMT) therapy, 242
Federated learning, 183
Fiducial points, 53
Flexible ureteroscopy, 142
Fluorescence microscopy, 243
Foetal development, 29
Follow-Up Consult(s), 40
Food and Drug Administration (FDA), 210
list of AI/ML-based medical technologies accepted by, 212
Fréchet Inception Distance (FID), 248
Functional MRI (fMRI), 243
Fuzzy inference systems, 30–31
Fuzzy logic system, 25–26, 27, 34–35, 45
for individual sensitivity, 36
for interpretation, 30–31

G

Gas chromatography analysis, 100–101
Gaussian filter, 137
GenBank, 183
Gene expression analysis, 185
General Data Protection Regulation (GDPR), 180, 184, 195
General Medical Council (GMC), 37
Generative adversarial networks (GANs), 191, 238, 247–249, 255
biogan, 249–251
classifications of, 249
types of, 248–249
Genetic algorithms (GAs), 26–27, 45
iterative nature of, 31
for optimization, 31, 34
Genetic data privacy
differential privacy, 265
equation of entropy, 265
equations for assessing privacy risks, 264, 265
ethical concerns related to, 264–265
K-anonymity, 265
mathematical representations of, 265
Genetic enhancement, mathematical concept of, 266–267
Genetic fructose intolerance, 191
Genetic information, 33, 35, 66, 78, 94, 179, 189, 193, 264, 265–266, 272

Genetic markers, 68, 103, 108, 185, 218
Genome editing, 256
　call to action for responsible, 273
　case studies, 271–272
　error analysis in, 261
　ethical considerations in cross-border genome editing
　　balancing local values with global standards, 270–271
　　mathematical approaches to, 270
　ethical implications of autonomous AI systems, 259
　ethical principles, 258
　ethical safeguards, 261
　explainability and transparency in
　　algorithmic transparency, 260
　　mathematical models for explaining AI decisions, 260–261
　　quantifying transparency levels, 261
　global governance and regulations on
　　equations for harmonizing international regulations, 269–270
　　ethical considerations in cross-border genome editing, 270–271
　　international perspectives on AI in CRISPR, 269–270
　　mathematical models for, 269
　human germline, manipulation of, 256, 258
　intersection with mathematics, 258
　mathematical equations for attributing responsibility in, 261–262
　　attribution models, 261
　　impact assessment, 261–262
　　international collaboration, 262
　　regulatory compliance, 262
　principle of autonomy in, 258–259
　probability and confidence scores, 260–261
　role of AI in ensuring accountability, 262
　tackling of moral issues related to, 271
Genomic data, AI solutions for securing
　balancing security with accessibility equations, 266
　mathematical encryption and decryption algorithms, 266
Genomic data analysis, use of neural networks in, 187
Genomic medicine, application in prenatal care, 44
Genomic research, use of AI in
　DNA sequencing, 188–189
　Hereditary Fructose Intolerance (HFI), 189–190
　predictive modeling, 189
Genomic sequencing, 106, 182, 265
　and analysis, 250
　personal genomics, 250
Germany, telemedicine licensing in, 37

Gestational diabetes mellitus (GDM), predictive models for early detection of
　confusion matrix, 124–126
　data modeling
　　mathematical description, 120–122
　　model building, 123
　future scope of, 127
　hybrid predictive modeling framework, 115
　machine learning algorithms, 117–118
　materials and methods, 117–118
　proposed work on
　　data collection, 118–119
　　data preprocessing, 119
　　feature selection, 119
　　flowchart of, 119
　Receiver Operating Characteristic (ROC) curve, 126
　related work, 115–116
　testing accuracy, 124
　training accuracy, 123
Gini coefficient, 267
GitHub, 238
Gleevec (Imatinib), 168, 171
Glide tool, 77, 100
Glycogen storage disease type III (GSDIII), 191
Google
　Cloud computing service, 183
　DeepMind, 100, 193
Gradient-based One-Side Sampling (GOSS), 118
Gradient Boosting Machine (GBM), 116, 183, 185, 189
GraphINVENT, 217
Graph neural networks (GNNs), 76, 238
Gray near co-occurrence matrix (GLCM), 136
GrayNet, 135
GreatCall, 15

H

Hardy-Weinberg equilibrium equation, 267
Healthcare delivery system, 14
　convergence of technological innovation and a patient-centric approach to, 21
　data-driven decision-making in, 196
　ethical ramifications of AI decision-making in, 180
　machine learning in
　　applications of, 94–100
　　disease diagnosis, 100–101
　　ethical considerations of, 107–108
　　future of, 108–109
　　personalized care plans, 22
　　principle of justice in, 108
Healthcare industry, 5, 7, 25, 105–106, 132, 178, 194–195, 230

Index

Healthcare-related big-data analysis, 4–5
   application of, 6
   linguamatics, 5
Healthcare sector, AI types relevant to
   challenges of, 238–239
   deep learning, 231–232
   machine learning, 229–230
Health data
   acquisition of, 30
   adaptive and learning algorithms in managing, 32–33
   pattern recognition, training neural networks for, 31
   use of recurrent neural networks in, 187–188
Health Fog framework, 10
Health Insurance Portability and Responsibility Act (HIPAA), 38, 108, 180, 184, 195
Health Level Seven International (HL7), 37
Health management, 19, 28, 93, 108–109, 180, 190, 192, 195
HealthMap, 192
Heart blockage, 49, 52
Heart diseases diagnosis, use of AI in, 9–10
Heart rhyme disorder, 51
Hereditary Fructose Intolerance (HFI), 189–190
Hereditary illnesses, 191
Hidden Markov models, 182
Hierarchical learning, 231
High-risk pregnancies, home monitoring of, 28
High-throughput screening (HTS), 98–99, 209
High-throughput sequencing technology, 183
Hit identification, 209
HIV protease inhibitors, 168
Holter monitor, 52
Home monitoring, of high-risk pregnancies, 28
Homomorphic encryption, 266
Human enhancement
   ethical limits on, 267–268
   mathematical concept of
      quantifying genetic enhancement through equations, 266–267
Human Genome Project (HGP), 182–183
Human germline, manipulation of, 256, 258
Human–machine interaction, 41
Human neurological system, 132
Hyperparathyroidism, 101
Hypertension disease detection, use of AI in, 10
Hypoglycemia, episodes of, 191

I

IBM Watson, 100, 193
ICM (MolSoft), 163
IEEE 11073 (Health informatics-particular health device communication), 38
Image analysis, AI-powered, 102
ImageNet, 135–136
Image segmentation, 9, 137, 242
Inception score, 248
India, telemedicine licensing in, 37
Infectious diseases
   outbreaks of, 184
   use of AI models in forecasting propagation of, 192
Influenza neuraminidase, 162, 165
Influenza outbreaks, 192
Information technology, 93, 108, 228
Informed consent, for data use, 107–108, 264, 272
Integrating the Healthcare Enterprise (IHE), 37
Intelligence technologies, 11
Intelligent agents, *see* Artificial intelligence (AI)
Interquartile range (IQR), 30
ISO 13485 (Medical Devices), 38

K

Kabsch algorithm, 253–254
Kidney diseases, 50
Kidney stones, 131, 134–135
   Extracorporeal Shockwave Lithotripsy (ESWL) for treatment of, 144
   Raman spectroscopy for analysis of, 152
K-Nearest Neighbors (KNN), 57, 100–101, 115, 137
KNIME software, 165, 170

L

Large Language Models (LLMs), 108
Lead optimization, 77, 78, 83, 164, 207–209
Learning models
   supervised, 75
   unsupervised, 75
Least absolute shrinkage and selection operator (LASSO) regression, 116
Licensing, in telemedicine practices, 36–37
LigandScout software, 163
LightGBM (Light Gradient Boosting Machine), 115, 118, 120, 141
Linguamatics, 5
Liver diseases
   factors for, 8
   liver disease severity algorithm, 8
Liver fibrosis diagnosis, 101
Logistic Regression (LR) model, 100
Long short-term memory (LSTM) networks, 187

M

Machine learning (ML), 4, 10, 29, 66, 77, 93, 101, 115, 151–152, 161, 168–170, 206
    algorithms, 33, 40, 135, 230
    black boxes, 40, 82
    future in healthcare, 108–109
    reinforcement learning (RL), 186
    supervised learning, 185, 230
    techniques in bioinformatics, 184–186
    as tool for early detection and diagnosis of diseases, 6–10
    unsupervised learning, 185–186, 230
    use in
        Chronic Disease (CD) diagnosis, 100
        healthcare systems, 94–101
        pharmaceutical industry, 11–12
        robot-assisted surgeries, 13–14
Machine vision, 231
Magnetic resonance imaging (MRI), 9, 101, 142, 242
Mass spectrometry (MS), 250–251
Maternal and foetal health, microbiome analysis for, 45
Maternal healthcare access, role of telemedicine in, 27–28
Maternal health indicators, 29, 34–35, 45
Maternal health monitoring, 20
Maternal perspectives, in accessing prenatal care, 23
Medical Expert system, 234
Medical imaging analysis, 96
    use of convolutional neural networks for, 187
Medical Nutrition Therapy (MNT), 116
Medical ultrasound imaging, 243–244
MedWhat, 15
Melanoma, 103
Metabolic disorders, 190
Metabolomics, 250
Microarray technology, 183
Microbiota-gut-brain axis, 242
Micro-computed tomography (MicroCT), 243
Microsoft Azure, 183
Microtomography, 152
Mobile health (mHealth), 19, 21
MoleAnalyser expert systems, 103
Molecular docking, 164–165
Molecular dynamics (MD), 162, 166–168
Molecular fingerprinting, 68
Molecular fragmentation, development advantages of, 74
Molecular Operating Environment (MOE), 163–165
Molecular property forecasting, 213
Molecular representation learning, 247
MolRNN model, 217
Multi-layer perception (MLP) classifier, 136
Multi-modal GANs, 249
Multimodal imaging technologies, 242
Multiphoton microscopy, 243
Mutations, 168
    case study on HIV protease inhibitors for, 168
    classification of, 169
Myocardial Perfusion Imaging (MPI), 103

N

Naive Bayes (NB) model, 100, 117, 121, 122
Nanopore sequencing, use of AI in, 189
Natural language processing (NLP), 5, 107, 132, 181, 190, 209, 244
    for biological text mining, 188
    and genomics, 188
    in telehealth communication, 45
Nearest-Neighbor Classifiers, 70
Negative predictive value (NPV), 136
Network Analysis Algorithms, 98
Neural networks (NNs), 26, 35, 45, 99, 233, 234
    artificial neural network (ANN), 27, 130, 134, 137, 141, 146
    convolutional neural networks (CNNs), 75, 100, 101, 131, 181, 187
    deep neural networks (DNNs), 71–72, 142, 187
    in genomic data analysis, 187
    graph neural networks (GNNs), 76
    incorporation of, 34
    methods for forecasting protein structures and gene sequences, 182
    for pattern recognition, 31
    recurrent neural networks (RNNs), 75, 100, 181, 187–188
    for treatment optimization, 36
Neurodegenerative diseases, 98
Neurodegenerative illnesses, risk factors for, 242
Non-invasive molecular imaging technologies, 242
Non-linear mathematical simulation systems, 131–132
Nuclear magnetic resonance (NMR) spectroscopy, 77, 215, 250–251
Nursing and Midwifery Council (NMC), 37

O

Omics research, use of deep learning in, 231
Optical coherence tomography (OCT), 146
Oseltamivir, *see* Tamiflu
Osteoporosis, 162

Index

P

Pacemaker, 53
Pain treatment, use of AI systems in, 218–219
Parkinson's disease, 99, 190
Particle Swarm Optimization (PSO), 27
Partition coefficient (logP), 70
Pathology, use of AI in, 102
    cardiac assessment, 103–105
    dermatology, 103
    personalized medicine, 105–106
    streamlining administrative tasks, 106–107
Patient empowerment, 22
Patient management, 40, 108
Patient privacy, 108
Pattern recognition, in health data, 191
Pattern recognition, neural networks for, 31
Pediatric urology, 133
Percutaneous nephrolithotomy (PCNL), 134, 142, 146
Personal genomics, 250
Personalized care plans, 31, 36, 42
    role of AI in development of, 191
Personalized medicinal decision support, in analgesic treatment, 218
Personalized medicine, 195, 257
    for individualized healthcare, 105–106
    treatment plans, 96
    use of AI and machine learning in diagnosis of, 11–12, 79
Pharmaceutical industry, 67
    use of AI and machine learning in, 11–12, 219
        databases and high-throughput screening, 68
        drug discovery and design, 11
        drug dosage, 12
        early development, 67
        historical context of, 67–68
        identifying clinical trial candidates, 12
        integration of genomics data, 68
        marketing, 12
        molecular modeling and docking, 67
        predicting treatment results, 12
        quantitative structure-activity relationship (QSAR), 67
        rare diseases and personalized medicine, 11–12
        research and development, 11
        rule-based systems and expert systems, 67
Pharmacogenomics, 191
Pharmacologically active compounds, 72
Pharmacological robots, 218
Pharmacophore modeling, in drug discovery, 163
    case study of
        raltegravir, 164
        zanamivir, 165
    softwares for, 163
    steps involved in the workflow of, 163
Pharmacovigilance (PV), 210
PharmaGist software, 163
Pharmit software, 163
Phase (Schrödinger) software, 163
Phenotypic screening, 97
Physicochemical properties, prediction of, 70–71
Physics-Informed Machine Learning (PIML), 237–238
Polycystic ovary syndrome (PCOS), 117
Polygenic Risk Scores, 250
Positive predictive value (PPV), 136
Positron Emission Tomography (PET), 244
Precision medicine, AI's role in, 190
Preclinical testing, use of AI in, 78
Predictive analytics
    application in the field of public health management, 192
    for early intervention, 43
    use AI to predict health outcomes, 191
Predictive modelling
    algorithms for enhanced, 97–98
    use of AI in, 189
    using soft computing, 34–35
Pregnancy, risk identification during, 31
Prenatal care
    application of genomic medicine to, 44
    biometric monitoring wearables for, 43–44
    challenges in accessing, 23
    decision support systems for, 33
    early identification and mitigation of risks, 22
    early risk assessment in, 35–36
    emerging technologies and potential areas for further research in, 43–45
    emotional and psychological support, 22
    importance of, 22
    optimal fatal development, 22
    role of telemedicine in, 28
        case study of, 24
        developments in, 20–22
        logistical hurdles, 24
        overcoming geographical barriers, 23–24
        in reducing the need to travel, 28
        remote monitoring, 24
    soft computing in remote prenatal care, 29
Pretrained molecular depiction, 213
Preventive healthcare
    importance of, 2
    in India, 1–2
    meaning of, 1

strategies of, 1
types of
  primary, 4
  primordial, 3
  quaternary, 4
  secondary, 4
  tertiary, 4
Primary prevention, 4
Primordial prevention, 3
Principal Component Analysis (PCA), 75, 186
Privacy and security, preservation of, 184
Privacy breaches, risk of, 36
Problem-solving, 14, 25, 29, 229
Prostate cancer
  digital pathology practice, 102
  Gleason grading system, 102
  use of AI in detection of, 102
Protein Data Bank (PDB), 183, 215, 248
Protein function prediction, 185
Protein-protein interaction networks, 209
Protein-protein interaction prediction, 245
Protein structure descriptors, 245
Proteomics, 250
Provider-patient dynamics, in accessing prenatal care, 23
PubChem, 162
Public health data, analysis of, 109
Public health initiatives, 192
Public health management, 180
  AI models for, 192
  disease prediction and, 192–194
Public health surveillance, 192
PubTator Central, 188
P waves, 53
PyTorch (software tool), 100

Q

Quality of life, 191
Quantitative structure-activity relationship (QSAR), 67, 161, 165–166, 170, 207
  Toolbox, 165
Quantitative structure-property relationship (QSPR), 71
Quaternary prevention, 4

R

Radiology diagnostics, AI solutions for, 100, 101–102
Raloxifene (Evista), 162
Raltegravir (Isentress), 162, 164
Raman spectroscopy, 152
Random Forest (RF) algorithm, 56, 62, 70, 77, 100, 115, 117–118, 120, 141, 170, 183, 185, 189, 212–213

Rare diseases, use of AI and machine learning in diagnosis of, 11–12
Real-time monitoring, 29
  revolution of, 78
Reasoning systems, rule-based, 25
Receptor-ligand complexes, 164
Recurrent neural networks (RNNs), 75, 100, 181, 187–188, 231, 238
Registered Medical Practitioners (RMPs), 39
  verification of, 39
Reinforcement Learning (RL), 83, 85, 182, 186
Remote consultation, role of telemedicine in, 27–28
Remote maternal mental health monitoring, technologies for, 45
Remote patient care, 19, 21
Remote patient monitoring
  adaptive algorithms for, 32
  for adjustments, 36
  AI-driven, 194
  technologies for, 20–21
  use of soft computing in
    data collection and preprocessing, 30
    data fusion and integration, 31–32
    fuzzy logic, 30–31
    genetic algorithms for optimization, 31
    neural networks for pattern recognition, 31
  use of wearable devices for, 42
Remote postpartum care, 43
Remote prenatal care, use of soft computing in, 29
Remote sensing, 118
Renal cell carcinoma, 79
Renal stones, 150
Renal transplant, 133
Representation learning, 238, 247
Reproductive urology, 133
Research and development, use of AI and machine learning in, 11
Resource allocation, 94–95, 192, 239, 263
  mathematical models for, 263–264
Res-U-Net network, 135
Retrosynthesis prediction, 213
Reverse pharmacology, 73
Robot-assisted surgeries, 96
  applications of, 14
  computer-controlled tool, 13
  laparoscopic surgery, 14
  role of, 13
  for treatment of cardiac illness and disease, 14
  use of AI and machine learning in, 13–14
Robotic medicine, 21
Robotic Process Automation (RPA), 107
Robotic surgical assistants (SRAs), 13
Root Mean Square Deviation (RMSD), 248, 253
Royal Free London NHS Foundation Trust, 219

Index

R—R gap, 53
RSA encryption, 266

S

Samuel, Arthur, 229
Saquinavir, 168
SARS-CoV-2, diagnosis of, 9
Scanning electron microscopy (SEM), 243
Schrödinger's QikProp, 165
Secondary prevention, 4
SegNet (ML algorithm), 135
Sensor-based devices, 94
Sequence alignment scores, 248
Shennong's information theory, 232
Shockwave lithotripsy (SWL), 134
SIDER database, 215
Simple logistic regression, 56
Singlecell RNA sequencing (scRNA-seq) data, 178, 248
Single-Photon Emission Computed Tomography (SPECT), 103, 244
Skin disease diagnosis, use of AI in, 10
Skin-to-stone distance (SSD), 141
SkinVision, 103
Sleep apnea, 132
Societal barriers, in accessing prenatal care, 23
Soft computing, in healthcare
    adaptive and learning algorithms in managing dynamic health data, 32–33
    for analysis of impact of telemedicine on pregnancy and postpartum care, 27–29
    in remote prenatal care, 29
    in supporting healthcare professionals, 33–34
    applications and techniques of, 27
    decision support systems for prenatal care, 33
    for early risk assessment in prenatal care, 35–36
    for enhancing healthcare quality, 27
    flow of, 26
    overview of, 25
    in predictive modelling, 34–35
    remote monitoring with
        data collection and preprocessing, 30
        data fusion and integration, 31–32
        fuzzy logic for interpretation, 30–31
        genetic algorithms for optimization, 31
        neural networks for pattern recognition, 31
        in prenatal care, 29
    role of, 25
    software informatics and, 25–27
    usage in medical for decision making, 25
Spontaneous stone passage (SSP), 139
Stacked GANs (StackGAN), 249
Stimulated Emission Depletion (STED) microscopy, 244
Stone disease
    CT-based detection of, 135, 150
    identification of
        use of ultrasonography (US) for, 136–137
        using cutting-edge techniques, 152
        via endoscopic pictures, 151
    uses in AI for diagnosing, 138–139
    uses of AI in understanding the chemical and compositional aspects of, 148–149
    X-ray-based setection of, 137–139
Stone-free rates (SFRs), 140–141
Stone-free status (SFS), 140–141
Structural hurdles, in accessing prenatal care, 23
Structure-activity relationships (SAR), 77, 161, 171, 209
Structure-based drug design (SBDD), 97, 99–100, 162, 217
Structured Illumination Microscopy (SIM), 244
Super-resolution GANs (SRGANs), 249
Super-resolution microscopy, 243–244
Super ventricular tachycardia (SVT), 51
Supervised learning, concept of, 185
Support Vector Machines (SVMs), 27, 55–56, 70, 77, 94, 98, 100, 136–137, 182, 185, 213
Support vectors, 55
SYBYL software, 163
Synthetic Minority Over-sampling Technique (SMOTE), 115
Systemic sclerosis (SSD), 141

T

Tamiflu, 162, 171
Target identification
    applications of AI for, 208–209
    building of databases for, 209
t-Distributed Stochastic Neighbor Embedding (t-SNE), 75, 186
Technological innovation, in healthcare delivery system, 21
Teledocs, 15
Teleeducation and information exchange, 21
Telegenetic counselling, 43
Telehealth
    empowering patients through, 22
    patient-centric, 21–22
Teleintervention, 21
Telemedicine, 19, 132
    challenges and future directions
        ethical and legal considerations, 36
        future prospects, 42–45

legal frameworks and guidelines governing telemedicine practices, 36–40
technological challenges, 40–44
evolution of, 19–22
historical development in the field of prenatal care, 20–22
  acceleration during the COVID-19 pandemic, 21
  early initiatives, 20
  empowering patients through telehealth, 22
  integration of mobile health (mHealth), 21
  integration with patient-centric telehealth, 21–22
  remote monitoring, 20–21
  teleeducation and information exchange, 21
  teleintervention and robotic medicine, 21
for home monitoring of high-risk pregnancies, 28
impact on pregnancy and postpartum care, 27–29
licensing requirements for, 36–37
progression in, 20
role in prenatal care
  case study of, 24
  logistical hurdles, 24
  in overcoming geographical barriers, 23–24
  remote monitoring, 24
use in prenatal and postnatal care, 28
user-friendly interfaces, 42
Telemedicine services, global expansion of, 43
Telenurses, 15
Telerobotic ultrasound technology, 45
TensorFlow (software tool), 100, 170
Terbinafine, 216
Tertiary prevention, 3–4
Text mining, 188, 209, 244
Thalassemia (blood disorder), 190–191
Therapeutic medication monitoring, 132
Three-dimensional (3D) GANs, 249
TorchDrug, 213
Trait distributions
  measurement of inequality in, 267
  prediction of, 267
Transfer learning, 183, 245, 247
Turing, Alan, 130
Turing test, 130
23andMe, 250

U

Ultrasonography (US), for the identification of stone illness, 136–137
U-Net, 135
UNETR (ML algorithm), 135
UniProt Knowledgebase, 215, 248
United Kingdom, telemedicine licensing in, 37
United Nations (UN), 271
United States
  Food and Drug Administration (FDA), 210
  Health Insurance Portability and Responsibility Act (HIPAA), 38, 108, 180
  telemedicine licensing in, 37
Unsupervised learning, 75, 130, 185–186, 230
Ureteral stones
  identification of, 134
  non-invasive treatment for, 140
Uric acid (UA) stones, 150
Urinary stone
  composition of, 152
  incidence, 131
Urography, 131
Urolithiasis, 131
  management of, 134
Urologic oncology, 133
Urology, AI applications in, 133–134
  in analysing relationship between blood, urine interaction, and other medical variables, 147–150
  enhancement of
    CT-based stone disease detection, 135–136
    endoscopic techniques, 146
    X-ray-based stone disease setection, 137–139
  in enhancing the efficiency of surgical procedures, 145
  forecasting results of
    endoscopic treatments, 142
    extracorporeal shockwave lithotripsy, 140–142
    overview of research on, 143
    prudent management, 139–140
  management of
    kidney stone, 134–135
    urolithiasis, 134
  by optimization of the surgical process
    by enhancement of the ESWL process, 144–146
  optimizing ultrasonography (US), 136–137
  use of augmented reality (AR), 133

V

Valvular disease, 103
Variational autoencoders (VAEs), 83, 238, 247
Vector-borne diseases, 193
Ventricular fibrillation, 51
Ventricular tachycardia, 51
Video conferencing, 19, 21, 24, 39

Index

Virtual assistants, AI-powered, 108, 179
Virtual healthcare practices, 21
Virtual nurses, driven by AI, 239
Virtual nursing assistants (VNAs), 14–15, 96
Virtual Reality (VR), 43
    for prenatal education, 45
Virtual screening (VS), 70, 76, 98, 162, 166, 246
    AI-powered, 98
Voice recognition, 14
Vrady cardia, 51
Vulnerable populations, prioritization of, 264

W

Wearable devices
    biometric monitoring wearables, 43
    use for remote monitoring, 42
Wearable technology, 15, 42, 239
World Health Organization (WHO), 270

X

XGBoost (eXtreme Gradient Boosting), 115, 118, 120, 122, 141
X-ray, 9, 92, 101
    crystallography, 99, 250–251
    diffraction, 77
    imaging, 141
    microscopy, 243
    stone disease setection based on, 137–139

Z

Zanamivir, 165
Zebra Medical Vision, 100–101
ZINC database, 215
ZINCPharmer MOE (Molecular Operating Environment) software, 163
Z-scores, 30

9781032832425